MW00964954

WIRELESS MULTIMEDIA NETWORK TECHNOLOGIES

THE KLUWER INTERNATIONAL SERIES
IN ENGINEERING AND COMPUTER SCIENCE

WIRELESS MULTIMEDIA NETWORK TECHNOLOGIES

edited by

Rajamani Ganesh
GTE Laboratories

Kaveh Pahlavan
Worcester Polytechnic Institute

Zoran Zvonar
Analog Devices

KLUWER ACADEMIC PUBLISHERS
Boston/London/Dordrecht

Distributors for North, Central and South America:
Kluwer Academic Publishers
101 Philip Drive
Assinippi Park
Norwell, Massachusetts 02061 USA
Tel: 781-871-6600
Fax: 781-871-6528
E-mail: kluwer@wkap.com

Distributors for all other countries:
Kluwer Academic Publishers Group
Distribution Centre
Post Office Box 322
3300 AH Dordrecht, THE NETHERLANDS
Tel: 31 78 6392 392
Fax: 31 78 6546 474
E-mail: orderdept@wkap.nl

 Electronic Services: http://www.wkap.nl

Library of Congress Cataloging-in-Publication Data

Wireless multimedia network technologies / edited by Rajamani Ganesh,
    Kaveh Pahlavan, Zoran Zvonar
        p.   cm. -- (Kluwer international series in engineering and
    computer science ; SECS 524)
    Papers presented at the 9th International Symposium on Personal,
Indoor, and Mobile Radio Communications, held Sept. 5-8, In
Boston, Mass.
    Includes bibliographical references (p.   ).
    ISBN 0-7923-8633-7
    1. Wireless communication systems.   2. Multimedia systems.
I. Ganesh , Rajamani.   II. Pahlavan, Kaveh , 1951-   .   III. Zvonar ,
Zoran.   IV. IEEE International Symposium on Personal, Indoor, and
Mobile Radio Communications (9th : 1998 : Boston, Mass.) V. Series.
TK5103.2.W5735   1999
621.382--dc21   99-40758
                                                                                     CIP

Printed on acid-free paper.
Printed in the United States of America

Contents

vi

PART III: ENABLING COMPONENTS OF WIRELESS NETWORKS

PREFACE

This book is a collection of invited papers that were presented at the Ninth IEEE International Symposium on Personal, Indoor and Mobile Radio Communications, September 5-8, 1998, Boston, MA. These papers are meant to provide a global view of the emerging third-generation wireless networks in the wake of the third millennium. Following the tradition of the PIMRC conferences, the papers are selected to strike a balance between the diverse interests of academia and industry by addressing issues of interest to the designers, manufacturers, and service providers involved in the wireless networking industry. The tradition of publishing a collection of the invited papers presented at the PIMRC started in PIMRC'97, Helsinki, Finland. There are two benefits to this tradition (1) it provides a shorter version of the proceedings of the conference that is more focused on a specific theme (2) the papers are comprehensive and are subject of a more careful review process to improve the contents as well as the presentation of the material, making it more appealing for archival as a reference book. The production costs of the book is subsidized by the conference and the editors have donated the royalty income of the book to the conference.

The main technical theme of the book is the multimedia technologies for wireless networks in the third millennium. It consists of eighteen chapters divided in three parts addressing evolving standards, systems, technologies and components. The authors are among the most respected individuals in the wireless networking industry and include senior executives of international organizations for development of technologies and standards, directors of research laboratories, and distinguished researchers in the industry and academia. This combination of diversified views results in a more thorough picture of the evolving multi-disciplinary wireless industry that relies on the standardization efforts, economical and political aspects as well as technological advances.

In the description, which follows, we used the name of the invited speaker at the PIMRC'98 conference, to refer to each chapter. We acknowledge, however, the fact that the invited speakers have received considerable help from their colleagues in preparing the chapters. A biography of all contributors is included at the end of this book.

The first part of the book includes five chapters on the evolving third-generation technologies. Robert Verrue, director general of the DG XIII in the European Commission, addresses critical regulatory issues ensuring successful development of the third generation systems from the European point of view. Lin-Nan Lee, VP of advanced development at Hughes Network Systems, discusses technical approaches to support reliable high-speed packet data delivery and improved voice capacity for the 3G networks. Tero Ojanpera, head of research, radio access systems at Nokia, presents an overall perspective of the technical aspects and trends in development of 3G systems. Jorge Pereira, scientific officer in the DG XIII of the European Commission, proposes a common architecture in the context of IP versus ATM that accommodates a variety of popular voice and data standards and yet stays open for innovative air interfaces. Antun Samukic, program manager of the ETSI's UMTS, describes formation and evolution of the Pan-European 3G standard UMTS.

The second part of the book includes seven papers on wideband multimedia wireless networks. Seiichi Sampei, associate professor of communications engineering at Osaka University, discusses flexible radio access and transmission technologies for support of multimedia services for both TDMA and CDMA systems. Dipankar Raychaudhuri, assistant general manager of the NEC USA C&C research laboratories, provides an authoritative view of the trends in wireless ATM for mobile applications and explains techniques for integration of IP based legacy services in wireless ATM. Krishnan Sabnani, head of networking software research department of Bell Laboratories, highlights challenges in providing Internet services to mobile phones and describes a prototype developed for this purpose. Ahmad Bahai, CTO of ALGOREX Inc., addresses complexities of the technologies for multi-mode, multi-band systems designed for high speed wireless Internet access. Prathima Agrawal, assistant vice president at Telcordia Technologies, relates power consumption in the terminal to the design of Medium Access Control protocols. Thomas La Porta, head of the networking techniques research department in Bell Laboratories, describes a novel modular software architecture for system design that facilitates migration from 2G to 3G networks. Kwang-Cheng Chen, professor of telecommunications at National Taiwan University, categorizes various approaches for providing quality of service (QoS) in wireless networks to satisfy end-to-end QoS requirements.

Part three of the book consists of six papers describing various aspects of implementation of wireless networks. Babak Daneshrad, assistant professor of EE at UCLA, provides the trends, challenges and options presented to designers of integrated circuits for wireless communication applications. Mohsen Kavehrad, professor and director of CICTR at Pennsylvania State

University, presents an overview of linearalization techniques for power amplifiers used in wideband wireless applications. Ramesh Rao, professor of EE at UCSD, explains two new methods to capture the behavior of wireless channels for higher layer error control protocols. David Goodman, professor and director of WINLAB at Rutgers University, introduces a new framework for distributed power control based on the economic principles of utility and pricing. Vijay Bhargava, professor of EE at University of Victoria, presents the framework for computing error rate of a broad class of communication systems with microdiversity reception. Savo Glisic, professor of Electrical Engineering at the University of Oulu, Finland, introduces a multitone MFSK signal for wireless ad hoc networking.

Collection of invited papers was a part of the overall agenda of the PIMRC'98 organization committee. As a result, other members of the organization committee have contributed in the selection of the topics and experts to present those topics. In particular we would like to express our thanks to Allen Levesque for his comments during the selection process and Jacques Beneat, Prashant Krishnamurthy and others at the Center for Wireless Information Network Studies, Worcester Polytechnic Institute who helped in communication with the authors.

Rajamani Ganesh

Kaveh Pahlavan

Zoran Zvonar

PART I

OVERVIEW OF EVOLVING

3G TECHNOLOGIES

Chapter 1

Ensuring the Success of Third Generation

ROBERT VERRUE*, JORGE PEREIRA♦
European Commission, DG XIII Rue de la Loi 200 B-1049 Brussels Belgium
Phone +32 2 296 1547, Fax +32 2 296 2178
Email: Jorge.Pereira@bxl.dg13.cec.be

Abstract: Starting from a sizeable investment in European R&D, always with a subsidiarity and consensus building perspective, resulting directly in crucial contributions to standardization both at the European (ETSI) and global (ITU) level, we build the case for the critical importance of the regulatory issues that make or brake a market (availability of spectrum, licensing, pan-European services, etc - and obviously the "single standard" versus "multiple standards" issue). Given the mobile explosion all over the world, and particularly in Europe, and the evolution towards mobile/wireless multimedia applications, we also analyze, in the context of ITU and WRC 2000, the anticipated need for additional spectrum for Third Generation in the medium to long term.

* Director General, Directorate-General XIII, Information Society: Telecommunications, Markets, Technologies - Innovation and Exploitation of Research
♦ Scientific Officer, Directorate-General XIII, Directorate F - Essential Information Society Technologies and Infrastructures, Unit F4 - Mobile and personal communications and systems, including satellite systems and services. On leave of absence from Instituto Superior Técnico, Lisbon Technical University, Portugal.

1 INTRODUCTION

The reader certainly shares with us the desire to make of Third Generation (3G) mobile communications a global success. A lot of effort has been put into the specification and standardization of 3G systems at national, regional and global levels (ANSI, ARIB/TTC, ETSI, ITU), even if the results are not the most encouraging in the latter. We need now to make sure that the right conditions are created to ensure the best results.

Europe brings to the table its experience with GSM (Global System for Mobile Communications), and the tremendous success it enjoys all over the world. And we want to make UMTS (Universal Mobile Telecommunications System), the European 3G submission to ITU as a member of the IMT-2000 (International Mobile Telecommunications-2000) family, even more successful than GSM, in fact targeting a mass market of mobile multimedia services [1]. Achieving that, however, requires more than the R&D that drives standardization; it requires the creation the right environment for the success of such advanced systems with capabilities well beyond those of Second Generation (2G).

In the following we will start by reviewing the accomplishments of European R&D in the area of Mobile and Personal Communications, and then revisit the tremendous success of GSM trying to identify what made it so successful *vis a vis* other systems. The lessons learned from GSM, together with the established principles of liberalization/de-regulation and competition, are then taken into account in establishing the right environment for the success of UMTS – and are suggested for other 3G systems.

Let us make it very clear that the success of UMTS/3G is not seen as implying the demise of GSM/2G; on the contrary, the two systems are seen as complementary, and the continued expansion and evolution of GSM is critical in the crucial phase of the launching of UMTS services. This not only because GSM will provide for the needed fallback network for the initial UMTS "islands", but mainly because GSM is now in the process of establishing the wireless data usage that will eventually lead to the much anticipated mass market of mobile multimedia services.

We conclude by restating the European Commission's commitment to support a successful preparation for and deployment of UMTS, not forgetting the need for continued R&D in this area. The industry is called to lead the development of UMTS from a global perspective, ensuring the universality of the system, and the (European) governments are invited to take the necessary measures and create the right environment for a successful and timely deployment of 3G systems.

2 R&D IN MOBILE AND PERSONAL COMMUNICATIONS

UMTS, we are proud to say, is a product of the European Commission (EC) initiative, conceived and developed in the scope of the RACE (Research into Advanced Communications for Europe – RACE I: 1985-1991; RACE II: 1991-1995) and ACTS (Advanced Communications Technologies and Services, 1995-1998) programs.

When GSM was still in the process of being launched in Europe, the EC started pushing the concept of 3G, and the first UMTS project was born in the scope of the RACE program. As a result, CEPT (European Conference of Postal and Telecommunications Administrations) was ready to propose and obtain at WARC 92 the allocation of 230 MHz of spectrum in the 2 GHz band, including both terrestrial and satellite components, on a world-wide basis, for what is now called IMT-2000 (previously FPLMTS, Future Public Land-Mobile Telecommunication Systems).

Table 1. European R&D in perspective

RACE I (1985-1991)
   "The" RACE Mobile Project
	1989 The RACE Mobile Project "invents" the
	   UMTS concept

RACE II (1991-1995) 12 mobile projects	1992 WARC allocates 230 MHz to FPLMTS
	1992 First GSM networks in operation in Finland: start of 2G
	1993 Creation of FAMOUS (trilateral US, Japan, EU on Mobile Communications)
	1994 Second wave of UMTS projects
	1995 RACE Vision of UMTS

ACTS program(1995-1998)
   32 trial-driven projects
   120 MECU (~$150 M) EC
   contribution
   Participants: operators,
     manufacturers, Universities,
     SMEs, users, content and
     service providers
 IST program (1999-2002)
	1996 Creation of UMTS Task Force
	 1996 Digital overcomes analog
	 1997 Establishment of the UMTS Forum
	 1992-1998 - Major contributions to ETSI and ITU
	   (GRAN, UTRA)
	 Jan 98 - ETSI landmark decision
	 Jan 99 - UMTS Decision

In the R&D front, two RACE projects [2], CODIT and ATDMA, started developing advanced high data rate systems, one CDMA- and the other TDMA-based. The test-beds produced in the scope of those projects were the main reason why such a tight schedule was proposed by Japan and followed by ITU for the specification and launching of 3G systems.

In 1996, the EC launched the UMTS Task Force with the mandate of identifying the requirements for 3G systems, conceived as a quantum leap relative to GSM [3]. As the concept matured, and industry became convinced that 3G systems had a role complementary to GSM, promising to bring about mobile multimedia, the UMTS Task Force evolved into the UMTS Forum, a full blown Industry Association with participation of manufacturers, operators and regulators, with more than 120 members from Europe, North America, Japan, Korea, Singapore, etc.

After the successful proof-of-concept, a whole series of projects started, in the scope of the ACTS Program [4], to look into the definition of UMTS. Two of them, FRAMES and RAINBOW, deserve particular mention.[1]

Table 2. FRAMES proposals

	FMA1 (TDMA with and without spreading)		FMA2 (W-CDMA)
	Without spreading	With spreading	
Multiple Access method	TDMA	TD/CDMA	W-CDMA
Carrier chip/bit rate	2.6 Mbps	2.167 Mcps	4.096 Mcps
Bandwidth	1.6 MHz		4.4 - 5 MHz
Duplex method	FDD and TDD		FDD, TDD for further study
Interference reduction	Joint detection supported	Joint detection	Multi-user detection supported
Spreading codes	N/A	Orthogonal of length 16 chips	Spreading factor 4 to 256, short codes
Multi-rate concept	Multi-slot	Multi-slot and multi-code	Variable spreading and multi-code
Detection	Coherent, based on mid-amble		Coherent (reference symbol or pilot)
Handover	Mobile assisted hard handover		Mobile controlled soft handover
Inter-frequency handover	Supported		Supported
Frequency hopping	Frame by frame / slot by slot		N/A

[1] Contributions from these and many other ACTS projects were showcased in four dedicated ACTS sessions organized in the context of PIMRC '98 [5], covering all aspects from terrestrial to satellite, from cellular to wireless indoor, from enabling technologies to advanced services.

FRAMES (Future Radio widebAnd Multiple accEss System), focusing on the Radio Interface, contributed 3 out of the 5 proposals considered in ETSI, and, in fact, is at the origin of the 2 proposals retained in ETSI's historic agreement of 29 January 1998 [6]: UTRA (UMTS Terrestrial Radio Access), a harmonized hybrid Wideband-CDMA plus TD/CDMA scheme.

Table 3. UTRA concept

	UTRA FDD	UTRA TDD
Multiple Access method	WCDMA	TD/CDMA
Carrier chip rate	4.096 Mcps	4.096 Mcps*
Pulse shape	root raised cosine, r = 0.22	root raised cosine, r = 0.22*
Carrier spacing	5 MHz	5 MHz*
Frame length	10 ms	10 ms*
Number of power control groups / time slots	16	16
Time slot duration	625 μs	625 μs*
Modulation	QPSK	QPSK*
Spreading factor	short codes, variable 4 - 256	1, 2, 4, 8 and 16

* same as FDD

On the other hand, RAINBOW (Radio Access INdependent Broadband On Wireless), looking at network issues, has been a major contributor of ITU-T. At the kernel of their contribution is an approach, GRAN (Generic Radio Access Network), that separates Radio Dependent and Radio Independent parts of the network, facilitating the design of the required interfaces, and the inter-working with 2G systems, namely GSM. The same approach makes it very simple to conceptualize the integration of the satellite component of UMTS (S-UMTS), an aspect pursued by the ACTS project SINUS (Satellite Integration into Network for UMTS Services).

It is important here to analyze the approach underlying R&D investments in the scope of the EC Framework Programs. EC contribution is reserved to those activities where there is an obvious European value added, in terms of achieving critical mass and/or assembling the right set of expertise dispersed throughout the Community and Associated States. The principle of *subsidiarity* applies whereby activities that are conducted at National level are not duplicated at Community level, on the contrary all R&D activities are coordinated to avoid wasting resources. As a consequence, by focusing in areas where impact at Community level is anticipated, important, even crucial contributions have been made with relatively small budgets.

But let us look now into the success of another pan-European initiative, GSM, which became truly a Global System for Mobile Communications.

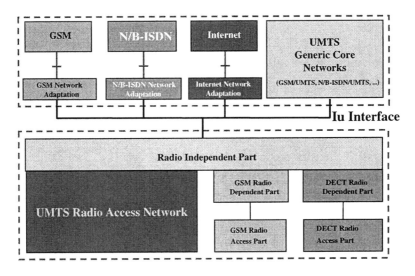

Figure 1. GRAN Model

Table 4. Cellular Subscribers at end of 1998 – expected ordering (*in italic cellular subscribers at end of 1997*) [Financial Times]

US (66.4 M)	France (11.2 M)
Japan (39.0 M)	*South Korea (6.81 M)*
China (13.9 M)	*Australia (5.45 M)*
Italy (20.2 M)	Spain (7.05 M)
Germany (14.0 M)	*Canada (4.31 M)*
UK (12.9 M)	Sweden (4.53 M)

3   THE SUCCESS OF GSM

The success of GSM, not only in Europe but also around the world, is unquestionable. At the end of 1998, in Europe, GSM systems at 900 and 1800 MHz already accounted for 91% of the cellular market, with 83.4 million subscribers (see Figure 2). Worldwide, there were 138 million subscribers, China being the country with most GSM subscribers. GSM has already been adopted by over 230 operators in more than 110 countries; 350

GSM networks in some 130 countries serving approximately 250 million users are expected by the year 2000.

In terms of penetration, at the end of 1998 Europe was still lagging with 23.8% penetration (versus 25.4% in the US and 30.8% in Japan), but the Nordic countries have penetrations above 50% (58% in Finland and 51% in Sweden), clearly leading the field (Figure 3b). In effect, eight countries already exceeded 30% penetration by the end of 1998, and are expected to be above 50% penetration before the end of the century. An important trend to note here is the fact that more and more people chose to have more than one subscription, namely to separate professional from private use. This will make it very likely that the 100% penetration mark will be achieved or even exceeded, even without taking into account cellular-equipped appliances.

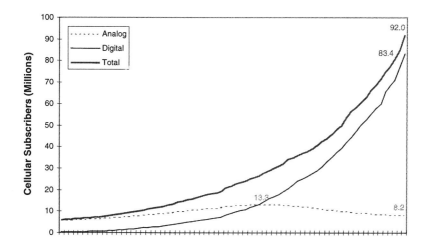

Figure 2. The explosive growth of GSM in Europe [Financial Times]

3.1 How did GSM become so successful?

The lackluster performance of a variety of incompatible analogue systems at 450 MHz and later at 900 MHz (NMT-450, NMT-900, TACS-900, RC2000, C-450) led to an effort led by the *Group Special Mobile* to develop a **single** digital system – cf. AMPS, the only analogue US standard – providing pan-European roaming. This required coordinated deployment across Europe, which was achieved through a GSM Directive [7] and an ERC Decision [8].

Figure 4 shows side by side the two growth curves. Note first the scale difference, and then the fact that the time scale has been halved for GSM. As a concrete example, while in 9 years the combined analogue systems approached only 6 million subscribers, in half that time GSM surpassed 30 million, a factor of more than 10 in take-up. Finally, from the figure it is easy to see that a better exponential fit exists for the digital growth, indicating a sustained better market uptake [9]. But a single digital standard is not, per se, a universal panacea.

The example of Japan confirms that **liberalization** and **competition** play a role the EC has long recognized: Figure 5 shows that, in spite of the single digital standard PDC, only after the first tentative steps at introducing competition, did the Japanese market seriously take up.

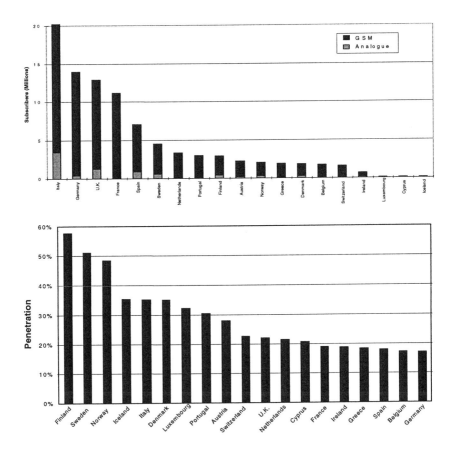

a) Cellular Subscribers,   b) Penetration
both analogue and digital

Figure 3. Cellular Subscribers and Penetration in Europe [Financial Times]

The situation in the EU at this moment is one of **full liberalization**, with most countries having at least two GSM 900 operators, a few with at least one GSM 1800 operator. To this, since recently, one must add joint GSM 900/1800 licenses attributed to old and new operators.

Figure 5 serves also to illustrate another common error: **artificially creating and sustaining a market**. The numbers show that the impressive growth of PHS halted as soon as the subsidies were suspended.

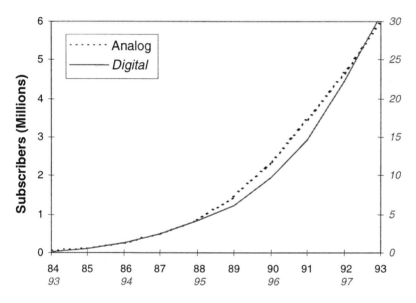

Figure 4. Growth of GSM versus various analog systems [Financial Times]

3.2   Situation in the US

It is easy to see from Figure 6 that the remarkable exponential growth of analogue cellular of a while back has slowed down to a linear growth with the introduction of digital. In fact, the performance of digital has been dismal, and only recently started to pick up. Taking into consideration that many of the new digital systems are being deployed in substitution of analog AMPS in the 800 MHz band, then PCS alone has faired even worst

It is in fact very meaningful that, in spite of the recognized advantages of digital, and of the ongoing substitution of AMPS with digital, analog is still growing; it shows that subscribers value the global roaming capabilities of AMPS, which comes at a premium with digital (**multi-mode, multi-band**

handsets – cf. SDR Forum's proposal of a quadruple-mode terminal as the Universal Terminal [10] – for the US only).

The difference in rate of growth is clear when we compare the growth of digital in Europe with that in the US (Figure 7), a single system versus many incompatible systems (some of them have in the meantime been "recalled").

As a result of this different growth – in 1998 alone, Europe added 37.8 million cellular subscribers, while the US added only 12.4 million and Japan 10.2 million – the US had already lost at the end of 1997 the lead it had maintained since the beginning (Figure 8).

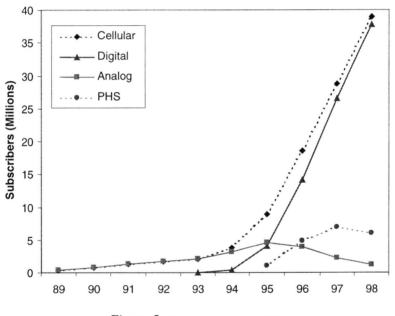

Figure 5. Situation in Japan [MPT]

Putting it candidly, the US is making now with 2G the same "mistake" Europe made by pursuing multiple incompatible 1G standards, even if now there is a solution in sight with "software radio", which comes at a cost however.

The fact that Calling Party Pays (CPP) is not practiced, does not help either. The CTIA is still pushing the FCC to address the concerns about consumer education, assurance that consumers will pay to place calls (beyond the wireline cost), and that CPP does not become a means of creating state regulation of wireless.

Table 5. Diversity of Cellular Systems in the US

800 MHz	AMPS, N-APMS, TDMA IS-136, CDMA IS-95
1900 MHz (PCS)	N-AMPS, GSM 1900, Omnipoint, TDMA IS-136, CDMA IS-95, PACS

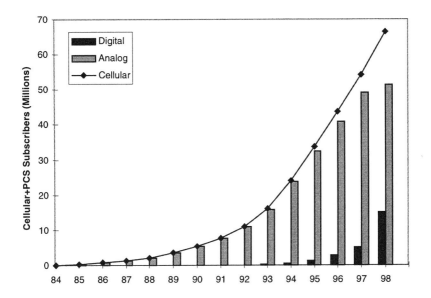

Figure 6. Cellular + PCS growth in the US [CTIA, CDG, UWC Consortium, Nextel[2]]

4   ENSURING THE SUCCESS OF UMTS

Even if it is understood that the GSM story cannot be repeated, as conditions are now very different, a few lessons can be extracted from the analysis above, which will be of help in preparing the launch of 3G.

History has repeatedly shown that the most careful system specification, even with unanimous support, is not enough guarantee of success. The conditions need to be there for the system to be deployed (namely spectrum needs to be available), the same for the "incentives" that make the system grow.

---

[2] Given the lack of discrimination in the CTIA figures from the semi-annual wireless industry survey, which include not only cellular and PCS but ESMR as well, and does not distinguish analog and cellular subscribers, we used information from the CDG and UWC Consortium for digital cellular + PCS subscribers, and subtracted Nextel's digital ESMR subscribers (2.79 million at the end of 1998).

14

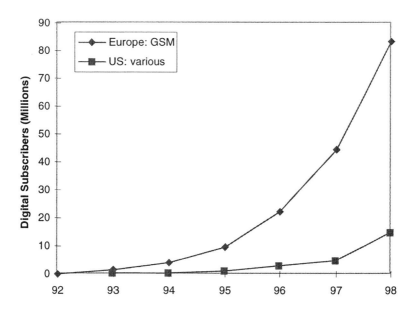

Figure 7. Comparing Digital Growth: Europe versus US [Financial Times, CTIA et al.]

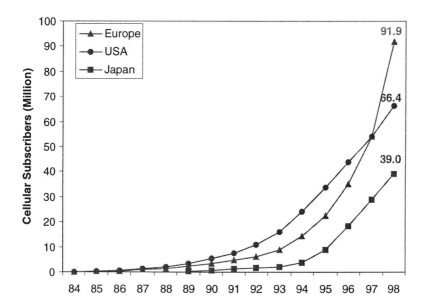

Figure 8. Comparative Growth [Financial Times, CTIA et al., MPT]

All starts (system definition) and then relies (system evolution) on R&D. In this area, the EC launched in January 1999 a new R&D program, the **Information Society Technologies** program, where all that concerns Mobile and Personal Communications will continue to be dealt with, building upon the work done in ACTS.

On the political side, to prepare the ground to make the success of UMTS a reality, the EC took a series of steps to establish the need for a coordinated approach to 3G systems in Europe, starting with a Green Paper on Mobile and Personal Communications [11] (cf. Table 1). A Council Decision [12] resulted, pointing towards the further development of the Mobile and Personal Communications sector. Two communications followed in 1997 identifying challenges and choices as well as strategy and policy orientations [13, 14].

It was soon recognized that the UMTS development and deployment should be a market driven, industry led process, and that the Commission, in close co-operation with Member States, was to promote UMTS by supporting R&D, establishing a favorable environment, and defending European interests from a Global perspective.

Furthermore, recognizing the risks associated with new technologies and the uncertainty inherent to new markets, and especially the Mobile Multimedia one, the need was identified to harness support of current players (incumbents), and to welcome new ones (new entrants), and even of re-inventing the playing field (new business models), keeping it open, fair and competitive.

Finally, a single, open system, compliant with the requirements for Third Generation, and globally competitive had to be standardized by ETSI. In that respect, the Commission committed to take all necessary measures, where appropriate in co-operation with ETSI, to promote a common and open standard for the provision of compatible UMTS services throughout Europe, in accordance with market requirements, taking into account the need to present a common standard to the ITU as an option for the worldwide IMT-2000 standard.

A joint UMTS Decision [15] was published in January 1999 establishing the conditions for the deployment of UMTS in Europe, with the first systems expected to become operational in 2002 leveraging existing infrastructure (building upon the GSM backbone), and relying upon GSM, at the initial stages of deployment, to have full coverage.

The UMTS Decision entails, namely identification of conditions which may be attached to licenses in order to ensure pan-European services; harmonization of frequency use and conditions associated to authorizations; harmonized provision of basic UMTS services by 1 January 2002; and that Member States will influence UMTS network operators to negotiate transnational roaming arrangements and take steps to ensure coverage of less populated areas.

UMTS is identified as one possible option for 3G, recognizing that others exist and are not per regulation excluded from the European market, even if regulation establishes that UMTS is to be used in the frequency bands designated to this effect. The UMTS Decision requires only that **at least one** of the 3G licensees deploy UMTS, to ensure pan-European service provision, a key element in the success of GSM.

CEPT, which published an ERC decision on UMTS [16], has been mandated to look into the harmonization of the conditions and procedures for authorization of UMTS networks to ensure the coordinated availability, licensing and use of the spectrum, the market driven and harmonized pan-European deployment of UMTS, ensuring pan-European services and roaming, and the right degree of competition, without diluting the market. CEPT has also to look into the need for additional spectrum in the longer-term.

Table 6. UMTS timeline

by February 99	CEPT mandated to allocate additional spectrum for UMTS with the objective of proposing those changes at the next WRC, and to harmonize the conditions attached to UMTS authorizations; CEPT will also look into re-farming the 900, 1800, 1900 MHz bands for UMTS
by the end of 99 [3]	Member States shall have established an one-stop-shopping licensing procedures where necessary
by 1 January 2002	provision of basic UMTS services
by 2005	full UMTS services will be available throughout the EU

Member States shall ensure that the provision of UMTS services is organized in frequency bands which are harmonized by CEPT, and pursuant to European standards developed by ETSI where available, including in particular a common, open and internationally competitive air-interface standard.

In what refers to licenses, where a Directive [17] already exists and is applicable, input has been provided, at national and EU level, mainly by the UMTS Forum, looking, in light of their perception of the regulatory

---

[3] Ahead of schedule, Finland became the first country in the world to attributed 3G licenses, by awarding 4 national licenses last 19 March 1999.

framework [18] into licensing conditions [19] and the minimum spectrum necessary for public UMTS service [20], which effectively conditions the number of licenses that can be accommodated in the finite spectrum available.

5 ITU: THE IMT-2000 PROCESS AND THE NEED FOR ADDITIONAL SPECTRUM

The ITU set out from the start to arrive at a global 3G system ensuring global roaming and interoperability. However, the different philosophies, as well as different degrees of market liberalization in different parts of the world, combined with the weight of legacy systems, lead to the concept of IMT-2000 "family" of standards, each satisfying agreed upon minimum requirements.

While the family concept ended up being reduced to no more than a list of possible parameter values, as the Fortaleza TG 8/1 final communication shows [21], detailed standards will be left to ANSI, ARIB/TTC, ETSI and other SGOs to define. In any case, the goal of ITU-R TG 8/1 is still "to enable world-wide compatibility of operation and equipment, including international and intercontinental roaming."

UMTS is the European IMT-2000 submission and will be a full member of the IMT-2000 family; it has been positioned from the outset as a global standard, not de facto as it was the case with GSM. That implies the UMTS must be specified by ETSI in close cooperation with other regional standards bodies, taking into account the input coming from ITU's evaluation process – here we should make reference to ETSI's recently launched 3GPP (3G Partnership Project).

We have already seen that, as a result of CEPT studies and proposals, based upon work performed in RACE, ITU designated at WRC '92 a total of 230 MHz of spectrum (see Figure 9) for terrestrial (200 MHz) and satellite (30 MHz) UMTS. CEPT immediately allocated that whole spectrum to UMTS, although effectively 35 MHz were already allocated to DECT. Japan followed suite, there the problem being with PHS, which occupies 20 MHz of the IMT-2000 band. The US, on the contrary, made different use of the IMT-2000 bands by auctioning part of it for PCS (2G), effectively preventing a common global IMT-2000 system.

In fact, in the US there is currently no spectrum available for 3G, which ends up not being as problematic as one would think a priori given the slow growth of 2G (see Figures 6 and 7).

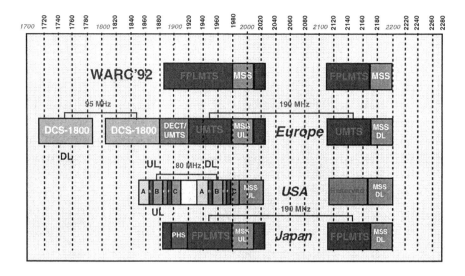

Figure 9. IMT-2000 Spectrum Allocation

In any case, the need was felt in all regions to start looking ahead to the (additional) spectrum needed for 3G (but taking into account the re-farming of 1G and 2G bands). In Europe, building upon an EC commissioned study [22], the UMTS Forum analyzed the UMTS market, and especially the Mobile Multimedia component, to identify the additional spectrum needed for 3G [23]. The result was the identification of the need for additional 185 MHz. At the recent Fortaleza meeting of ITU TG 8/1, consensus emerged around the value of additional 160 MHz in all regions [24], but none upon the bands where UMTS will be accommodated.

This decision, provided the necessary follow-up, is seen as essential to establish the right climate of confidence that will enable the huge investments required in deploying 3G systems.

6   CONCLUSIONS

The European Commission is committed to support a successful preparation and deployment of UMTS, not forgetting the need for continued R&D in this area. Politically, the UMTS Decision helps establish the right climate for the required investment in UMTS development and deployment; in terms of R&D, Mobile and Personal Communications will remain a major area of interest, and the area is expected to continue contributing actively to ETSI.

The Commission is aware of the essential and critical role of ETSI in this process: ETSI has the mandate to specify UMTS, and will do so in cooperation with other non-European standardization bodies (ETSI's 3GPP).

The Industry, in particular through the UMTS Forum, is called to lead the development of UMTS from a Global perspective. The UMTS Forum has provided to the Commission Industry feedback on the political, spectrum and R&D requirements for the success of UMTS, and has established links with non-European fora. The involvement of the GSM Association (ex-GSM MoU) is also seen as extremely important, as it brings in non-European input (and considerable non-European interests).

Finally, the Member States are called upon to transpose into national law the UMTS Decision in all its consequences, and to free the necessary spectrum and establish the licensing procedures in a harmonized manner. Only this cooperation will drive and ensure the success of UMTS.

It is our sincere conviction that, by ensuring the availability of spectrum, creating the right licensing and regulatory conditions, and making it possible pan-European/continent-wide service offerings, UMTS/3G will indeed experience the success we all hope for.

7 AKNOWLEDGEMENTS

In my capacity of Director General of DG XIII, I would like to take this opportunity to convey my appreciation to the very dedicated people that pull together our R&D and Regulatory Framework activities in the area of Mobile and Personal Communications.

The ACTS program was until the end of 1998 the focal point of R&D activity in the area of Mobile and Personal Communications. I would like to thank the Director a.i., Spyros Konidaris and his predecessors who led the ACTS program, and want to single out the Mobile and Personal Communications unit, namely João da Silva, Bartolomé Arroyo-Fernández, Bernard Barani, Jorge Pereira and Demosthenes Ikonomou. Besides their R&D-related tasks, they also support the Commission's participation at WRC and associated preparatory work, and provide the necessary technical assistance on regulation-related issues such as spectrum planning and management, standards, technology and market trends, etc.

Playing a key role in the definition of regulation and policy for the sector of Mobile and Personal Communications, I would like to thank Nicholas Argyris, Director responsible for the Regulatory Framework for Telecommunications and his colleagues, notably Jean-Eric de Cockborne, Ruprecht Niepold, Leo Koolen, and Huibert van Wagensveld. *[RV]*

20

The authors also wish to thank all the colleagues, who in the scope of the RACE and ACTS programs have so decisively contributed to the advancement of the state-of-the-art in Mobile and Personal Communications, and provided such critical input to ETSI and ITU in the area of Third Generation.

REFERENCES

[1] "From Wireless Data to Mobile Multimedia: R&D Perspectives in Europe", J. Pereira, J. da Silva, B. Arroyo-Fernández, B. Barani, D. Ikonomou, European Commission, DG XIII, in *Insights into Mobile Multimedia Communications*, D. R. Bull, C. N. Canagarajah and A. R. Nix, Eds, Academic Press, 1999

[2] 10 Years of RACE, European Commission, DG XIII, CD-97-96-887-EN-Z

[3] UMTS Task Force Report - The Road to UMTS, European Commission, DG XIII, March 1996.

[4] *ACTS 98*, European Commission, DG XIII, AC 980872-CD

[5] *Proceedings* of PIMRC '98, Boston, Sep 98

[6] *Consensus decision on the UTRA concept to be refined by ETSI SMG 2*, Tdoc SMG 39/98, SMG No. 24bis, ETSI SMG, 28-29 January 1998

[7] *GSM Frequency Directive*, Council Directive 87/372/EEC, 25 June 1987 (OJ 17 July 1987)

[8] ERC Decision *on the frequency bands to be designated for the coordinated introduction of the GSM digital pan-European communications system*, CEPT/ERC/DEC(94)01, 24 October 1994

[9] "Indoor Wireless Broadband Communications: R&D Perspectives in Europe", Jorge M. Pereira, European Commission, Colloquium on Indoor Communications, TU Delft, Delft, The Netherlands, Oct 97

[10] "The impact of Step 7 Key Characteristics on Wireless Devices for IMT-2000", SDR Forum, Document 8-1/INFO/25-E, 11 Nov 98, 15th Meeting of ITU Task Group 8/1, 9 - 20 November 1998, Jersey, Channel Islands

[11] Green Paper on Mobile and Personal Communications, *Towards the Personal Communications Environment,* European Commission, DG XIII, April 94

[12] Council Decision, *On the Further Development of Mobile and Personal Communications Sector in the European Union*, June 95

[13] Communication to the EP and Council, *On the Further Development of Mobile and Wireless Communication - Challenges and Choices*, COM(97)217, May 97

[14] Communication to the EP and Council, *Strategy and Policy Orientations with Regard to the Further Development of Mobile and Wireless Communications (UMTS)*, COM(97)513, October 97

[15] Decision 128/1999/EC of the European Parliament and of the Council of 14 December 1998 *On the coordinated introduction of a third-generation mobile and wireless communications system (UMTS) in the Community* (OJ 22 January 1999)

[16] ERC Decision *On the introduction of UMTS*, CEPT ERC/DEC/(97)07, 30 June 1997

[17] *European Parliament and Council Directive 97/13/EC of 10 April 1997: On a common framework for general authorizations and individual licenses in the field of telecommunication services* (OJ 7 May 1997)

[18] *A Regulatory Framework for UMTS*, Report No. 1, UMTS Forum, June 1997

[19] *Considerations of Licensing Conditions for UMTS Network Operations*, Report No. 4, UMTS Forum, 1998

[20] *Minimum spectrum demand per public terrestrial UMTS operator in the initial phase*, Report No. 5, UMTS Forum, 1998

[21] *Key Characteristics for the IMT-2000 Radio Interfaces*, Document 8-1/TEMP/168, 17 March 1999, 16th Meeting of TG 8/1, Fortaleza, Brazil

[22] *UMTS Market Forecast Study*, Final Report to the European Commission, DG XIII, Analysis/Internal Report No. 97043, February 1997

[23] *Spectrum for IMT-2000*, SAG Report, UMTS Forum, 1997

[24] *Spectrum Requirements for IMT-2000*, Draft new report, Document 8-1/TEMP/140(Rev.1), 17 March 1999, 16th Meeting of TG 8/1, Fortaleza, Brazil.

Chapter 2

Third Generation Wireless Communication Technologies

LIN-NAN LEE, FENG-WEN SUN, KHALID KARIMULLAH, MUSTAFA
EROZ, AND ROGER HAMMONS
Hughes Network System, 11717 Exploration Lane Germantown MD U.S.A.

Abstract: The debate on the third generation wireless system is not about the selection of the core technologies, such as CDMA versus TDMA. Rather, it is on whether the design objectives properly address the market needs and whether there are sufficient technical innovations to achieve these objectives. This chapter discusses the essential design objectives for a successful third generation wireless system. A number of key innovations intended to meet these objectives are presented. We will briefly describe the cdma2000 forward and reverse link design. In the forward link an alternative approach for the 3X mode that exhibits advantages of two approaches currently included in the cdma2000 standard is outlined. In the reverse link, we will present a media access scheme that combines PRMA for large packets and ISMA with capture message for short packets to significantly improve the efficiency of channel utilisation for high-speed data transmission. In both the forward and reverse links, Turbo FEC codes further increase the underlying channel capacity by 50-60 percent. By combining an efficient reverse link multiple access architecture, and optimised FEC coding techniques, third generation wireless systems can reach an unprecedented level of performance in high-speed packet data delivery and improved voice capacity with minimum impact on implementation complexity.

1   BACKGROUND

The third generation wireless communications standards debate has been centered on the IMT-2000 standardisation activities at the International Telecommunications Union (ITU) during the last couple of years. A quick review on the proposed air-interface standards around the world reveals similarities in terms of the fundamental technologies behind the proposals. The "mainstream" proposals [1,2] heavily leverage the technologies of a second-generation mobile system known as the IS-95 Code Division Multiple Access (CDMA) system currently being used in the United States and several countries around the world. Unlike the development of the second generation system, the debate on the third generation system is not about the selection of the core technologies, such as CDMA versus TDMA. Rather, it is on whether the design objectives properly address the market needs and whether there are sufficient technical innovations to achieve these objectives.

When proposals were solicited by ITU for the third generation wireless standards, the objectives set by ITU were global roaming and integrated data and voice services. Serious doubts were cast on the ITU initiatives by some of the U.S. wireless service providers and manufacturers. From the voice service perspective, it was a time that IS-95 based CDMA system just started to be deployed in the field. Service providers and manufacturers needed to recoup their investment. It was not clear whether there is a market need for the third generation system. From the data services perspective, the wireless industry has yet created a profitable data business, let alone high-speed data. Even worse, the ITU requirements on the third generation, formulated more than ten years ago, was rather outdated. It heavily emphasised circuit-switched data. Actually, packet-switched data was not even part of the requirements to be supported. Neither was the Quality of Service (QoS) requirement as defined by ITU geared towards the wireless environment. Thus, each of the third generation proposals was left to define its own vision and technologies to achieve their objectives.

In order to define the role of third generation in the marketplace and the technologies needed to achieve the goal, it would be helpful to review the wireless industry from a historic perspective.

The Advanced Mobile Phone Systems (AMPS) and its variation are often considered as the first generation wireless technologies. As two-way FM radio was not new, the major breakthrough for AMPS is the cellular concept itself. By organizing the frequencies in groups and establishing a frequency reuse pattern in cellular clusters as well as the capability to hand over a call from one cell to another as the mobile travels through them brings to the birth of today's wireless industry. The cellular concept was conceived and developed in late 1970s, and the deployment of the AMPS network started in early 1980s. Following the AMPS' lead, other parts of the world developed their respective first generation wireless standards. Whereas they generally differ by selecting a different channel bandwidth and/or different frequency band, the use of FM radio, and the basic cellular concept is identical. First generation wireless system is revolutionary in terms of defining a complete new type of service, mobile telephony for the general public.

Global Systems for Mobile (GSM) in Europe, the IS-136 (or D-AMPS) and IS-95 CDMA in the US, and Pacific Digital Cellular (PDC) in Japan are often considered as the second generation wireless technologies, of which GSM has been the acknowledged leader while the IS-95 CDMA plays catch-up. Generally, GSM, D-AMPS, and PDC digital technologies introduced low-rate voice coding, forward error correction (FEC) coding, digital modulation, channel equalization technologies and Time-Division Multiple Access (TDMA). From fundamental technology stand point, the differences between IS-136 or PDC and GSM are small. But, GSM is undoubtedly the most successful among the three. The reason for GSM success in the market place, is primarily due to its open architecture and standards, as well as international agreements among most of the European countries to support it, as it is the first wireless standard designed with international roaming in mind.

IS-95 CDMA has many technological breakthroughs in terms of use of Direct Sequence Spread Spectrum, soft handoff, and tight power control to achieve frequency reuse in every cell and every sector. It also uses a variable rate vocoder to reduce the amount of data transmitted over silence periods occurred during natural speech. Due to the uses of these newer technologies, much greater voice capacity than previously thought possible was achieved at the expense of implementation complexity. From technology standpoint, the success of IS-95 technology is based on the continued progress in digital solid state circuit technology during the 1990s, which eventually made implementation of this technology cost effective.

From service standpoint, second generation system is evolutionary since its objective is still mobile telephony. It advances to digital technology to enhance the capacity of first generation wireless system. In terms of deployment effort, second generation is more revolution than evolution since it is meant to replace the first generation equipment, though in reality, it is happening much slower than anticipated initially.

It is by necessity that all the successful technologies in each generation offered significantly greater capabilities than the previous generation to enjoy the market success. The same will most certainly be applicable to the third generation as well. If we look at the second generation systems being offered today, we can conclude that both GSM and IS-95 CDMA with the latest enhanced vocoders offer fairly good voice services. IS-95, in addition, offers unprecedented capacity for voice services. Even though both GSM and IS-95 have room for further capacity improvements, and the voice coding technology will continue to progress, it is apparent that better voice service alone will not differentiate a third generation system from successful second generation systems.

On the other hand, if we look at the present data service capability of the second generation wireless systems, we immediately noticed the discrepancy between speeds offered by wireless and wire-line systems. It is therefore concluded that the third generation wireless systems must offer substantially increased data speed. In this sense, the ITU requirement on an integrated voice and data network for the third generation system hit the right target, though it failed to address the need for packet-switched data sufficiently.

All the second generation digital wireless systems were designed and optimized for circuit-switched voice. Packet data provided by these systems are essentially based on the circuit-switched concepts. Therefore, third generation circuit-switched data can heavily leverage second generation technologies. However, this is no longer adequate for multimedia third generation networks based on packet-switching. Given the explosion of Internet Web applications, successful data network architecture must be based on packet-switching technology.

Packet data presents different characteristics for optimization. It is of a very bursty nature and low duty cycle. However, it is less stringent in terms of delay requirements. In typical applications, packet data is also highly asymmetrical in terms of traffic directions. The forward link often requires much larger capacity than the reverse link.

For the service providers who have just completed the deployment of the second generation networks, a third generation wireless system can be cost effective only if it offers backward compatibility with the second generation networks from the infrastructure perspective. Backward compatibility is defined as the capability to maximally preserve the value of second generation system, to provide value-added services to existing network, and to increase the system capacity with *incremental* upgrades. Therefore, third generation has to be evolutionary rather than revolutionary.

In conclusion, the wireless industry faces two challenges: first designing such a backward compatible system without sacrificing performance and future capabilities; secondly, identifying issues and solutions associated with hybrid wireless networks that supports both circuit and packet-switched traffic.

As further discussed in the following sections, the first challenge can be largely answered by the capability of overlaying data services to the current system without reducing the performance of either the data or the voice services. The second challenge requires us to re-examine the design of current circuit-switched wireless networks and push the technologies one step further to address the technical issues arisen from wireless packet data networks.

2 FORWARD LINK DESIGN

One reason that the forward link design is critical lies in the fact that for data application, forward link typically has much heavier load than in the reverse link. For Internet Web applications, it is also desirable to maximize the peak data rate for the forward link.

A less obvious reason is that forward link design is also critical for a very desirable backward compatible capability: spectral overlay.

Figure 1 illustrates the differences between two approaches of overlaying a high-speed data network on a second generation network: spectral overlay versus spectral coexistence. As shown in

Figure 1(a), coexistence means that the third generation system is deployed in a different part of the spectrum from that occupied by the second generation signals, whereas in spectral overlay the third generation system shares at least part of the second generation spectrum. Spectral overlay can be achieved without loss of efficiency only with properly designed forward link.

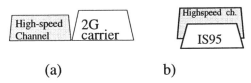

(a)   b)

Figure 1. (a) Spectral coexistence (b) spectral overlay

As previously mentioned, second generation CDMA system is adequate for voice application for most of the service providers for the foreseeable future. Therefore, third generation should provide value-added services. A simple, low cost and minimum risk transition from second generation to third generation can be realised by spectrally overlaying a high speed data layer to an existing second generation voice system. The cost saving of transition from existing networks is achieved by continuing to use the existing voice infrastructure for voice service until such a time it becomes economical to change over to more efficient next generation equipment. With this architecture, the high-speed data channel and the voice communication system can be optimized independently with their own evolution paths. Hence, we can fully exploit the packet data characteristics in the design to maximize the total throughput and performance. When properly designed, overlay capability can be realised without performance compromise [3,4].

For the reverse link of CDMA system, spectral overlay is straightforward since users appear to be random to each other. It does not make much difference in the system design whether the interference comes from underlay or overlay users. However, for the forward link, orthogonality between the overlay and underlay signals is very much desired to minimize the mutual interference. Orthogonality is also very important for effective fast forward link power control.

cdma2000 is a third generation wireless communications air-interface standard being developed under the Telecommunication Industry Association (TIA), which is also known as IS-2000. The cdma2000 standard allows system to be deployed in the bandwidth of multiple of 1.25 MHz. A system of n times IS-95 bandwidth is commonly referred to as a nX system. In other words, 1X system has the same bandwidth as IS-95; and 3X system has three times of the IS-95 bandwidth. For 1.25 MHz overlay, it is a relative simple task to maintain the orthogonality between the overlay system and the underlay IS-95 system. But, when the third generation bandwidth is larger than 1.25 MHz, it is very challenging to maintain the forward link orthogonality between the wider bandwidth overlay signal and the narrower band underlay signal. The following subsections elaborate the forward link design for different bandwidths.

2.1   1.25 MHz mode of cdma2000

The 1.25 MHz mode of the cdma2000 has the same chip rate and bandwidth occupancy as the IS-95. The fundamental differences between IS-95 and this mode are:

1. QPSK modulation with complex spreading. Third generation system employs Quaternary Phase-Shifted-Keying (QPSK) modulation in the forward link. The incoming bits are converted into I and Q data streams. The in-phase pseudo-random number (I-PN) and the quadrature pseudo-random number (Q-PN) forms a complex spreading sequence. The complex spreading sequence is multiplied with the incoming data stream. In contrast, IS-95 uses Binary Phase-Shifted-Keying (BPSK) modulation in the forward link. The same data stream goes to in-phase and quadrature phase and are spread by different pseudo-random sequence. Note that IS-95 signal constellation is also four-phase signal due to the spreading

sequences. However, each modulation symbol carries only one FEC coded bit rather than 2 FEC coded bits as in the 'true' QPSK case.

2. The QPSK modulation symbol rate (1.2288Mbits/s) is determined by the channel bandwidth. In order to achieve the same information data rate, the spreading gain between the information sequence and the QPSK modulator is twice of IS-95. The total power gain however is the same as IS-95 since the IS-95 BPSK modulator regains the 3 dB difference over QPSK modulator. This extra spreading gain between the information and the QPSK modulator can be distributed by FEC coding and orthogonal spreading. If we allocate the extra spreading gain for FEC, a rate 3/8 convolutional code instead of the rate ¾ can be used for information rate of 14.4 kbps and multiples. Similarly, a rate ¼ convolutional code instead of the rate ½ can be used for information rate of 9.6 kbps and multiples.

3. Similarly, length 128 Walsh codes can be used to channelize 8 kbps voice channels instead the length 64 Walsh codes. As a result, the Walsh code space can be doubled.

Figure 2 depicts the structure of the cdma2000 1X transmitter. By incorporating fast power control in the forward link along with these changes, cdma2000 manages to achieve 2dB gain over IS-95A/B on the forward link. The system capacity is increased accordingly. Since the 1X cdma2000 has the same R.F. characteristics and PN codes as IS-95, we can overlay the signal directly on the IS-95A/B signal in the same spectrum, as long as the two signals are chip synchronized and they do not unse overlapping Walsh codes. *Table 1* is a summary of the estimate gain in E_b/N_o over IS-95A/B.

Table 1. Summary of Forward Link Gain of cdma2000 1X mode vs. IS95

Gain Vs IS-95A/B	9600 bps	14400 bps
Power Control	1.5 dB	0.5 dB
FEC Coding	0.5 dB	2.0 dB
Forward Link Total Gain	2.0 dB	2.5 dB

2.2   Forward Link Options for 3.75 MHz Systems

The 1X mode is limited in terms of the maximum data rate it can support. To achieve higher data rate, a wider bandwidth design that utilizes 3.75 MHz has been considered by the cdma2000 developers. For the 3.75 MHz, or 3X system, cdma2000 has a direct sequence spreading mode and a multi-carrier mode. The direct sequence spreading mode is similar to the 1X mode except that all the sequences are clocked at the 3X rate. In the multi-carrier mode, a transmitter puts out three 1X carriers at the same time in the forward link, with the coded data demultiplexed into three separate streams, one for each

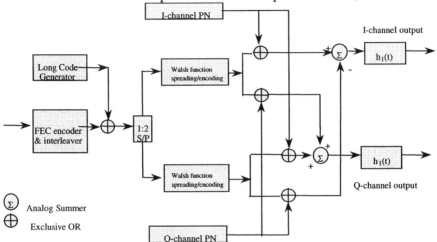

Figure 2. Structure of cdma2000 1X mode transmitter.

1X carrier. The direct sequence speading mode cannot be overlaid directly to IS-95A/B signals, whereas the multi-carrier mode is designed for spectral overlay to the IS-95A/B signals. In addition to the overlay capability, the multi-carrier mode also offers possibility to reuse the basestation R.F. transmit equipment from IS-95A/B, eliminating the need to develop and deploy a new wideband amplifier, and is potentially easier to deploy than the direct sequence spreading mode.

However, the direct spreading does take advantage of the true wideband signal to optimize the performance in a multipath environment. The subscriber unit is simpler to implement, having to de-spread and demodulate only one carrier. Similarly, implementation is simpler for the direct sequence spreading approach in terms of power control and handoff, as we will never have to resolve conflicting requirements between different carriers as a result of fading or load imbalance in the underlay carriers in an spectral overlay situation. To prevent sudden surge of traffic load in one of the underlay carriers from causing significant degradation on the overlay high-speed data channel, the multi-carrier approach requires sophisticated load balancing among the three carriers in the underlay IS-95 channels. Arguments were made by direct sequence spreading proponents that the spectral overlay with the multi-carrier approach will not provide sufficient capacity and performance. As such, spectral overlay should not be used.

Given the tradeoffs between direct spreading and multi-carrier design, the ideal solution is to design a direct spreading scheme capable of overlay to the IS-95 signals. Not to compromise performance, overlaid signal of wider bandwidth must be orthogonal to the underlay signal of narrower bandwidth in the forward link, as mentioned previously. The orthogonality must be achieved without any change to modulation and demodulation procedure of the underlay signal since it is impossible to replace all the second generation handsets already in the field. Note that IS-95 subscriber units will filter down the incoming signal with existing narrow band matched filters. Such narrow band filters will change the structure of the wide band signal. Furthermore, the carrier frequency of the wider band signal may not be the same as the narrower band signal. It was not obvious whether such orthogonal overlay technique can be designed. Until the following scheme was conceived, the common wisdom in the industry considered it to be an impossible task. To provide insights to the solution, we will first define the problem systematically.

The overlay system can be modelled by two data streams: one represents the wider band signal and the other represents the narrower band signal. In our particular case, the narrower band signal corresponds to IS-95 signal and the wider band signal corresponds to the M-X system with M larger than 1. We denote the wider band signal by

$$s_1(t) = \sum_{i=-\infty}^{\infty} a_i h_1(t - iT) e^{j2\pi f_1 t}$$

where T is the signalling interval and f_1 is the carrier frequency and $h_1(t)$ is the pulse-shaping filter. With a larger signalling interval for the narrower band signal, we can assume that the signalling interval is an integer multiple of T, MT for M-X system. Therefore, we can denote the narrower band signal by

$$s_2(t) = \sum_{i=-\infty}^{\infty} b_i h2(t - iMT) e^{j2\pi f_2 t}$$

where $h_2(t)$ is the response of the pulse-shaping filter and f_2 is the carrier frequency. A timing diagram for $s_1(t)$ and $s_2(t)$ are illustrated in *Figure 3*, for the case of M=3.

When both $s_1(t)$ and $s_2(t)$ are transmitted, the received signal is equal to $s_1(t) + s_2(t)$. Additive noise is omitted since it does not play any role when analyzing the mutual interference between the wider band and narrower band signals.

Consider the demodulation of the narrower band signal. In this case, the received signal must first be down-converted by multiplying $e^{-j2\pi f_2 t}$. The down-converted signal passes a filter matched to $h_2(t)$ and then sampled at time instance iMT. Let $F_i(t)$ denote the convolution of $h_1(t) e^{j2\pi f_1 t}$ and $h_2(MT - t)$, i.e.,

$$F_1(t) = \int_{-\infty}^{\infty} h_1(\tau)e^{j2\pi\Delta f\tau}h_2(\tau-(t-MT))d\tau,$$

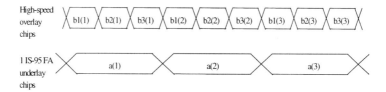

Figure 3. Timing diagram: relationship between the Walsh chips from the high speed data overlay (b1,b2,b3) and an IS-95 underlay signal

where $\Delta f = f_1 - f_2$. Then the convolution of $[h_1(-iT)e^{2\pi\Delta ft}] * h_2(MT-t)$ is equal to $e^{j2\pi\Delta f(iT)}F_1(t-iT)$. Thus, the wider band signal filtered by the narrower band filter is equal to

$$g_1(t) = \sum_{i=-\infty}^{\infty} a_i e^{j2\pi\Delta f(iT)} F_1(t-iT)$$

Sampling this signal at $(l+1)MT$ for an integer l, we obtain

$$g_1((l+1)MT) = e^{j2\pi\Delta flMT} \sum_{i=-\infty}^{\infty} a_{lM+u}e^{j2\pi\Delta f(uT)}F_1(MT-uT)$$

Assuming $F_1(iT)$ is negligible for i<0 or i>M. Note that this condition can be met if both $h_1(t)$ and $h_2(t)$ are strictly time-limited respectively to [0,T] and [0,MT]. Based on this assumption, the right hand side of above equation reduces to

$$\sum_{i=-\infty}^{\infty} a_i e^{j2\pi\Delta f(iT)} F_1((l+1)MT-iT)$$

The interference of the wider band signal to the narrower band signal is determined by the samples at time instants $g_1(iMT)$. For a transmitted narrower band sequence $(b_0,b_1,...,b_{N-1})$, it is desirable to make the correlation

$$\sum_{l=0}^{N-1} b_l g_1((l+1)MT)$$

equal to zero in order to minimize interference. Substituting the formula for $g_1((l+1)T)$ into the above equation, it yields

$$\sum_{u=0}^{M-1} e^{j2\pi\Delta fuT} F_1(MT-uT) \sum_{l=0}^{N-1} b_l a_{lM+u}e^{j2\pi\Delta flMT}$$

Therefore, orthogonality can be achieved as long as

$$\sum_{l=0}^{N-1} b_l a_{lM+u} e^{j2\pi\Delta flMT} = 0$$

for all $0 \le u < M$. We can similarly show that the condition for achieving orthogonality for the wider band receiver is

$$\sum_{l=0}^{N-1} b_l a_{lM+u} e^{-j2\pi\Delta flMT} = 0$$

If the difference of carrier frequencies is equal to an even integer multiple of $1/MT$. The orthogonality condition becomes

$$\sum_{l=0}^{N-1} b_l a_{lM+u} (-1)^l = 0$$

If the difference of carrier frequencies is equal to an odd integer multiple of $1/(2MT)$, the orthogonality condition becomes

$$\sum_{l=0}^{N-1} b_l a_{lM+u} = 0$$

Therefore, when properly designed, the overlaid direct-sequence spreading can maintain the forward link orthogonality with the underlaid IS-95 signal.

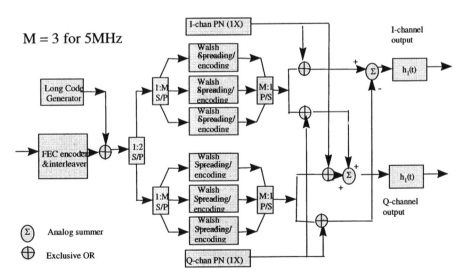

Figure 4. Functional Block Diagram for Forward Link Orthogonal Direct Sequence Spreading.

Figure 4 illustrates the design for M=3. As in the two other approaches discussed previously, the high-speed data is first encoded by a forward error correcting (FEC) code and interleaver. The coded bits are then masked by a pseudo random number (PN) sequence, which can be the same as the long code in IS-95 system. The resultant randomized bits are then demultiplexed into M separate data streams, where M is the number of IS-95 underlay frequencies to be accommodated in the overlay spectrum. When M is odd, each of the M data streams are then encoded and spread by the Walsh function and multiplexed back on a chip by chip basis. For even M, the M data streams are encoded and spread by a sequence obtained by alternating a Walsh code. These chips are then further masked by the same in-phase and quadrature PN sequence used by the IS-95 underlay at the IS-95 chip rate, and modulated accordingly. As illustrated by the resultant chips can be seen as a time multiplexed version of M IS-95 spreading codes, each generated by the same (or separate) Walsh code which is orthogonal to the underlay IS-95 Walsh codes.

The frequency separations between the IS-95 carriers are integer multiple of its chip rate. The orthogonality condition holds for both odd and even M. For odd M, the center frequency for the wideband overlay carrier is the same as the center IS-95 carrier for. For even M, the carrier frequency of the underlaid IS-95 is at multiple of $1/(2MT)$ from the carrier frequency of the overlaid carrier frequency. Figure 5 illustrates the case for M=2,3.

Figure 5. Spectral relationships between the high-speed overlay and IS-95 channels.

Even though the carriers are not spaced exactly integer multiple of the chip rate in the current IS-95 standards, it can be demonstrated so that the residual error signal generated by the small frequency offset is negligible.

It should be mentioned that this design can be considered as the time-domain dual of the multi-carrier design, the receiver can take full advantages of the true wideband signal to optimize the performance in a multipath environment. It also possesses an inherent multi-chip multipath fade resistance when the delay spread is less than the duration of one IS-95 chip, or 0.8 μsec. Also, in this approach all the multiplexing and demultiplexing is done in the time domain, which lend themselves to digital implementation, whereas the multi-carrier multiplexing is done in the frequency domain, and tends to be more difficult to implement.

Figure 6 shows the performance of the proposed direct spread forward link with the overly capability for a 3X system without underlay signals. The performance of conventional direct spreading method gives precise the same results when simulated under the identical assumptions. This shows that modulation scheme with the overlay capability will not compromise performance when deployed in a new spectrum. Along with the analysis presented above, we demonstrated that the orthogonal direct sequence overlay forward link design combines the best properties of both direct sequence spreading and multi-carrier approaches in the cdma2000 standard.

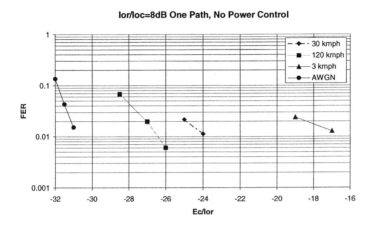

Figure 6 . Performance of the proposed forward link.

2.3   Extension to 6X, 9X and beyond

Obviously, both direct sequence spreading and multi-carrier approaches can be extended to even wider bandwidth. In the case of direct spreading, it can be done by simply increasing the clock rate to the highest multiple of the IS-95 chip rate so that the signal still maintains its desired R.F. characteristics. Similarly for the multi-carrier, additional carriers can be added until they nearly fill the band. However, it becomes increasingly more difficult to resolve multipath at chip rates higher than 3X from implementation point of view for the direct sequence spreading approach. Therefore, the ability of resolving multipath is best tapped in the 3X configuration. Similarly, for the multi-carrier approach, it will be difficult for the subscriber units to add more and more rake receivers. To extend to 6X and 9X bandwidths, an approach to use the 3X direct sequence as the basic building blocks is proposed. Multiple 3X carriers are used to fill the band, as illustrated in Figure 7. This approach can avoid most of the implementation difficulties that either direct spread or multi-carrier alone will not be able to solve easily.

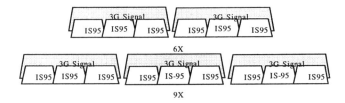

Figure 7. Extending to 6X and 9X using the 3X direct sequence spreading

3   ACCESS CHANNEL

IS-95 uses slotted ALOHA for initial access. For voice applications, access channel efficiency is of relative small impact on the system capacity since access time is typically much shorter than the duration of a call. Slotted ALOHA provides a satisfactory solution for IS-95 system. However, this is no longer true for multi-media applications. Packet data transmissions are typically very bursty and of low duty factor. For very short packets, it may be more efficient to send the packet directly into the access channel or 'access-like channels'. Furthermore, it is often not capacity efficient to maintain power control and synchronisation loops during periods of inactivity. However, if power control and synchronisation loops are turned off, mobile has to perform 'access' type of procedures to get back to resume transmissions. When a separate access attempt is needed, a packet reservation multiple access (PRMA) scheme becomes a natural approach to handle this process.

With PRMA, when a user terminal needs to transmit a data packet, it first transmits a reservation request through the access channel; and then bursts out its data at the scheduled time after the request was granted. As transmission from different high-speed data users are separated by time, mutual interference among users in the same sector or cell is eliminated. The traffic channel still utilises the spread-spectrum technology so that other cell interference is minimised. The overall link capacity is maximized. Users transmit at separate time can use the same code channel, conserving the number of Walsh codes needed. The PRMA procedure does require a reservation from the mobile terminal to be transmitted and granted, which introduces some delay. This delay is generally negligible for large packets. For shorter packets, this reservation step can be bypassed and delay eliminated with the Inhibit Sense Multiple Access (ISMA).

ISMA is a random access protocol for Frequency Division Duplexed (FDD) wireless channel, modified from the well-known Carrier Sensed Multiple Access (CSMA) protocol used by the Ethernet. Instead of having all mobiles sense the presence or absence of a carrier from other mobile terminals, the basestation performs this function and transmits an "BUSY/IDLE (B/I) bit" periodically to indicate whether or not the channel is busy. All mobile terminals check the B/I bit from the basestation and transmit data on the random access channel only if the channel is idle, thus achieving the same objective as CSMA. Like in CSMA, contention resolution is based on a p-persistent concept. Once a mobile finds the B/I status idle, it sends its packet with a probability p. To minimize probability of contention, the period of B/I bit is kept very small compared to the duration of the access packet in the reverse link. It could be as small as the time it takes to detect the preamble of the access packet and setting/resetting of the B/I bit in the forward link control channel. The ISMA protocol has been implemented in Cellular Digital Packet Data (CDPD) and proved to be effective. The ISMA protocol can be further improved in a CDMA environment by taking advantage of the capture capability of the CDMA

Note that for the access channel, the same spreading and scrambling code is shared by all the users in any given cell. Therefore collision may occur when two users attempt to access the system. In general, when two packets from two different users are transmitted at the same time, one of the packets may be demodulated correctly if the received power from this packet is much greater than the other packet. In CDMA reverse link, the mobile terminals are generally not synchronised to the chip level. Both packets can theoretically be captured, if their transmit timing is separated by more than one chip time. However, because a single receiver is typically assigned to one particular code channel, only the first packet to arrive is captured. Packets transmitted by other mobile terminals are essentially interference to the active users, and would be suppressed.

Cdma2000 adopted this approach in which the captured user is notified by sending a "Capture Message" from the basestation over a forward common control channel. This control channel is monitored by all accessing mobiles. Mobile terminals accessing the same random access channel cease transmission as soon as mismatch is detected. This not only reduces interference but also allows a forward link common channel to power control the captured access signal, further improves the system efficiency. Capture Message can also provide other information to perform scheduled packet data transfer or redirect data packets to other code channels. Along or after the capture message, the basestation can also instruct the mobile to change its data rate and transmit power.

To avoid incorrectly combining signals from two different mobile terminals on the same code channel, the basestation would generally be unable to activate the multipath rake receiver in a common access channel. Using ISMA with capture, transmissions by other terminals in the channel are turned off after the receipt of the capture message. The rest of data burst is free from interference. The basestation can then take advantage of maximum ratio combining capability of the rake receiver to improve the link performance.

To identify the mobile terminal accessing the system, a shortened or hashed version of mobile identification number (MIN) or electronic series number (ESN) could be included in the header of each packet. The turn-around time of the capture message has a significant impact on performance. One possible approach to facilitate faster capture is to exploit the characteristics of the CDMA signal.

The mobile's selected PN-offset of the access long code in the preamble can be used as a short address in capture message. This approach takes advantage that the mobiles, attempting access on each access long code, use some level of PN randomization which facilitates capture capability in the receiver. Each mobile may select this offset randomly or use a MIN/ESN based hashing procedure. In either case the information about the offset used by the mobile is available to the base right after detecting the mobile's preamble and comparing it with its own PN-code phase. In this approach, the number of possible PN offsets available on each access long code should be large enough to provide sufficient addressing capacity without ambiguity. Depending on the cell size and access pattern, a number between 8-12 bit may be appropriate. (The resolution should be sufficient to account for propagation delay variations.) Note that in this case there is no need for sending a short address explicitly and the decoding process is eliminated. The basestation will transmit the detected offset as part of the capture message in the forward control channel. Mobiles monitoring forward control channel will compare the offset value received to the one used for access to determine capture success.

In summary, ISMA performance superiority is accomplished by the following factors. First, the collision probability and interference created by unsuccessful access attempts is reduced. This results in better throughput and delay performance for each channel and improved system and cell capacity. Second, other mobiles whose transmission collides with the successful mobile are forced to terminate their transmission immediately upon receipt of the capture message. Thereby, a major part of each packet is free from interference generated by contention. Third, power control can be exercised on the captured data burst. Fourth, better link performance is achieved by using multipath combining of the rake receiver, instead of path selection, right after capture is completed.

Figure 8 and Figure 9 are comparisons of performance of the ISMA with capture and conventional Slotted-Aloha access technique in terms of the throughput vs. offered load and delay vs. offered load[5]. The performance is derived based on a reverse link which has 141 simultaneous voice users and 100 data terminals operating at transmission probability of 0.004. A packet is in error when access traffic plus the background traffic causes total interference to exceed a threshold. In this case the entire packet is re-transmitted. The following parameters were used in the analysis and simulation. Spreading rate is 3X (3.6864 Mchips/sec). Eb/No of 4.4 dB with a standard deviation of 1.5 dB, and voice activity of 0.4 are used for voice users. Eb/No of 7.5 dB is used for data traffic with a standard deviation of 4.0 dB to account for lack of power control. The data rate on voice and data traffic is 9.6 Kbps. In the figures, identifiers 1,2 and 4 represent packet length of one, two or four frames. Also the curve marked 'Sim' represents simulation results of four-frame packets while all others represent analytical results. In the simulation, the capture message reception delay is included, which causes the simulated data to be slightly worse than the analytical results which assume no delay. Also the BUSY/IDLE bit are transmitted repeatedly at every quarter frame interval and received with a quarter frame delay. Evidently, the throughput of PR-ISMA is significantly better than the slotted Aloha protocol for moderate to high values of the offered load, from both throughput and delay performance point of view.

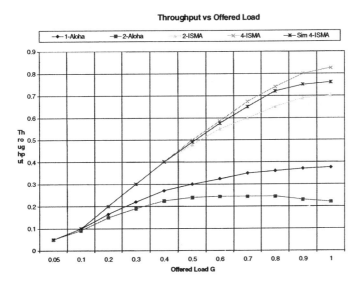

Figure 8. Comparison between ISMA with capture and conventional slotted-aloha: throughput vs. offered load.

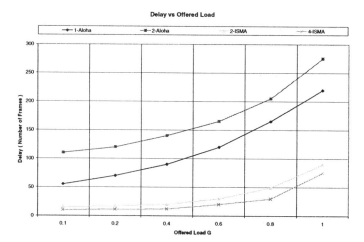

Figure 9. Comparison between ISMA with capture and conventional slotted aloha: delay vs. offered load.

4 TURBO CODES

Since the capacity of a CDMA system is highly dependent on the operating E_b/N_o, turbo codes are very appealing because of its capability of achieving near Shannon limit performance [6,7]. Turbo codes are not suitable for voice and low-speed data applications because it requires a relatively long interleaver and deinterleaver, which introduce delay. But, we found it possible to reduce the interleaver size to as low as 500 information bits while still gaining significant coding improvements over conventional convolutional codes of similar complexity. For high-speed data, the resulting interleaving delay appears quite acceptable. The Turbo encoder employs two systematic recursive convolutional codes connected in parallel, with an interleaver (the "Turbo interleaver") preceding the second recursive convolutional encoder. The two-recursive convolutional codes are called the constituent codes of the Turbo code. While it was not obvious, turbo codes are particularly suited for the high-speed CDMA data service applications. Fast power control in a CDMA system serves to mitigate Raleigh fading and make the mobile radio channel closer to an additive white Gaussian noise (AWGN) channel, the type of channel for which Turbo codes have been demonstrated to provide performance near the Shannon limit. Increased forward error correction (FEC) performance at the expense of a modest increase in decoder complexity is an attractive trade-off for CDMA systems, since every E_b/N_o gain translates directly into increased system capacity. At higher speed, the longer interleaver frame sizes (and associated delay) required by Turbo codes to achieve their exceptional performance appear to be a very acceptable.

Extensive trade-offs regarding generator polynomials and puncturing patterns to identify universal Turbo codes offering uniformly optimal or near-optimal performance across code rates and interleaver depths have been performed [8][9]. The specific Turbo codes accepted into the cdma2000 proposal submitted by the U.S. Telecommunication Industry Association (TIA) are described below. For all high-speed data channels, the Turbo codes of rate ½, 1/3 and ¼ share a common constituent code specified by the following generator polynomials:

$$d(D) = 1 + D^2 + D^3$$
$$n_1(D) = 1 + D + D^3$$
$$n_2(D) = 1 + D + D^2 + D^3$$

For the rate ½ Turbo code, the parity bits $n_2(D)$ and every other parity bit $n_1(D)$ from both constituent encoders are punctured. For the rate 1/3 Turbo code, all the parity bits $n_2(D)$ are punctured. For the rate ¼ Turbo code, the parity bits $n_1(D)$ from one of the encoders and $n_2(D)$ from the other encoder are alternately punctured. Three flush bits are used per constituent encoder to force the constituent encoder to the all-zero state. This is accomplished by setting the flush bits equal to the corresponding feedback values of the encoder. Thus, the flush bits for the two constituent encoders are independent, and a total of six flush bits are required.

The frame error rate performance of this Turbo code in the AWGN versus standard constraint length K=9 convolutional codes [10] are shown in Figure 10. Eight iterations were used for the turbo code. The figure demonstrates an interesting difference between Turbo codes and convolutional codes with respect to frame error rate statistics. As the information frame size increases from 512 to 3072, the frame error rate for Turbo codes decreases sharply, while that of the convolutional code increases significantly. Ignoring edge effects, the bit error rate for convolutional codes is essentially uniform across the frame and is a constant independent of frame size. Thus, for a given E_b/N_o, the expected number of bit errors increases with frame size, and the frame error rate worsens. For Turbo codes, however, the power of the code increases significantly as the frame size increases. This increase in power is sufficient to overcome the burden of protecting a larger frame of data.

Coding gain on fading channels without power control is typically reduced. The use of fast power control can restore the performance advantage to that achieved on the AWGN channel. The fading channel model that we implement is the Vehicular Test Environment Channel A (V-A) specified by ITU. In this model, there are six delayed Rayleigh fading paths of diminishing power. The simulation results are shown in Figure 11. In the simulation, other-cell interference is kept 8 dB below the intended base station. The data rate is 76.8 kbps, and the frame duration is 20 ms.

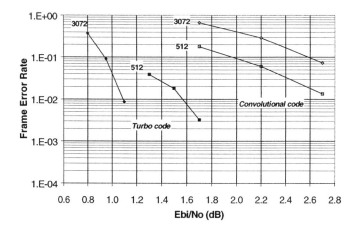

Figure 10. FER Performance for Rate ½, in AWGN channel.

Each power control bit is sent every 1.25 ms with a delay of 0.715 ms and an error rate of 4%. To simulate real systems accurately, interference from other users is obtained by actually encoding their bits and sending them over the channel. A pulse shaping filter at the transmitter and a matched filter at the receiver is also implemented. As in the AWGN, the number of turbo decoding iterations is eight. While only the 30 kmphr case was shown, Turbo codes provide substantial gain over conventional convolutional codes in all speeds, with more dramatic gain at higher vehicle speeds.

5   SUMMARY

We have briefly described the cdma2000 forward and reverse link design and described an alternative approach for the forward link for the 3X mode that retains the advantage of both the direct sequence spreading and multi-carrier approaches currently included in the cdma2000 standard. We introduced a reverse link media access scheme that combines PRMA for large packets and ISMA with capture message for short packets to significantly improve the efficiency of channel utilisation for high-speed data transmission. The underlying physical layer channel capacity is further increased by an additional 50-60 percent using properly selected Turbo code family with 8-state constituent codes. By combining an efficient reverse link multiple access architecture, and optimised FEC, we expect to provide third generation wireless systems with an unprecedented level of performance in high-speed packet data delivery and improved voice capacity with minimum impact on implementation complexity.

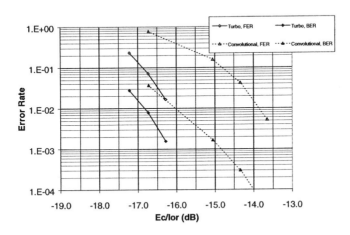

Figure 11. Ior/Ioc=8 dB, vehicle speed=30 kmph,V-A power controlled.

REFERENCES

[1] "The ETSI UMTS Terrestrial Radio Access (UTRA) ITU-R RTT Candidate Submission," Proposal Submitted to the International Telecommunications Union (ITU), by ETSI-SMG2.

[2] "The cdma2000 ITU-R RTT Candidate Submission," Proposal Submitted to the International Telecommunications Union (ITU), by TIA/TR45.5.

[3] L. Lee, K. Karimullah, F.W. Sun, and M. Eroz, "Third Generation Wireless Technologies - Expectations and Realities," (Invited Paper) Proceedings, the 9th IEEE International Symposium on Personal, Indoor, and Mobile Radio Communications (PRMIC'98), Boston, Mass, USA, September, 8-11, pp.

[4] L. Lee, M. Eroz, R. Hammons, K. Karimullah, and F.W. Sun, "Third Generation Mobile Telephone Systems and Turbo Codes," (Invited Paper) Proceedings, the 3rd International Symposium on Multi-Dimensional Mobile Communications (MDMC'98), Menlo Park, CA, USA, September 21-22, 1998, pp.

[5] W. Chan, E. Geraniotis, K. Etemad, "ISMA-based Packet Reservation Scheme for 3rd Generation CDMA Networks," paper submitted to *IEEE Journal for Special Areas of Communications,* September 1998.

[6] C. Berrou, A. Galvieux, and P. Thitimajshima, "Near Shannon Limit Error Correcting Coding and Decoding: Turbo Codes," *Proceedings of ICC* (Geneva, Switzerland), May 1993.

[7] S. Benedetto and G. Montorsi, "Design of Parallel Concatenated Convolutional Codes," *IEEE Transactions on Communications*, May 1996, vol. COM-44, pp. 591-600.

[8] M. Eroz, A.R. Hammons Jr, and L. Lee, "Recommendation for Universal Turbo Codes for CDMA 3G Data", Contribution # TR45.5.4 Turbo/98.01.26.02.

[9] M. Eroz, A.R. Hammons Jr, and L. Lee, "Recommendation for Rate ¼ Turbo Code for CDMA 3G Data", Contribution # TR45.5.4 Turbo/98.02.18.02.

[10] S. Lin and D.J. Costello, *Error Control Coding: Fundamentals and Applications*, Prentice-Hall, 1983.

Chapter 3

Perspectives on 3G System Development

TERO OJANPERÄ
Nokia Telecommunications

Abstract: This chapter presents different perspectives on the development of third generation (3G) mobile communications. Historical, technical, market, regulatory, and future perspectives are presented.

1 INTRODUCTION

Emerging requirements for higher rate data services and better spectrum efficiency are the main drivers identified for the third generation mobile radio systems. At ITU (International Telecommunication Union) level, third generation is known as IMT-2000 (International Mobile Telecommunications by 2000)[4] [1]. In Europe, the third-generation of wireless personal communications is known as Universal Mobile Telecommunication Systems (UMTS). The main objectives for the IMT-2000 air interface can be summarized as

- Full coverage and mobility for 144 kbps, preferably 384 kbps;
- Limited coverage and mobility for 2 Mbps;
- High spectrum efficiency compared to existing systems;
- High flexibility to introduce new services.

2 HISTORICAL PERSPECTIVE

The development of IMT-2000 started in ITU already in 1986. It began as an initiative to provide people with a hand-held device facilitating worldwide roaming, driven by a common frequency allocation. In Europe, the driving force behind the third generation development has been the European Commission funded research programs RACE (Research of Advanced Communication Technologies in Europe) from 1988 to 1995 and ACTS (Advanced Communication Technologies and Services) from 1995 to 1998 [2]-[4]. In ETSI, UMTS standardization started in 1990 when the subtechnical committee SMG5 was established. Japan initiated their third generation effort in the beginning of the 1990s. The IMT-2000 Study Committee in ARIB was established in April 1993 to coordinate the Japanese research and development activities for IMT-2000. Korea also started its IMT-2000 activities in the beginning of the 1990s. In the United States, third generation gained wider attention in 1997.

The transition from first generation to second generation systems happened in different ways in different countries. The U.S. moved from a single analog system, AMPS, into multiple digital systems, TDMA, GSM, and IS-95. In contrast, Europe moved from multiple analog systems into one digital standard, GSM, driven by operators.

Since the world is changing so fast, it is hard to use the lessons learned from the past in the future. However, as the infrastructure still requires

---

[4] Previously IMT-2000 has been called FPLMTS (Future Public Land Mobile Telecommunication Systems)

extensive investments, operators need be sure that the standard they adopt will have wide support to make their business viable. Therefore, well-standardized, widely recognized systems are essential for third generation success.

CDMA is the youngest second generation mobile radio technology. Furthermore, since wideband CDMA seems to gain the most momentum as a third generation radio technology [5], we review its history shortly. The origins of spread spectrum are in the military field and navigation systems. Techniques developed to counteract intentional jamming have also been proved suitable for communication through dispersive channels in cellular applications. In the following we highlight the milestones for CDMA development starting from the 1950s after the invention of the Shannon theorem [6]. An extensive overview of spread spectrum history is given in [7][8].

In 1949, John Pierce wrote a technical memorandum where he described a multiplexing system in which a common medium carries coded signals that need not to be synchronized. This system can be classified as a time hopping spread spectrum multiple access system [7]. Claude Shannon and Robert Pierce introduced the basic ideas of CDMA in 1949 by describing the interference averaging effect and the graceful degradation of CDMA [9]. In 1950, De Rosa-Rogoff proposed a direct sequence spread spectrum system, and introduced the processing gain equation and noise multiplexing idea [7]. In 1956, Price and Green filed for the anti-multipath "RAKE" patent [7]. Signals arriving over different propagation paths can be resolved by a wideband spread spectrum signal and combined by the Rake receiver. The near-far problem (i.e., a high interference overwhelming a weaker spread spectrum signal) was first mentioned in 1961 by Magnuski [7]. For cellular application, Cooper and Nettleton suggested spread spectrum in 1978 [10]. During the 1980s DS-CDMA techniques were investigated and this led to the commercialization of cellular spread spectrum communications in the form of the narrowband CDMA IS-95 standard in July 1993. Commercial operation of IS-95 systems started in 1996. Multiuser detection (MUD) has been subject to extensive research since 1986 when Verdu formulated an optimum multiuser detection for the AWGN channel, maximum likelihood sequence estimator (MLSE) [11]. During the 1990s wideband CDMA techniques with a bandwidth of 5 MHz or more have been studied intensively throughout the world and several trial systems have been built and tested [12]. These include FRAMES FMA2 (FRAMES Multiple Access) in Europe, Core-A in Japan, the European/Japanese harmonized WCDMA scheme, cdma2000 in the U.S., and the TTA I and TTA II (Telecommunication Technology Association) schemes in Korea. Introduction of third generation wireless communication systems using wideband CDMA is expected by 2001/2002.

Based on the above description, the CDMA era is divided into three periods: (a) pioneer CDMA era, (b) narrowband CDMA era, and (c) wideband CDMA era, as shown in Table 1 [5].

2.1 Radio Access

After extensive research, the main regional standards bodies have already defined candidate technologies for IMT-2000 [12]-[14]. The main parameters of the third generation air interface proposals are presented in Table 2. The third generation air interface standardization for the schemes based on CDMA seem to focus on two main types of wideband CDMA: network asynchronous and network synchronous. In network asynchronous schemes the base stations are not synchronized, while in network synchronous schemes the base stations are synchronized to each other within a few microseconds. In all wideband CDMA schemes, coherent detection is performed using either time or code multiplexed pilot. In the uplink each user has its own pilot for coherent detection (user dedicated pilot). In the downlink a common pilot for all users is available. In addition, WCDMA and cdma2000 schemes have a user/group specific pilot in the downlink.

Table 1. CDMA Era.

(a) Pioneer Era

1949	John Pierce: time hopping spread spectrum
1949	Claude Shannon and Robert Pierce: basic ideas of CDMA
1950	De Rosa-Rogoff: direct sequence spread spectrum
1956	Price and Green: anti-multipath "RAKE" patent
1961	Magnuski: near-far problem
1970s	Several development for military field and navigation systems

(b) Narrowband CDMA Era

1978	Cooper and Nettleton: Cellular application of spread spectrum
1980s	Investigation of narrowband CDMA techniques for cellular applications
1986	Formulation of optimum multiuser detection by Verdu
1993	IS-95 standard

(c) Wideband CDMA Era

1995-	Europe	: FRAMES FMA2	
	Japan	: Core-A	WCDMA
	USA	: cdma2000	
	Korea	: TTA I, TTA II	
2000s	Commercialization of wideband CDMA systems		

3 TECHNICAL PERSPECTIVE

The fast third generation development during recent years has been due to Japanese initiative. In the beginning of 1997, Association for Radio Industry and Business (ARIB), a standardization body responsible for Japan's radio standardization, decided to proceed with detailed standardization of wideband CDMA. The technology push from Japan accelerated standardization in Europe and the US. During 1997 joint parameters for Japanese and European wideband CDMA proposals were agreed. The air interface is now commonly referred as WCDMA. In January 1998, strong support behind wideband CDMA led to the selection of WCDMA as the UMTS terrestrial access (UTRA) scheme for paired (FDD) and TD/CDMA for unpaired (TDD) frequency bands in ETSI. The selection of the UTRA concept was also backed by Asian and American GSM operators. In the US, the TIA (Telecommunications Industry Association) TR45.5 committee, responsible for IS-95 standardization, adopted a framework for wideband CDMA backward compatible to IS-95, cdma2000, in December 1997. TR45.3, responsible for IS-136 standardization, adopted TDMA based third generation proposal, UWC-136 (Universal Wireless Communications), based on the recommendation from the UWCC (Universal Wireless Communications Consortia) in February 1998. The 200 kHz carrier of the UWC-136 proposal is harmonized with the EDGE (Enhanced Data Rates for GSM Evolution) concept of GSM. The 200 kHz carrier is especially suitable for operators with limited third generation spectrum. Korea has proposed two wideband CDMA technologies, one being similar to WCDMA and the other similar to cdma2000. These have been harmonized with respective technologies.

In 1998 the different proposals have been submitted into the ITU RTT selection process, which is aiming to harmonize the different proposals. In March 1999, a high level grouping into CDMA and TDMA was defined leading into the definition of key characteristics of IMT-2000 radio access.

Table 2. Main parameters of third generation air interface proposal

	WCDMA	cdma2000	UWC-136/136HS
Multiple Access	CDMA	CDMA	TDMA
Chip rate/carrier bit rate	(1.024)/4.096/8.1 92/ 16.384 Mcps	1.2288/3.6864/7.3728/ 11.0593/14.7456 Mcps for direct spread n×1.2288 Mcps (n=1,3,6,9,12) for multicarrier	200 kHz carrier: - 812.5 kbps (8-PSK) - 270.8 kbit/s (GMSK)
Carrier spacing	5, 10, 20 MHz	1.25, 5, 10, 15, 20 MHz	200 kHz, 1.6 MHz
Frame length	10 ms	20 ms	4.615 ms, 8 slots/frame (200 kHz) 16/64 slots/frame, (1.6 MHz)
Inter base station synchronization	Asynchronous	Synchronous	Asynchronous
Coherent detection	User dedicated time multiplexed pilot (downlink and uplink), Common pilot in downlink	DL: Common continuous pilot channel and auxiliary pilot UL: Pilot time multiplexed with PC and EIB	Training sequence

3.1 Core Network

While wideband CDMA has been identified as a mainstream technology for third generation radio, GSM core network is gaining similar momentum in the network side. Currently, an extensive work to specify the GSM based third generation core network including WCDMA based radio access is on-going into the third generation partnership project (3GPP), with a wide participation from different regions. 3GPP was established in December 1998. Figure 1 describes the evolution of the GSM platform. New capabilities are developed that facilitate timely adoption of appropriate radio technology based on operator and customer needs.

It should also be noted that all emerging satellite personal communication networks (S-PCN) Iridium, Globalstar, and ICO utilize the GSM core network. Thus also the IMT-2000 satellite component can be readily integrated into the evolved GSM platform.

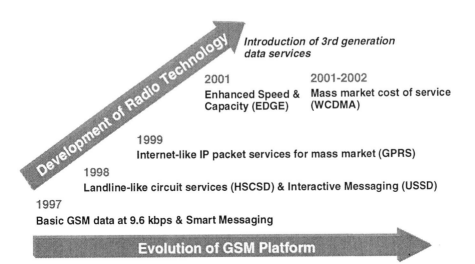

Figure 1. Evolution of GSM platform

4 MARKET PERSPECTIVE

Most important success factor for third generation wireless systems will be the existence of mass market. The basis for the third generation market will be the existing second generation market. Second generation systems will offer speech, low, and medium bit rate data services. Therefore, the third generation market, especially in the beginning of its life cycle, will concentrate into high bit rate multimedia and packet data services, complementing the services offered by second generation systems such as GSM. Possible spectrum congestion in second generation networks can be relieved by migrating heavy data users into third generation systems.

The development of the third generation market depends on several factors including coverage, tariffing, lifestyle changes, and terminal offerings. Table 3 presents world subscriber growth for different compounded annual growth percentages for 1998-2001. According to [15] growth of 40 to 45 percent is considered most plausible. Between 1996 and 1997, the annual growth worldwide averaged 51.6 percent and since 1990 compounded annual growth has averaged 50.9 percent.

Table 3. Forecast world subscriber growth (millions). (Source: Hershel Shosteck Associates, Ltd [15])

	Assumed Compounded Annual Growth				
	25%	30%	35%	40%	45%
1998	258	268	279	289	299
1990	322	349	376	404	434
2000	403	453	508	566	629
2001	504	589	685	793	912

The market analysis group in the UMTS Forum has conducted a study that analyzed four different market scenarios for mobile multimedia market [16]. These scenarios give an indication of the factors that will determine the third generation market development. Table 4 shows the number of mobile users and multimedia users year 2005 for the different scenarios.

Table 4. Number of mobile and multimedia users in Europe 2005 [16].

Scenario	Mobile users (penetration)	Multimedia users
1. Slow Evolution	82 M (22 %)	7.5 M
2. Business Centric	82 M (22 %)	9 M
3. Sophisticated Mass Market	123 M (35 %)	19 M
4. Commoditised Mass Market	140 M (40 %)	27 M

In Scenario 1 (Slow Evolution), mobile multimedia development is slow, characterized by limited applications, and high service and terminal prices due to unsuccessful liberalization and fragmented standards.

In Scenario 2 (Business Centric), mobile multimedia takes off in the business sector but not in the consumer sector, as there is a lack of innovation in consumer applications.

In Scenario 3 (Sophisticated Mass Market), mass market for mobile multimedia has emerged. Terminals and applications have large set of features and can be customized for personal needs. However, user interfaces are still rather complicated being suitable for IT-literate users.

In Scenario 4 (Commoditised Mass Market), a real mobile multimedia mass market has emerged fast and comprises both business and consumer users. Main difference to the previous scenario is development of simple user interfaces, and cheaper spectrum.

5 REGULATORY PERSPECTIVE

Fundamentally, regulatory matters are national issues. However, in some cases regional and global harmonization of regulations and policy issues is desirable [17]. Areas where common regulations might be required are licensing, services, provisioning, interconnection, infrastructure, frequencies, numbering, security and policies [17]. With regard to a radio access system, the most important regulatory areas are spectrum coordination and licensing.

The process of obtaining frequencies can be divided into three phases: identification of spectrum, allocating the spectrum for specific purposes, and licensing the spectrum. Since the identified spectrum is usually already used for some other purposes, it needs to be cleaned from existing users before it can be used for the new purpose. Therefore, a proactive long term planning is required to ensure the availability of spectrum when required.

ITU plays an important role in spectrum regulation. The Radio Regulations (RR) of ITU are updated in WARCs (World Administrative Radio Conference). The next WARC is scheduled for 1999. The national regulators are not bounded to follow the ITU guidelines for spectrum allocation. However, the ITU RRs form a tool to encourage the national regulators to do that, in order to achieve global harmonization of spectrum [17]. The IMT-2000 spectrum was identified in the year 1992 by WARC92 as a result of ITU studies on IMT-2000. These studies indicated that the minimum spectrum for IMT-2000 should be 230 MHz.

In general, Europe follows ITU recommendations for spectrum issues. The European wide harmonization is carried out by the Conference of European Post and Telecommunications (CEPT). The European Commission can issue a directive to create harmonized frequency allocations for specific technologies. Examples of such directives are GSM in 1987 and DECT 1991. Recently, the Commission has published its proposed UMTS Decision, which sets in place time-scales and actions for national licenses and spectrum harmonization by the year 2000 [18]. The European Radiocommunications Committee (ERC) of CEPT makes decisions, which usually form the basis for the harmonized spectrum designations. CEPT has designated most of the IMT-2000 spectrum for UMTS with the adoption of the ERC Decision on UMTS, which identifies a total of 155 MHz of spectrum for terrestrial UMTS services with an additional 60 MHz set aside for UMTS satellite services [19].

6 FUTURE PERSPECTIVE

"Regardless of whether the market reaches 400 million, 500 million, or more subscribers, the greater risk to manufacturers and carriers is to delay investing in capacity, component sources, and sales/distribution infrastructure. A conservative position of delaying investment until the size of market is certain, virtually assures loss of market position and profits" [15].

Based on that statement and the above presented market perspective we can conclude that the success of third generation depends on the investment into second generation systems and their evolution. The third generation multimedia market will develop with the help of second generation systems. Moreover, timely development of new third generation standards will allow seamless migration into third generation when user requirements cannot be met with existing systems. Winners will select their system, not based on single-minded focus on radio, but based on network architecture that will provide the widest selection of features and a future-proof platform for new radio access technologies.

REFERENCES

[1] Special Issue on IMT-2000: Standards Efforts of the ITU, *IEEE Pers. Commun.*, Vol. 4, No. 4, August 1997.

[2] "The European Path Towards UMTS", *IEEE Pers. Commun.*, Special Issue, Vol.2, No. 1, Feb. 1995.

[3] J. S. DaSilva, B. Arroyo, B. Barani, and D. Ikonomou, "European Third-Generation Mobile Systems", *IEEE Communications Magazine*, Vol. 34, No.10, Oct. 1996 pp. 68-83.

[4] R. Prasad, J. S. DaSilva, and B. Arroyo, "ACTS Mobile Programme in Europe", *Guest Editorial in IEEE Communications Magazine*, Vol. 36, No. 2, Feb. 1998.

[5] T. Ojanperä and R. Prasad, *Wideband CDMA for Third Generation Mobile Communications*, Artech House, 1998.

[6] C. E. Shannon, "A Mathematical Theory of Communication", Bell System Technical Journal, Vol. 27, 1948, pp. 379-423 and 623-656.

[7] R. A. Scholtz, "The Evolution of Spread-Spectrum Multiple-Access Communications", in *Code Division Multiple Access Communications* (S. G. Glisic and P. A. Leppänen, eds.), Kluwer Academic Publishers, 1995.

[8] Simon, M. K., J. K. Omura, R. A. Scholtz, and B. K. Levitt, *Spread-Spectrum Communications*, McGraw Hill, 1994.

[9] "A conversation with Claude Shannon", *IEEE Communications Magazine*, Vol. 22, No. 5, May 1984, pp.123-126.

[10] G. R. Cooper and R. W. Nettleton, "A spread-spectrum techniques for high-capacity mobile communications", *IEEE Trans. Veh. Tech.*, Vol. 27, No. 4, Nov. 1978, pp. 264-275.

[11] S. Verdu, "Minimum Probability of Error for Asynchronous Gaussian Multiple Access", *IEEE Trans. on IT.,* Vol. IT-32, No. 1, January 1986, pp. 85-96.

[12] T. Ojanperä, "Overview of Research Activities for Third Generation Mobile Communication", in *Wireless Communications TDMA vs. CDMA* (S.G. Glisic and P.A. Leppanen, eds), Kluwer Academic Publishers, 1997, pp. 415-446.

[13] T. Ojanperä and R. Prasad, "Overview of Air Interface Multiple Access for IMT-2000/UMTS", *IEEE Communications Magazine,* 1998.

[14] T. Ojanperä and R. Prasad, "Overview of Wireless Personal Communications – European Perspective", *IEEE Personal Communications,* December 1998.

[15] H. Shosteck, "The Explosion in World Cellular Growth", The 2nd Annual UWC Global Summit, Vancouver, BC, Canada, April 15-17, 1998.

[16] The UMTS Market Aspects Group, UMTS Market Forecast Study, 1997.

[17] F. Leite, R. Engelman, S. Kodama, H. Mennenga and S. Towaij, "Regulatory Considerations Relating to IMT-2000", IEEE Personal Communications Magazine, pp. 14-19, August 1997.

[18] European Commission, "Proposal by the Commission for a Decision of the European Parliament and Council on the co-ordinated introduction of mobile and wireless communications (UMTS) in the Community", 1997.

[19] EUROPEAN RADIOCOMMUNICATIONS COMMITTEE ERC Decision of 30 June 1997 on the frequency bands for the introduction of the Universal Mobile Telecommunications System (UMTS), ERC/DEC/(97)07, can be found from ERO website http://www.ero.dk/.

Chapter 4

An Unified, Open Architecture to Deliver Mobile Multimedia

JORGE M. PEREIRA
European Commission*, DG XIII, BU-9 04/61, Rue de la Loi 200, B-1049 Brussels, Belgium
Phone: +32 2 296 1547, Fax: +32 2 296 2178; Email: Jorge.PEREIRA@bxl.dg13.cec.be

Abstract: Since 1985, European Commission-funded R&D projects have contributed significantly to the development of data capabilities in GSM. In the context of the ACTS program (1995-98), projects working towards the development of the next generations of mobile communication concepts, systems and networks, have been looking into many of the technological and service engineering issues involved in the provision of data services, both outdoors and indoors. To leverage existing investments in infrastructure, while at the same time preserving a maximum of flexibility to offer more opportunities and easier access to new entrants, a common architecture capable of accommodating the whole range of GSM/DECT/UMTS/MBS/W-LAN and still open to allow for the introduction of innovative radio access techniques, is proposed and discussed in the context of now somewhat academic discussion IP versus ATM.

* The views expressed herein are those of the author, and do not necessarily reflect the views of the European Commission. Any mention of companies or products should not be construed as an endorsement. The author is on leave of absence from Instituto Superior Técnico, Lisbon Technical University, Lisboa, Portugal.

1 INTRODUCTION

Mobile and Personal Communications are recognized as a major driving force of socio-economic progress and are crucial for fostering the European industrial competitiveness and for its sustained economic growth, as well as for its balanced social and cultural development. Their impact extends well beyond the industries directly involved, in fact enabling a totally new way of life and a wealth of new ways of working and doing business.

The emergence of GSM (Global System for Mobile Communications) as a world standard shows the potential of a concerted action at a European level, covering all areas from R&D to political and regulatory to spectrum availability. Resulting notably from extensive research in the scope of the European Commission RACE (Research into Advanced Communications for Europe – RACE I: 1985-1991; RACE II: 1991-1995) program, GSM is a demonstration of the European know-how and the basis for the European competitiveness in the world telecommunications scene.

The unprecedented growth of worldwide mobile and wireless markets, coupled with advances in communications technology and the accelerated development of services taking place in fixed networks, points now to the urgent introduction of a flexible and cost-effective Third Generation Mobile Communication System. In this context, UMTS (Universal Mobile Telecommunications System), as such system is commonly referred to in Europe, has been the subject of extensive research carried out primarily in the context of the RACE and ACTS (Advanced Communications Technologies and Services, 1995-1998) R&D programs.

MBS/W-LAN issues are also under investigation in ACTS, with MBS dating back to the RACE program. More recently, W-LAN work has taken an impetus of its own, with a clear standardization focus.

In this contribution we analyze the present GSM data offerings and their evolution towards higher data rates, trying to anticipate how data could be provided over UMTS. We also discuss the trends in Wireless Broadband data beyond UMTS, including MBS and W-LANs, and identify an underlying common architecture open enough to accommodate innovative radio interfaces. Finally, the IP versus ATM debate is analyzed in that perspective.

2 DATA OVER GSM

Short Message Service (SMS) and circuit switched data at 9.6 kbps (allowing e-mail and fax), already enjoy considerable success[5] in spite of their inherent limitations.

From this incipient start, the data capabilities of GSM are being enhanced, starting with direct connection to the ISDN network. With high-speed circuit-switched data (HSCSD) and the introduction of packet-switched data (GPRS, General Packet Radio Service), all with multi-slot capability, data over GSM will accommodate a considerable range of applications, this even if HSCSD and GPRS are not expected to be integrated any time soon. Finally, through higher level modulation, EDGE (Enhanced Data rate for GSM Evolution) will further extend the data rates at least under high link quality conditions.

2.1 SMS

SMS is an inherent capability of GSM; it operates much like two-way paging, but has the potential and is rapidly evolving into an electronic messaging system. When GSM phones with PDA-like functionality show up, SMS is expected to take off as a platform for communication applications with low data-transfer requirements. Some projections [1] in fact suggest that SMS might account for more that 10% of all cellular network revenues by 2002, representing at that time more than half of the expected revenues from all data services.

Work on SMS is ongoing in ETSI (European Telecommunications Standards Institute) SMG (Special Mobile Group), with SMS Inter-working Extensions, Second SMS Broadcast Channel and Concurrent SMS/CB and Data Transfer included in GSM 2+ Release 96. Simultaneously, new applications are being devised to make use of the SMS capabilities. With the appropriate software, SMS messages can be managed like e-mail: users can send, receive, and forward messages from the laptop to a single user or groups of users.

Value-added applications gateways are also being developed that facilitate message transfer between corporate LANs and the network

---

[5] The GSM Association announced one billion GSM text messages were carried over by global GSM operators in April 1999, and that traffic is expected to grow at a rate of 40-50% in 1999. Services available from many of the GSM networks include, besides the user generated text messages, news, sport, financial, and location-based services, as well as mobile e-commerce services like stock and share price information/alert, mobile banking, and travel/leisure booking.

operator's SMS center (SMSC). The gateway acts as an intelligent two-way multiplexing system between regular messaging protocols and the SMSC.

2.2 GSM Circuit Switched Data

By virtue of being a digital system, GSM was conceived from the start for wireless data – all the way to ISDN rates. However, only in March 1994 was data first sent over GSM from a notebook PC linked to a GSM phone via a PCMCIA card (now PC Card). Today, several vendors offer PC Cards that provide a 9.6 kbps interface to the GSM network. A communications application on the notebook/mobile computer sees the card as if it were a regular data/fax modem, enabling e-mail, file transfer, and faxing to be done with regular Windows programs. Unfortunately, all the current notebook-GSM phone interfaces are proprietary, but vendors are working their way toward a standard.

The PC Card provides a communication link to the GSM network, but the modulation/demodulation facility is part of the network infrastructure. The Mobile Switching Center (MSC) of the network interfaces the Public Switched Telephone Network (PSTN), and the modem pool of the GSM network operator pumps the data over the phone network. Over the air, data is conveyed by the GSM radio link protocol (RLP) that handles traffic from the data card to the MSC, while V.42-compliant communication takes over for the modem-to-modem part over the PSTN. What was needed, and is already available, is a standard that implements V.42bis on GSM networks. It has already been approved[6] by ETSI but has not yet been implemented by every network operator, although V.42bis enabled PC Cards are already available.

An all-digital link from a GSM notebook client to a corporate LAN or Web server over ISDN was demonstrated in the summer of 1996. The PC Card ISDN controller, which complies with the Common ISDN API (CAPI), compresses the data to allow for effective data-transfer rates of up to 40 kbps. It uses the standard ISDN rate adapter of the MSC to connect to the ISDN network.

The advantages of having an all-digital mobile ISDN link are significant, particularly for short communications. Modems need time to negotiate with each other (i.e., find the highest common speed, as well as the compression and error-correction capabilities): it takes 30 to 40 seconds to set up a connection over GSM. This set-up time imposes a pretty steep penalty when sending just short email messages. A mobile ISDN connection, however,

---

[6] It must be noted here that, with V.42bis being an openly available ITU standard, it took still an inordinate amount of time for it to be incorporated into the GSM specification.

takes only 5 to 7 seconds. Together with the high data throughput, this solution makes a number of heavy-duty applications more cost-effective[7]. A wireless ISDN user could connect, send a short message and disconnect in less time than it takes to simply set up a connection via a traditional circuit switched channel.

Higher data throughput via V.42bis compression, coupled with the fast set-up times of all-digital connections, will boost applications such as mobile access to the Internet, particularly when this data network is used to access a corporate LAN or host.

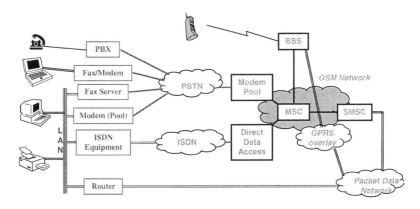

Figure 1. How the GSM Network connects to a LAN

The difference between ISDN-enabled applications and systems that employ the PSTN is usage of the transport layers of the communication protocol. Regular AT-commands send data to the PSTN modem facility in the GSM infrastructure. In an all-digital scenario, a wireless AT-command extension directs the call over ISDN.

A more elegant solution would use CAPI and so-called service indicators. ETSI is currently in the process of extending CAPI to include GSM specifications. The service indicators are control instructions, which are sent over a separate channel. In this case, data and instructions are independent and can be sent simultaneously. Service indicators also enable much quicker set-up times. Thus, when a CAPI-compliant application is loaded, file transfer can start almost immediately. If the instruction is to use ISDN but the called party has an analogue connection, then the modem pool in the digital network is used. The set-up time will still be quicker, but the end-to-end transfer rate will be determined by the weakest link in the chain –

---

[7] The cost and time savings are obvious once you consider that voice and data connections are billed in some countries (e.g., UK) in one-second increments.

the analogue modem. The separate control channel enables GSM to send a fax and receive, for example, a short message via SMS at the same time.

The availability of compression, allowing faster effective transmission, and of direct links from the GSM network to data applications, reducing significantly the handshake period, together with the development of software that allows an application to appear to reside in the background while the GSM call is actually disconnected, are expected to significantly impact the circuit-switched data take-up.

Another interesting development which will certainly contribute to increased take-up is the recent offer of a cheaper and simpler software alternative to the PC Card. Integrated software packages have recently become available which enable the mobile computer user to send data and faxes using a standard serial cable connected to a mobile phone handset. The software includes all the functionality associated with traditional PC Cards, being both cheaper and less complex. The software, which works with most GSM handsets, senses the type of handset being used and loads the appropriate driver.

2.3   High-Speed Data over GSM

We have already identified the limitations of present GSM data offerings as one of the reasons why wireless data has not took-off. This has in fact been widely acknowledged, and as a result more efficient, higher data rate services are being standardized in ETSI SMG, namely High-Speed Circuit-Switched Data (HSCSD) and General Packet Radio Service (GPRS).

While some have questioned the need for two advanced data services on GSM, it should be clear that the focus of these developments is different, and that both will prove valuable additions to the operator's arsenal. However, the point remains that we are not dealing with an integrated high speed data offering, i.e., that we cannot switch between HSCSD and GPRS, each of them requiring a dedicated, specialized radio.

2.3.1   High-Speed Circuit-Switched Data (HSCSD)

Higher data rates will come as a result of compression, including compression at the application level[8]. But multi-slot services can take it all the way up to 64 kbps and thereby facilitate inter-networking with ISDN.

The recently standardized HSCSD (GSM Release 96, March 97), allows for the combination of multiple time slots. By using up to four time slots in

---

[8] We can foresee intelligent agents and other technologies being used to automate, simplify, and otherwise enhance the mobile communications process, by filtering out unnecessary information.

each direction (uplink *and* downlink), the channels can be multiplexed together to offer a *raw* data rate of up to 64 kbps (38.4 kbps *user* data rate, or up to 153.6 kbps with data compression). However, because each time slot could carry a conventional conversation, the use of multiple slots restricts the capacity for speech traffic, forcing the user to specify a minimum acceptable data rate and a preferred (and usually higher) data rate. The network will then attempt to provide as much bandwidth as required, without compromising the capacity for voice traffic. The only limitation will be the price subscribers will pay for the extra bandwidth.

Trials are already ongoing and HSCSD is expected to be commercially available in the 99 time-frame, initially offering 19.2 kbps (two slots), the targeted service being fax and file transfer at 14.4 kbps (up to 56 kbps with data compression). It will prove particularly useful for applications with high-speed data requirements, such as large file transfers, advanced fax services and mobile video communications.

2.3.2 General Packet Radio Service (GPRS)

As its name suggests, GPRS is based on the transportation and routing of packetized data, reducing greatly the time spent setting up and taking down connections. Capacity limitation is hence in terms of the amount of data being transmitted rather than the number of connections. At the same time, charges will be based on the amount of data actually transferred and on the actual data rate achieved[9] and no longer on the connection time. Multi-slot services will likely be marketed as a bandwidth-on-demand facility, making it ideal for high-speed file transfers and mobile video communications.

GPRS will work with public data networks using Internet Protocol and with legacy X.25 networks, and it is likely to prove very successful in bursty applications such as email, telemetry and some Intelligent Transportation Systems services. Perhaps its greatest asset, however, is that it offers a perfect medium for Internet access. The requirements for short bursts of high-speed transmission (downloading) interspersed with longer periods of inactivity (perusing) are better suited to this connectionless approach than a circuit-switched environment in which precious airtime must be charged for all the time a connection is open.

GPRS requires significant modifications to the GSM system architecture, with the need to introduce a data overlay, shown bold in Figure 2.

---

[9] The whole issue of wireless data billing is still far from being resolved, and the need to reconcile "best effort" with QoS (implying data rate) guarantees only makes the solution more elusive.

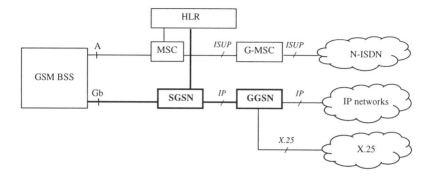

SGSN: Serving GPRS Support Node; GGSN: Gateway GPRS Support Node

Figure 2. GPRS overlay on the current GSM network

A first specification of GPRS was made available in 1998 as part of GSM Release 97, focusing on point-to-point connections, and based upon a frame relay (FR) implementation of the overlay network. Release 98 is expected to include the specification of the ATM implementation, and full IPv4/IPv6 capabilities. GPRS trials[10] are expected to start late in 1999-early 2000, and commercial availability late in 2000.

2.4 EDGE (Enhanced Data rates for GSM Evolution)

EDGE, also known as Evolved GSM, will enable higher data rates using the same frequency bands in use today (900 and 1800 MHz in Europe, and 1900 MHz in the USA) with relatively small additional hardware and software upgrades, and keeping full GSM backward compatibility.

This will be achieved through higher level modulation: instead of the conventional GMSK of GSM, EDGE will rely on O-16QAM. As a result, instead of today's 14.4 kbps of GSM, 48 kbps per time slot will be possible. One has to understand that for the higher data rate (which can in principle be taken as high as 64 kbps per time slot in indoor environments) higher radio signal quality is required than for GSM. The system, therefore, will have to automatically adapt to radio conditions (link adaptation), dropping back to GSM data rates when necessary.

Furthermore, channels with EDGE functionality will be able to operate in either GSM or EDGE mode, allowing the two types of channels to co-exist in the same network, facilitating the step-by-step introduction of EDGE.

---

[10] Trials are scheduled in Europe but also in the US, with Omnipoint having already signed an agreement for the first trial and launch.

Table 1. GSM versus EDGE

	GSM	**EDGE**
Modulation	GMSK	16-QAM
Carrier Spacing	200 kHz	200 kHz
Modulation Bit Rate	270 kbps	640 kbps
User Data Rate per Time Slot	9.6/14.4 kbps	48 kbps
Total User Data Rate*	76.8/115.2 kbps	384 kbps

*no necessarily what would be available to a single user

EDGE will make it possible to explore the full advantages of GPRS (and HSCSD as well): fast set-up, higher data rates, and the fact that many users can share the same channel will result in a highly improved utilization of the network, especially for bursty applications.

3 UMTS DATA

What is unique about data over UMTS is that in the same network, as a function of the environment and the user/terminal mobility, the maximum peak data rate available to the user will vary from 144 kbps to 2048 kbps. This will imply the need for dynamic adjustment of the maximum data rate available to the applications as the user moves about the network. Moreover, as UMTS will be progressively deployed and is not expected to provide full coverage, not even in urban areas, from day one, the user will be required to drop back to GSM data when outside coverage. Then, a more complicated handoff mechanism beyond rate adaptation will be required, especially if it needs to be seamless.

The final implementation of data over UMTS will certainly result from the same quest for improved efficiency relying upon data rate (i.e. bandwidth) on demand, dynamic resource allocation, and QoS negotiations based upon user profile. It is also clear that both circuit- and packet-switched data will have to be provided over a single radio, allowing the user to select on of the modes at will, unlike it is the case for GSM, where GPRS and HSCSD each requires a dedicated radio implementation.

The ACTS project FRAMES has contributed significantly to the definition of the UMTS Terrestrial Radio Access (UTRA), proposing the two modes retained by ETSI in its historic January 1998 decision. As for the UMTS backbone, at this point in time there seems to be consensus that it will evolve from the GSM backbone and its GPRS overlay. In fact, ETSI's

GMM Report [2] singles out GPRS as a key link of GSM to UMTS, both in terms of the packetized transfer made possible, and the mere existence of an overlay. Where the perspectives differ is in the way that evolution is expected to occur.

Figure 3 shows the approach proposed by the ACTS project RAINBOW, where the interaction of the new UTRA Network and Core Network (CN) with the existing GSM network is done through Inter-working Function (IWF) units. It is obvious from the figure that the UMTS network can thus be looked at as yet another overlay on the GSM network. This conceptual image does not however spell out the exact implementation of the UMTS Core Network, nor its connection with existing networks.

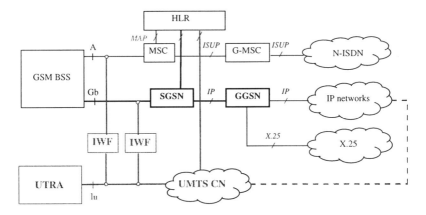

Figure 3. UMTS as an overlay on the evolved GSM network [Source: ACTS Project RAINBOW]

In the author's opinion, the UMTS CN, certainly IP-based[11], be it Mobile-IP and/or IPv6, might come about in many formats: e.g., IP over ATM, native IP or a hybrid of the two. Furthermore, the overlay nature of this network will make it even possible for operators (especially new entrants) to skip the deployment of the GPRS overlay, jumping directly into UMTS, while at the same time allowing incumbents to maximize the return on investment on existing networks.

---

[11] Project RAINBOW, on the contrary, always advocated a purely ATM solution, and only recently started considering a full integration of UMTS and IP contexts.

4 BEYOND UMTS: BROADBAND WIRELESS

The never-ending demand for higher and higher data rates, even while on the move, will not be satisfied with the 2 Mbps made possible by UMTS. Higher data rates, up to 155 Mbps, are being pursued to provide truly broadband wireless access at par with the broadband wired access. The flexibility and mobility offered by wireless solutions, not forgetting the significant cost savings in wiring and re-wiring buildings, plus the dropping costs of portable computers and the associated proliferation of portable devices, are motivating serious interest towards the deployment of private and public wireless broadband networks, even if the market perspectives are not that reassuring. Wireless-LANs and Mobile Broadband Systems (MBS), a concept born in the EC-funded RACE Program, are analyzed below.

4.1 W-LANs

The W-LAN market is currently dominated by proprietary systems operating in the 2 GHz ISM band, with data rates ranging from 1 to 10 Mbps. In order to improve customer acceptance and market penetration, interoperability becomes strategically important. The IEEE 802.11 standard defines an air interface that facilitates interoperability between wireless-LAN products from many different suppliers. The Ethernet-like approach of IEEE 802.11 enables operation in two modes: independent (allowing for ad hoc networking) and infrastructure-based. Data rates are 1 or 2 Mbps, depending on the PHY (Physical layer) implemented (Frequency Hopping Spread Spectrum Radio, Direct Sequence Spread Spectrum Radio, Infrared). Recently, a new study group has been established to address higher data rates (up to 25 Mbps).

In opposition to this "pragmatic" approach, the attention of other standardization bodies seems to focus now solely on Wireless-ATM. The ATM Forum and the ETSI project BRAN (Broadband Radio Access Networks) lead, and in fact co-ordinate together with the Japanese MMAC (Mobile Multimedia Access Communication) organization, the work in the Wireless-ATM area.

A whole family of Wireless-ATM-based protocols (HiperLAN Type 2, HiperAccess and HiperLink) is under preparation in BRAN for W-LAN, remote access/WLL and point-to-point usage, covering data rates from 25 up to 155 Mbps. Interestingly enough, the MMAC concept consists of two systems: Ultra-High Speed Radio LAN (fixed access, up to 156 Mbps) and High-Speed Wireless Access (pedestrian speeds, up to 25 Mbps).

Table 2. European HiperLAN "family" and associated ERC Decision/ CEPT Recommendation specifying spectrum and maximum power

HiperLAN Type 1 Wireless-LAN		HiperLAN Type 2 Wireless-ATM	HiperAccess Wireless Remote Access	HiperLink Wireless Point-to-Point
MAC: CSMA/CA		W-ATM DLC	W-ATM DLC	W-ATM DLC
PHY: 5 GHz	PHY: 17 GHz	PHY: 5 GHz	PHY: 5 GHz	PHY: 17 GHz
up to 25 Mbps		up to 25 Mbps	up to 25 Mbps	up to 155 Mbps
Range: ~50 m				
100 MHz common to all CEPT countries plus 50 MHz on a national basis according to market demand	200 MHz common to all CEPT countries			
1 W transmitter	100 mW transmitter			
ERC/DEC(96)03 [3]		CEPT Rec. T/R 22-06		

HiperLAN Type 1 has already been standardized in ETSI RES10, with first products promised before the end of 1999. The HiperLAN Type 1 protocol, contrary to the others, is Ethernet-like, reflecting its computer communications background (Apple, INRIA, Symbionics). The standard only defines *part* of the lower two layers of the OSI model (Physical and DLC). Within the DLC (Data Link Control) layer, only the MAC (Medium Access Control) sub-layer (CSMA/CA[12]) is specific to HiperLAN. The organization of the MAC sub-layer provides a fully decentralized subsystem which does not require any central control point to operate, enabling ad hoc networking.

ACTS projects [4] are contributing to the BRAN work: The Magic WAND project is involved in the definition of HiperLAN Type 2 and is also involved in HiperLink, while project AWACS is contributing to HiperLink

---

[12] Carrier Sense Multiple Access with Collision Avoidance is a variation of CSMA/CD, CSMA with Collision Detection, or IEEE 802.3, usually referred to as the Ethernet protocol, although some minor differences exist.

and HiperAccess. Wireless customer premises (local area) networks (WCPN/WLAN) systems are also under study: the ACTS project MEDIAN is looking at providing 20-155 Mbps at 60 GHz with a W-ATM approach.

4.2 Mobile Broadband Systems (MBS)

For an outdoor, cellular scenario, Mobile Broadband Systems (MBS), a concept born in the RACE Program, are under consideration in the ACTS project SAMBA, targeting data rates up to 155 Mbps with high mobility. Here again, the solution adopted in Europe is Wireless-ATM-based. No standardization activity is so far foreseen, although the ERO has already considered the spectrum needs of this new service [5].

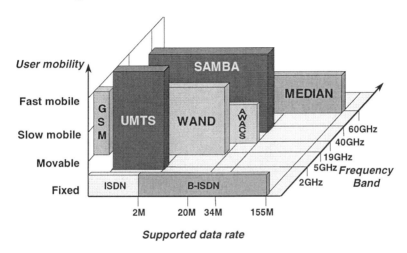

Figure 4. Broadband Wireless ACTS Projects relative to GSM and UMTS [Source: The Magic WAND]

4.3 An Unified, Open Architecture

As W-LANs and MBS are seen as higher data rate extensions of UMTS, providing "hot spot" coverage, an architecture that leverages existing infrastructure and maximizes flexibility in the access to existing networks is essential. Adding W-LAN/MBS connectivity to the GSM/UMTS networks should correspond to implementing yet another (inexpensive) modular overlay, both in terms of functionalities and coverage (to be added on demand).

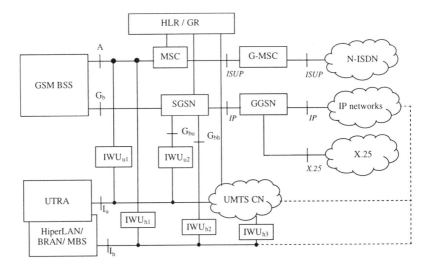

Figure 5. A Common Architecture for GSM/UMTS/MBS/W-LANs building upon the evolved GSM network

In many cases, a voice connection will be reduced to a simple voice gateway to the PSTN (possibly IP-telephony-based, not shown in Figure 5), while data will be pumped directly into an IP-based packet data network.

Another way to view this flexibility is to refer to the GRAN (Generic Radio Access Network) approach proposed by the ACTS project RAINBOW [6] By separating radio-dependent from radio-independent parts, it becomes much easier to introduce new air interfaces, starting with UMTS (and DECT for that matter), and adding MBS and W-LANs where and when appropriate, nothing but different instantiations of air interface (Figure 6).

Another advantage of this open architecture is that it is intrinsically able to accommodate any innovative air interface proposals that might be proposed. As for the Core Network, it is easy to see that all possible implementations can be accommodated, and can co-exist, through inter-working functions; the overlay nature of a multi-network implementation is also suggested.

The question remains what are the "standard" voice and data core networks. For voice, it is almost unanimous that it will be the evolved GSM CN, while for data, as discussed before, the options remain open (by design). The impressive growth of the Internet, and the most likely implementation of GPRS as an IP-based overlay, make IP the most likely to dominate.

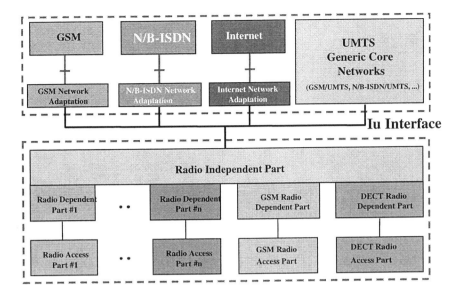

Figure 6. Generic Architectural Model [Source: ACTS project RAINBOW]

5   IP VERSUS ATM

The IP versus ATM debate has captured, perhaps beyond the reasonable, a whole generation of engineers. We discuss here the role of ATM and IP in providing access to multimedia to the mobile user, in the context of the distinct approaches presented in Section 3 for UMTS and in Section 4 for MBS/W-LANs.

The approach in Europe seems to be quite distinct when discussing UMTS or beyond. While Third Generation developments seems to be all geared towards providing mobile multimedia over IP at up to 2 Mbps, on top of an evolved GSM core network building upon the GPRS (IP-based) overlay, Wireless-ATM is currently the only solution under consideration for data rates of 25 Mbps or more. The Japanese approach seems to be more coherent, very Wireless-ATM-oriented, reflecting their (more conventional) B-ISDN tradition.

It is true that in the fixed network ATM offers bandwidth-on-demand and is capable of simultaneously supporting voice, data and video, which is important at a time when operators, carriers and ISPs are all vying to deploy an efficient, economic infrastructure, allowing them to bundle services, and be able to carry good quality traffic. These capabilities, along with the required mobility management functionalities, are attractive also for wireless service provision [7].

However, against the more traditional "circuit-switched" Wireless-ATM approach, one cannot ignore the ongoing efforts to extend the Ethernet-like IEEE 802.11 standard to higher data rates, nor the fast paced development at IETF of real time/time constrained protocols (RTP, RSVP), in parallel with the development of Mobile IP and IPv6, as well as of versions of TCP suitable for mobile environments (e.g., M-TCP [8]) as well as the recent development of concepts such as IP Switching [9].

The same way Wireless-ATM is seen as a natural extension of the ATM backbone, Ethernet-like W-LAN protocols fit naturally with the ubiquitous, ever faster (from 10 Mbps to 100 Mbps to 1 Gbps) Ethernet LANs. And here, as already pointed out in [10], the trend is not very favorable to Wireless-ATM.

Although ATM offers clear and appealing advantages from a technological point of view, its commercialization still faces many problems. Ovum sees a scenario where the first barrier lies in investments already carried out by network operators to build other high-speed data transmission networks, which they would like to recover before boosting alternative infrastructures. A further issue concerns billing: a fair and appealing method has not yet been agreed upon, although some interesting, innovative work is being done in this area in the scope of the ACTS program. Until network operators are willing to offer ATM at attractive tariffs, network managers will also be reluctant to move to ATM in the LAN. If there is little or no likelihood of connecting individual LANs running ATM, ATM LANs just cannot compete with other LAN technologies.

The ATM component of the LAN market is not expected to exceed by much the 10% mark: IDC forecasts that ATM will not significantly impact LAN hubs sales over the next five years because of issues such as incomplete standards, higher prices, limited application support and immature technology.

Recent inroads in IP Switching, which anticipate the deployment of fully IP-based networks, will certainly help IP-based W-LAN solutions, well matched to Ethernet-like protocols, dominate any Wireless-ATM alternatives.

In a recent study [11], Ovum predicts that as early as 2002 the usage of IP will become dominant, surpassing that of other (non-IP and non-ATM) protocols, and certainly above ATM. ATM usage will still grow in importance but less rapidly than IP, even in Europe and Japan where the influence of traditional operators is greater. At the anticipated rates of growth, we foresee that before 2005 more than half of the world traffic will be carried by IP, a phenomenon bound to extend onto the mobile/wireless arena.

As a consequence, it seems certain that IP-based W-LAN solutions will be given more attention in Europe in the future, helped by the expected availability of HiperLAN Type 1.

However, in the Core Network the winning strategy will certainly be, in the short to medium term, (a hybrid) one that musters the best of both (ATM and IP) worlds in an optimal combination that takes into account the advantages and disadvantages of each implementation, but... it just might be pragmatically dictated by availability.

6   CONCLUSIONS

The objective of leveraging existing investment in mobile network infrastructure while simultaneously providing flexible access to existing networks has led to the approach of implementing current developments of GSM (like GPRS) and soon UMTS as overlay to an evolved GSM network. A similar approach is suggested herein for accommodating MBS/W-LANs now under consideration to provide higher data rates, as well as any future innovative radio interface that might be proposed. The objective is also to enable access to GSM and UMTS data via inexpensive "gateways" or directly via IP (native, over ATM or both), allowing full connectivity even in "hot spots" and private/corporate environments.

7   AKNOWLEDGEMENTS

The author would like to thank all the colleagues, who in the scope of the ACTS program have contributed to the around network evolution around, and particularly those in project RAINBOW: their GRAN model is critical to the successful deployment of future systems.

REFERENCES

[1] Strategy Analytics, The European and US Markets for Wireless Data Services, Nov 97
[2] ETSI SMG 3GIG, GMM Report – A Standardization Framework, Jun 96
[3] ERO, Wireless LANs, Apr 96
[4] *ACTS 98*, European Commission, DG XIII, AC 980872-CD; also please check regularly the ACTS Mobile home page at
http://www.infowin.org/ACTS/ANALYSYS/CONCERTATION/MOBILITY

74

[5] ERO, Mobile Broadband Systems (MBS), Jul 97

[6] RAINBOW home page: http://www.cselt.it/sonah/RAINBOW/

[7] D. Raychauduri, N. Wilson, "ATM-based Transport Architecture for Multiservices Wireless Personal Communication Networks", IEEE Journal on Selected Areas in Communications, Oct 94

[8] K. Brown, S. Singh, "M-TCP: TCP for Mobile Cellular Networks", ACM Computer Communication Review, Oct 97

[9] Ipsilon Networks, IP Switching: The Intelligence of Routing, the Performance of Switching, white paper

[10] Jorge M. Pereira, "Indoor Wireless Broadband Communications: R&D Perspectives in Europe", in Proceedings of the Colloquium on Indoor Communications, TU Delft, Delft, The Netherlands, Oct 97

[11] Ovum, The Future of Broadband Networking: ATM vs. IP, white paper, 1997

Chapter 5

UMTS/IMT-2000 Standardisation[*]

3 GPP Third Generation Partnership Project: Development of Standards for the New Millennium

ANTUN SAMUKIC

Ericsson Radio Systems AB (Project Manager UMTS Architecture,
ETSI, MCC Mobile Competence Centre[13])
Tel:+33 4 92944352, Fax:+33 4 93652817, E-mail: antun.samukic@etsi.fr)

Abstract   UMTS is a set of standards aimed at the global market and belongs to the IMT-2000 family. Today UMTS/IMT-2000 technical specifications are being developed in a partnership 3GPP (3rd Generation Partnership Project) consisting of ARIB and TTC from Japan, TTA from Korea, T1 from USA and ETSI. 3GPP partnership project has been started in December 1998 and the full working meetings have already dominated the standardization scene at the beginning of 1999. 3GPP's decisions made during the 1999 shall shape a mobile telecommunication solutions for the beginning of the New Millennium. The article intends to give a picture of the current and planned work within 3GPP but also some historical information related to ground work performed by ETSI SMG. This article attempts to give an objective view of the UMTS standardization. The article does not necessarily express the opinions of any group or company I am affiliated with, neither of 3GPP, ETSI SMG Special Mobile Group, 3GPP, ETSI MCC Mobile Competence Centre or Ericsson.

---

[*] Portions reprinted, with permission, from (IEEE Transaction on Vehicular Technology, volume 47, number 4, page 1099 - 1104; November 1998). © 1998 IEEE.

[13] The work of the 3GPP is supported by a project support team MCC Mobile Competence Centre working full time together, located at ETSI Head Quarter,

1 INTRODUCTION

1.1 User's choice and importance of flexibility of telecommunications options

The development trend in telecommunications is driven by user service requirements asking for access to a diversified range of personalized set of services to anyone, anywhere, anytime but not at any price. Borders between telecommunications, information technology and entertainment services are disappearing and users can combine service offerings from various operators. Full coverage and availability of mobile telecommunication services and increase of Internet services influences development and usage of services. Deregulated world-wide market and rapid introduction of mobile services of the second generation, specifically GSM, Global System for Mobile Communications, has lead to the conclusion that one "ultimate mobile solution ", one radio access network and one single core network standardized to a very detailed level is not realistic. Flexibility of telecommunication services and opportunities of user's choice within mobile communications are, to a large extent, already available today. The choice is available for users, service operators, network operators and manufacturers. Consequently UMTS development, developed by partnership, takes into account the opportunity of the choice for users, network and service operators and the multiplicity of existing and future fixed and mobile telecommunication networks and services. These existing 2nd generation networks and services can evolve towards UMTS but with their own starting points, targets and own pace. Such facts influence the current standardization process at national, regional and ITU level. Such a flexible approach created a ground for active collaboration between the major players covering the global mobile telecommunication needs. There are opportunities of choice for each standardization partner also. The 3GPP partners have choice and opportunity to select only a part of all deliverables and convert them to regional and national standards. At the same time standardization through partnership offers possibilities for all players to suit development according to their own judgement of market needs in order to achieve their time to market.

1.2 The second generation - a platform for the third generation

Initially UMTS has been considered as a system that should offer services for a mass market. Indeed a mass market has been already achieved through the 2nd generation networks and services, reaching high terminal penetration in some markets, already passed 50% penetration in Scandinavia. The second generation system in general and GSM specifically has been well introduced to the global market. GSM, being also used as an evolution platform for the UMTS, has already reached global acceptance and mass market. At the end of December 1998 there were more then 300 members of Association GSM from 112 countries world-wide that are committed to GSM system. There were 311 GSM networks in operation serving 135 million subscribers [14]. It is expected around 250 millions subscribers should be served by GSM networks at the end of the year 2000[15]. GSM serves 65 % of all digital mobile users.

The mass market development is being further strengthened through the multi-band terminals (i.e. GSM/DCS1800/PCS1900) or multi-mode multi-band terminals covering various combination of the 2nd generation terminals (e.g. DECT/GSM, PHS/GSM, GSM/satellite). The expected penetration for mobile communication in the developed countries is expected to rise to 50 - 80 % within the time frame for the introduction of UMTS, accentuating importance of an evolution approach towards UMTS. The UMTS/GSM standards family offering dual-band, and multi-band/mode terminals and networks will become the leading digital standard for mobile communications according the predictions of the leading players within mobile communications. The merging of the GSM family networks in different bands (900, 1800, 1900 MHz) and UMTS results in much higher capacity network, cheaper and better equipment optimized for various environments and larger choice for users. It is important to note that regulatory bodies also support such trend by assignment of mixed bands to one network operator. In addition efficient access to Internet has also been achieved by the second generation making possible to use Internet services by circuit switched or packet switched mode.

---

[14] GSM Association, June 1998
[15] Various market reports including GSM Association

2 THE AGREED KEY FEATURES OF UMTS RELEASE 1999[16]

Operators and manufacturers objectives can be fulfilled only if they serve customers. The operators, other interested parties and manufacturers, set UMTS key features and requirements. Market driven development of standards ensures that UMTS standards fulfil objectives of users and operators. High level requirements and key features of UMTS have been developed in close co-operation with GSM Association 3GIG (3rd Generation Interest Group) including GSM operators worldwide, UMTS Forum and PSOs. Operators and manufacturers from other regions that also helped to create 3GPP also favored such key features.

Key features of the UMTS are the base for the development of standards for the UMTS Release 1999 (initially planned as UMTS Phase 1) making commercial operation possible by 2002. Further development of the UMTS standards should be on a yearly basis introducing new functions as a timely response to the users and operators requirements. It should be noted that UMTS Release 1999 standards should be developed in a way making possible even earlier introduction of UMTS/IMT2000 in Japan.

	UMTS Release 1999 standards (operation 2002 possible)

Services	• multimedia services phase 1:
	    • at least 2 Mbps with maximal speed of 10 km/h
	    • at least 384 kbps in suburban outdoor with maximal speed of 120 km/h
	    • at least 144 kbps in rural with maximal speed of 500 km/h
	 • high quality speech (like fixed networks) using low bit rates
	 • packet and circuit switched services for the different bit rates and the different radio environments
	 • service creation and measurement toolkit
	 • services portability, when roaming into other networks
	 • advanced addressing mechanisms, e.g. personal, Internet-style
	 • new charging mechanisms (e.g. volume)
	 • dual band/mode of operation UMTS/GSM incl. roaming between UMTS "islands"
	 • roaming between UMTS and GSM networks
	 • operate in any suitable band that becomes available, e.g. GSM, DCS1800, PCS1900

---

[16] UMTS 30.01 UMTS Baseline document, ETSI SMG#23,
13-17 October 1997, Budapest

Terminals	• mobiles and SIM with downloading capabilities over the air for e.g. data and applications (feasibility in phase 1 needs further study) • multi media terminals • dual mode/band GSM/UMTS terminals • adaptive terminals

Access network	• New UMTS BSS
	   • flexible bearer
	   • rates <= 2 Mbit/s
	   • fast, self adapting interface
	• high capacity
	• support of variable bit rates and of mixed traffic types
	• high spectrum efficiency for multimedia and low bit rate speech

Core transport network	• evolution of the GSM NSS (Network Switching Subsystem) and ISDN/IN (Intelligent Network) • new charging and accounting mechanisms • support of service mobility across networks - VHE (Virtual Home Network) • support of variable bit rates and of mixed traffic types • mobile fixed convergence elements • support of packet data by Internet protocols
Security	• protection of network use • provision of security services to the user • control of misuse and/or abuse of the network
Operation & Mainten-ance	• automatic establishment of roaming relations • support of multi-vendor networks

Table 1: The content of the UMTS Release 1999

The requirements on the UMTS/IMT2000 coming from the all 3GPP partners have been further evaluated at the beginning of 1999. The similarities of the requirements have been noted while having some variations related to the market introduction.

3   EVOLUTIONARY APPROACH

Evolution studies started both within ITU and ETSI as an expression of market needs to continue service offering to the very large and increasing number of customers of the second generation mobile networks. Several international groups, including FAMOUS (Future Advanced Mobile Universal Systems, trilateral group USA, Japan and Europe), RAST (Radio Standardisation, meeting of standardization bodies dealing with radio standardization) and ETSI GMM (Global Multimedia Mobility) have recognized that there are large number of the cellular networks in operation with capabilities of further development. ITU-R TG8/1, GMM Global Multimedia Mobility group and SMG committee played the key role for the introduction of the UMTS/IMT-2000 evolutionary path within the standardization process, as requested by the telecommunications industry. The customer must have choice and opportunities to use new services governed only by own needs without being pushed for change for the sake of technology or regulation. The acceptance of UMTS on the market should depend on its capabilities to provide new multimedia and data services, access to Internet and Intranet and enhanced speech quality. The early ETSI study of evolution towards UMTS was initiated in SMG5, being adopted by ETSI, reflected in much broader GMM study and opened the door for UMTS evolutionary approach:

"The third generation mobile communications systems will be introduced in the early years of the 21st century. They will consist of radio interfaces, supporting infrastructure and connections to networks. In this frame, the third generation systems are likely to evolve from existing (also called pre-UMTS) mobile systems and at the same time integrate new system components and concepts. The form of the evolution will be strongly influenced by market considerations. UMTS should be standardized by a managed evolution process starting from GSM and N/B-ISDN, using a generic access part"[17].

The ETSI GMM (Global Multimedia Mobility) report[18] clearly concluded: "The UMTS standardization process requires:

- Standards for new terminals
- Standards for a new access network
- Enhanced or new network capabilities in existing core transport networks".

The evolution paths from GSM and narrowband ISDN networks to support UMTS and other future mobile services were accentuated by the GMM report explicitly. Dedication to UMTS and development of services for customers have also led to further work on GMM companion document

---

[17] ETR 312 "Scenarios and considerations for the introduction of the UMTS"
[18] ETSI GMM Global Multimedia Mobility report

covering the service aspects of global mobile multimedia. The companion document should be available during the middle of 1999.

4 FOCUS OF THE WORK ON THE UMTS CONCEPT

During 1997 - 1998 ETSI SMG has put great emphasis on the UMTS concept development [19]. UMTS concept development has been paired with the other elements of equal importance for UMTS introduction to the market in the year 2002. These elements are availability and allocation of the UMTS spectrum[20] and regulative and licensing framework[21] and elaboration of operators' commitment. The UMTS standardization process has been accompanied by the activities of the other bodies relevant for the introduction of UMTS to the market place. For those reasons an early regulatory framework and licensing of the UMTS frequencies to operators is a prerequisite. Operators' commitment plays important role for the acceptance of the UMTS, just as the GSM Association has played the role at the introduction of GSM. The work on the selection of the UTRA UMTS Terrestrial Radio Access technologies already broadened work on UMTS through inputs and participation of T1P1 and ARIB.

During 1997 - 1998 the first contacts between ETSI and other standardization bodies were established, strengthened and finally standardization partnership has been signed in December 1998.

During 1999, and later, further work on UMTS specifications are reflected within 3rd GPP Agreement[22]:

---

[19] UMTS 30.01 UMTS Baseline document: Collections of ETSI SMG's positions on the UMTS

UMTS 22.01 UMTS Service principles

UMTS 22.70 Virtual Home Environment

UMTS 30.04 Definition of the limited number of UTRA concepts

UMTS 30.06 UTRA Concept evaluation reports

UMTS 23.20 Evolution of the GSM platform towards UMTS

ETSI SMG Tdoc 39/98, ETSI SMG#24bis,
28-29 January 1998, Paris

[20] European Radiocommunications Committee ERC Decision of 30 June 1997 on the frequency bands for the introduction of the UMTS, (ERC/DEC/(97)07)

European Radio Office ERO , UMTS (Frequency Allocation Report), September 1996

[21] European Commission, "Proposal by the Commission for a Decision of the European Parliament and Council on the co-ordinated introduction of mobile and wireless communications (UMTS) in the Community", 1998

UMTS Forum, A Regulatory Framework for UMTS, Report no. 1 from the UMTS Forum, 26 June 1997

[22] Third Generation Partnership Project Agreement, December 1999

"The purpose of 3GPP is to prepare, approve and maintain globally applicable Technical Specifications and Technical Reports for a 3rd Generation Mobile System based on the evolved GSM core networks, and the radio access technologies supported by the partners (i.e., UTRA both FDD and TDD modes), to be transposed by the Organizational Partners into appropriate deliverables (e.g., standards)."

Development of detailed specifications, to be transposed to standards including detailed protocols specifications, is the main task for the year 1999 and 2000.

5 UMTS TIME SCHEDULE[23]

This timetable was established during the earlier time of ETSI standardization of UMTS. The timetable has not been updated later but it has been agreed that suitable phasing is needed in order to make possible to introduce UMTS services on commercial bases already in 2001. Depending on operator's strategy and availability of frequencies it should be possible to start commercial services already in 2001. It should be noted that the first release of the UMTS standard specification shall be approved at the end of 1999 as a Release 1999. The first set of 3GPP deliverables dealing with radio access should be approved at April/May 1999.

6 3GPP SPECIFICATIONS

6.1 Responsibilities for 3GPP specifications [24]

It must be pointed out that majority of work continuos within 3GPP and its TSGs. Each TSG shall prepare, approve and maintain the specifications within its own terms of reference. "Officially recognized Standardization Organizations have agreed to work collaboratively for the production of Third Generation System specifications. The means for this collaborative activity has been provided in the form of a partnership Project.

---

[23] SMG UMTS Work Programme, ETSI SMG#23, 13-17 October 1997, Budapest
[24] 3GPP Agreement, December 1998

Table 2: Overall schedule for the UMTS Release 1999 development

Task Name	1996	1997	1998	1999	2000	2001	2002
GSM900 Phase 2+ implementation		▓	▓	▓	▓	▓	▓
UMTS Vision		▓	▓				
Co-operative research: ACTS	▓	▓					
Regulation: framework (report UMTS Forum)		▓	▓				
Regulation: CEC, ECTRA, ERC decisions			▓				
Regulation: national license conditions			▓				
Regulation: license awards				▓			
Operators' commitment: elaboration of draft		▓					
Operators' commitment: signature			▓				
ETSI: basic UMTS standards studies	▓	▓					
ETSI: freezing basic parameters of standard			▓	▓			
ETSI: UMTS Phase 1 standards			▓	▓			
Regulation: conformity assessment conditions			▓	▓			
Pre-operational trials						▓	
UMTS Phase 1: commercial operation							▓

The Partners have agreed to co-operate in the produc ion in globally applicable Technical Specifications and Technical Reports for a Third Generation Mobile System based on evolved GSM core networks and the radio access technologies that they support (i.e., Universal Terrestrial Radio Access (UTRA) both Frequency Division Duplex (FDD) and Time Division Duplex (TDD) modes). 3GPP is established for the preparation, approval and maintenance of the above mentioned Technical Specifications and Technical Reports. Initially, 3GPP shall prepare, approve and maintain the necessary set of Technical Specifications and Technical Reports for the first phase of a 3rd Generation Mobile System including:

- UTRAN (including UTRA: W-CDMA in Frequency Division Duplex (FDD) mode and TD-CDMA in Time Division Duplex (TDD) mode);

- 3GPP Core Network (Third Generation networking capabilities evolved from GSM. These capabilities include mobility management and global roaming);

- Terminals for access to the above (including specifications for a UIM; and

- System and service aspects. " [25]

The UMTS standards are based on UTRAN and GSM network platform. The strength of the concept has been further enhanced due to commitment of several key players in the market and standardization arena to work together on technical specifications. The standards are developed within 3GPP partnership while formal approval is within responsibility of each PSO Partner Standardization Organization. The deliverables of 3GPP are common property of 3GPP partners while each standardization bodies select a set of deliverables to be transformed to standards.

The partners within 3GPP are ARIB and TTC from Japan, TTA from Korea, T1 from USA and ETSI. 3GPP shall cover the complete, all aspects of the standardization. The work is carried in TSG Technical Specification Groups and WG Working Groups. Formal approval of 3GPP deliverables as ETSI standards is within responsibility of TC SMG. In addition EP UMTS has been created within ETSI dealing with UMTS long term aspects and all other aspects not covered by 3GPPP. In addition PCG Project Co-ordination Group has been established with a task to monitor TSGs work and give them a strategic guidance when needed. The consists of TSGs officials and high representatives of 3GPP partners. During its first year of existence ETSI DG Director General chairs the PCG.

The following structure has been agreed for 3GPP (see Figure 1):

---

[25] Third Generation Partnership Project Agreement, December 1999

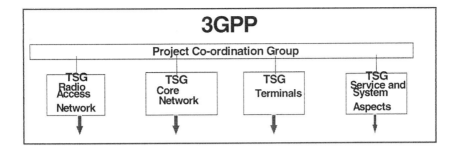

Figure 1. The 3GPP structure

6.1.1   TSG SA   Services and Architecture

The Terms of Reference for TSG Services and System Aspects says: "The TSG Services and System Aspects (TSG-SA) is responsible for the overall architecture and service capabilities of systems based on 3GPP specifications and, as such, has a responsibility for cross TSG co-ordination. Any difficulty that may appear in this role shall be reported to the PCG:
The following Working Groups were agreed:
WG1   Services (including value added services)
WG2   Architecture
WG3   Security
WG4   CODEC
WG5   Telecom Management

6.1.2   TSG RAN   Radio Access Network

The Terms of Reference for TSG Radio Access Network says: "The TSG Radio Access Network (TSG-R) is responsible for the radio access part, including its internal structure, of systems based on 3GPP specifications".
The following WG Working Groups were agreed :
WG1   Radio layer 1 specifications
WG2   Radio layer 2 and layer 3 specifications
WG3   Iub, Iur, Iu specifications, UTRAN and O&M Requirements
WG4   Radio performance and protocol aspects, RF parameters and BS conformance
Ad-hoc group on ITU co-ordination

6.1.3   TSG CN   Core Network

The Terms of Reference for TSG Core Network says: "The TSG Core Network (TSG-N) is responsible for the specifications of the Core network part of systems based on 3GPP specifications".

The following WG Working Groups were agreed :

WG1   MM, CC, SM (Iu)

WG2   CAMEL/MAP

WG3   Inter-working with external network

6.1.4   TSG Terminals

The Terms of Reference for TSG Terminals says: "The TSG Terminals (TSG-T) is responsible for specifying the Terminal Equipment interfaces ensuring that terminals based on the relevant 3GPP specifications meet the 3GPP objectives".

The following WG Working Groups were agreed :

WG1   Mobile terminal conformance testing

WG2   Mobile Terminal services and capabilities

WG3   USIM

6.1.5   PCG Project Co-ordination Group

"The Project Co-ordination group (PCG) is responsible for overall time-frame and management of technical work to ensure that the 3GPP specifications are produced in a timely manner as required by the market place according to the principles and rules contained in the Project reference documentation (Partnership Project Description, Partnership Project Agreement, Partnership Project Working Procedures)".

6.2   UTRA (UMTS Terrestrial Radio Access) definition and time table

6.2.1   UTRA Procedure and time table

The originally planned selection procedure , "beauty contest" type selection, has been replaced with well experienced ETSI contributions and consensus driven process. The procedure is open for a large number of participants without requesting a complete system. Such a procedure maximizes the benefits by using expertise from all players within a large number of the areas of radio access technology The utilization of contributions from all competent players has resulted in a UTRA with an overall performance superior to that obtainable by a traditional "beauty contest" type of selection. The new defined procedure ensures a timely definition of the UMTS Terrestrial Radio Access (UTRA). The time frame set by the market demand for the introduction of UMTS at the year 2002 should be fulfilled. Another important aim of the procedure was achieved by submitting UTRAN as a candidate for IMT-2000 to ITU (International Telecommunication Union) before 30 June 1998. Several UTRA documents were approved by SMG#28 (February 1999) and all UTRA documents and WI Work Items have been transferred to 3GPPP.

The UTRAN milestones consisting of three phases were all achieved according to original time plan.

SMG#22, June 97 (milestone M1)
Approval of the definition of a limited number of UTRA concepts
SMG#24, Dec. 97 (milestone M2)
Approval of the selection of one UMTS terrestrial radio access concept
SMG#26, June 98 (milestone M3)
Approval of the specification of key technical aspects of the UMTS terrestrial radio access and submission to

ITU-R. Evaluation report to ITU-R in September (SMG2 has in its role of ITU-R evaluation group to prepare the UTRA proposal to ITU-R by the end of June 1998. The evaluation of the proposal should be completed by September 1998)
SMG#28, February 1999
Elaboration of technical description of UTRA and transfer of work to 3GPP.
Further milestones to be finished by 3GPP include:
- elaboration of specifications completion ultimo 99
- correction based on experiences
- UMTS phase 2 standardization start primo 2002.

6.2.2   Consensus Decision on the UTRA Concept [26]

ETSI SMG group responsible for the development of UTRA for the entire ETSI has achieved a historic milestones by selecting a new radio interface for UMTS in January 1998. This unanimous consensus decision has far reaching consequences influencing all development in the next 10-15 years. This decision opens flexible solutions for radio access network and

[26] ETSI SMG document Tdoc 39/98, ETSI SMG#24bis,
   28-29 January 1998, Paris

terminals capable of multimedia and data services and high speed access to Internet. It also opens possibility of defining a full global solutions for the radio access, covering all applications and cell sizes. Already this decision has involved participation of 3GPP partners even if without their formal voting rights. The UTRA decision received a broad support of very large number of non-ETSI members. NTT DoCoMo participated as an observer at SMG#24bis meeting and expressed support for ETSI SMG decision. A number of Asian Pacific operators did the same support paving a way for a global solution.

This consensus agreement contains the key elements and advantages of both WCDMA and TD/CDMA, and contains the following elements:

1. In the paired band we adopt the radio access technique proposed by the ETSI Alpha group, that is WCDMA (Wideband Code Division Multiple Access)[27]

2. In the unpaired band, we adopt the radio access technique proposed by the ETSI Delta group, that is TD/CDMA (Time Division/Code Division Multiple Access)[28]

3. In the process in selecting the technical parameters the following shall be the objectives:

-     Low cost terminal
-     Harmonization with GSM
-     FDD/TDD dual mode operation
-     Fit into 2 * 5 MHz spectrum allocation

6.2.3   Submission of UTRA as IMT-2000 RTT Radio Transmission Technology candidate and further work on UTRA

At its Helsinki meeting at the end of June 1998 ETSI SMG approved UTRA description to be submitted as a proposal to ITU-R. The approval of UTRA as a ITU-R IMT-2000 candidate was unanimous without any objections or abstentions. ETSI SMG indicated also its willingness to be an active participant in the ITU-R consensus building group. Also ETSI SMG submitted a report including references of IPR (Intellectual Property Rights) status which have been notified to ETSI as being essential to UMTS.

Harmonization of the various ETSI SMG proposals leading to UTRA was based on the UMTS requirements. Harmonization with other technologies outside of ETSI SMG has been already achieved to a much larger degree then it was expected some time ago. UTRA descriptions are

[27] ETSI SMG document Tdoc SMG 903/97
[28] ETSI SMG document Tdoc SMG 897/97

still subject to refinements and improvements, but any changes in UTRA description must lead to increased performance. UTRA technical description and performance will be elaborated until the end of 1998 and the technical standards and specifications should be ready by the end of 1999. Further improvement of UTRA standards should be introduced during the years 2000-2001 by the process of CR change request to the specifications.

6.2.4 Approval of UTRAN documents and transfer of work to 3GPP[29]

SMG#28 (February 1999) has approved a complete set of existing UTRA documents covering areas of UTRAN Architecture, UTRAN Layer1, Layer 2 and Layer 3. All these documents were transferred to 3GPP mainly 3GPP TSG RAN for further work..

6.3 UMTS Network principles

The principles of the evolving GSM core network supporting UMTS services have been studied and accepted by ETSI SMG and all partners in 3GPP, TTC from Japan, TTA from Korea and T1 from USA. . The work on UMTS Concept Architecture specifications was carried by SMG12 System Architecture committee combining both GSM and UMTS network experts. The work continues in WG2 of the TSG SA Technical Specification Group Service and Architecture of the 3GPPP. The current work concentrates on the UMTS network principles, general UMTS architecture, evolution of the GSM platform, principles for the Iu Interface, UMTS access stratum and framework of network functions for support of multimedia services in UMTS. Some requirements already being accentuated UMTS architecture includes:

- High data rate and asymmetric data transmission
- Importance and relationship to the Internet
- VHE Virtual Home Environment

Core issues are handling of mobility management, transport technologies and their relationship to higher protocol layers, the split of functionality and the definition of the Iu interface between core networks and a radio access network Initially ETSI SMG has decided to evolve GSM/GPRS core network in order to support both UMTS and GSM radio access by the

---

[29] ETSI SMG#28 Meeting Report, February 1999

evolved GSM-UMTS network. The UMTS network principles and network architecture were formally approved at the beginning of 1999. The work has been transferred to 3GPP but a large number of WI work items are of the joined responsibility of 3GPP and TC SMG due to fact that the evolved GSM/GPRS core network shall be capable to support both GSM and UMTS services. Basic concept documents were approved during the first half of 1999 while detailed specifications including protocol MAP updates should be developed by TSG CN during 1999. The key document UMTS23.20 Evolution of the GSM platform towards UMTS shall be used with 3GPP as a document for technical co-ordination among 3GPP TSGs and Wags. .

6.4   UMTS Service principles

Development of standards for UMTS has been always characterized by putting great emphasis's on services to be offered to end users. Standardization concentrates on UMTS service capabilities rather than the services themselves.   Standardization of service capabilities includes standardization of bearers , Quality of Service (QoS) parameters and additional mechanisms needed to realize the required services. These additional mechanisms include, the service creation functionality of various network elements and the communication between element and the storage of associated data. Such standardized service capabilities should provide a defined platform that will enable support of speech, video, multi-media, access to Internet, messaging, data, other teleservices, user applications and supplementary services. A set of service capabilities should enable users, service providers and network operators to define services themselves according to their needs. The very high interest to direct UMTS standardization from the service point of view have resulted in a large number of the UMTS services reports and specifications:
- UMTS phase 1 capabilities
- UMTS service principles
- Services and service capabilities
- UMTS terminal and smart card concepts
- Charging and accounting mechanism
- VHE Virtual Home Environment making possible to offer user specific services across range of various networks
- Quality of service and network performance
- Mobile multimedia services including mobile Intranet and Internet services
- Advanced addressing
- UMTS service scenarios
- Automatic establishment of roaming relations

The document UMTS 20.00 UMTS phase 1 capabilities has been agreed as another key document for technical co-ordination among 3GPPs TSGs and WGs.

7 THIRD GENERATION STANDARDISATION POLICY AND UMTS AND IMT-2000 RELATIONS

7.1 Focused co-operation of standardization bodies

Already work on GSM standards requested a new spirit of co-operation. ETSI SMG started working together with the North American standards organization ANSI T1P1 (American National Standards Institute) on GSM standardization. GSM is implemented in the North America in the 1900 MHz band. Such co-operation has been very successful strengthening market leadership of SMG. T1 and ETSI SMG continued co-operation also in UMTS field.

During 1998 ETSI SMG with ETSI support initiated a dialogue with Japanese standardization bodies, ARIB (Association of Radio Industry and Businesses), and TTC (Telecommunication Technology Committee). ARIB is responsible within Japan for the IMT-2000 radio standardization and TTC is responsible within Japan for IMT-2000 network standardization. As a first step an exchange of technical information was agreed and further co-operation was discussed between ETSI and ARIB/TTC within the scope of the UGG (UMTS Globalization Group). Korean TTA also favors strength of co-operative work concentrating on well defined targets, as well strength of asynchrony W-CDMA radio access and great developing capabilities of GSM network platform and joined 3GPP consequently.

Common interest and willingness to create a system for the new millennium and wish to create this new system in a quite new co-operative spirit, these standardization organization have agreed to sign co-operation agreement known as 3GPP 3rd Generation Partnership Project.

The areas of common interest cover work on the initial phase of a complete 3rd generation system including both network aspects based on GSM evolution and radio aspects based on UTRA.

7.2 UMTS relationship with IMT-2000

It has been also recognized that 3rd generation systems are expected to evolve from the existing systems. Therefore, there may be several compatible systems belonging to IMT-2000 family but with full global roaming and global service support. Relationship between UMTS and IMT-2000 are of great importance for entire communication community. As proposed by ETSI SMG, all 3rd generation systems, that fulfil essential requirements on the 3rd generation, should be seen as a IMT-2000 family member leaving implementation options to operators. The concept of "IMT-2000 Family" is recognized by ITU as a suitable approach to accommodate the need for evolution/migration toward IMT-2000 in the wireless communications industry. All ITU study groups and entirely telecommunication community have positively accepted these proposals.

IMT-2000 family concept offers needed flexibility as required by users and operators. The concept offers full flexibility while at the same time a large number of considerable commonality have been achieved without forcing a single solution and degrading overall system performance. All efforts have been made and should be made to enable roaming between IMT-2000 family members and enable a free circulation of IMT-2000/UMTS terminals..

8 CONCLUSIONS

Basing developments of UMTS standards on the requirements developed together with the known customers provides an opportunity for a global success. Second generation's users are aware that they have access to advanced services already today and that they should have a choice and option to use new advanced services as it suits them through GSM, UMTS or their combinations. Operators shall have options of introducing new services according to their business plans and strategies. UMTS strength comes also from the fact that all other activities like spectrum allocation, regulative framework and licensing are being developed in parallel with the standardization process. UTRA UMTS Terrestrial Radio Access has received tremendous level of interest and support from very large number of operators and manufacturers. Dual mode operations , UMTS in combination with GSM and D-AMPS and PDC offers an introductory path for 3rd generation mobile services.

Very high levels of harmonization have been achieved between UTRA and other radio transmission technologies elaborated by T1P1 and ARIB. UMTS standardization process ensures technical capabilities for introduction of the full UMTS services in the year 2002 while suitable phasing of the UMTS standardization makes possible to offer some services even earlier as requested by some Japanese operators. Creation of 3GPP reflects a new spirit of co-operation that applies a leading edge of mobile technology and covers global market needs.

Acceptance of the IMT-2000 family concept has opened further possibilities for creating a global system with enough flexibility while achieving needed level of commonality as required by market forces. 3GPP partnership including standardization bodies like ARIB and TTC from Japan, TTA in Korea, T1 and ETSI are paving a way for a new 3rd generation system building on the common interest and flexibility of the system to cater even for different requirements. The IMT-2000 family concept opens way for recognition of differences while exploiting similarities expressed by the motto:

"Getting together without being the same"

PART II

TOPICS IN WIRELESS
MULTIMEDIA NETWORKS

Chapter 6

Intelligent and Flexible Radio Access/Transmission Technologies for Wireless Multimedia Communication Systems in the Software Defined Radio Era

SEIICHI SAMPEI AND NORIHIKO MORINAGA
Department of Communication Engineering, Graduate School of Engineering, Osaka Universit, Japan

Abstract: This paper discusses intelligent and flexible radio access/transmission technologies necessary to smoothly upgrade supported services as well as to economically extend coverage and operational environments. After brief discussion on classification of key technologies for intelligent and flexible radio access/transmission technologies this paper will discuss the adaptive transmission parameter control techniques for both time division multiple access (TDMA) based and code division multiple access (CDMA) based systems. Finally, this paper will propose a new time division multiplexing (TDM) based adaptive access control technique to further enhance system capacity as well as to introduce higher-grade services for CDMA-based wireless multimedia communication systems.

1   INTRODUCTION

In the past 10 years, very extensive studies have been made on the development of digital wireless communication technologies. During this period, although increase in voice capacity for cellular systems was the main driving force in the early stage, the main driving force at present is the support of multimedia services rather than the higher voice capacity. This change is mainly due to the rapid growth of multimedia communications via the Internet as well as the emerging note-sized and pocket-sized computers.

The International Mobile Telecommunication 2000 (IMT-2000) [1], [2] is being standardized to support multimedia services with its bit rate of up to 2 Mbit/s. Among many requirements for the IMT-2000, the followings are unique requirements compared to the 2nd generation systems [1];

– seamless operability between picocell, microcell, macrocell and/or megacell environments
– smooth migration/evolution from the currently operated system to the next generation system
– ease of introduction of technology advancements and different applications

and they will be getting more important when the IMT-2000 is upgraded or migrated to the next generation systems.

A solution for these requirements could be a multi-mode terminal. Thanks to the recent device technologies, dual mode terminal is practical even at present. Therefore, it would be possible in the near future to make small-sized multi-mode terminals having three or more modes, if hardware commonality between modes is maximized and device technologies are further improved.

When we consider a multimode terminal, we usually imagine that a mode is selected by a hardware switch or a software switch according to the operational environments. However, if the terminal has several transmission modes, we can also have a chance to adaptively select transmission modes even during communication according to, for example, the fast variation of the propagation path characteristics.

Other than the adaptive transmission mode control, flexible selection of access schemes would also be necessary in the future because service availability for the TDMA-based and CDMA-based systems greatly depend on the operational environments as well as any other constraint conditions such as the assigned system bandwidth. Moreover, such an adaptive transmission mode and access scheme control would be an essential technology when software defined radio technologies become core technologies for wireless multimedia communication systems.

Therefore, this paper will discuss intelligent and flexible radio transmission/access technologies for wireless multimedia communication system in the software defined radio era.

2   CONCEPT FOR INTELLIGENT AND FLEXIBLE RADIO ACCESS/TRANSMISSION SYSTEMS

To design wireless multimedia communication systems, we have to take into account the following requirements [3]:

– mitigating dynamic variation of multimedia traffic
– controlling quality of service (QoS) according to the type of information
– compensating for multipath fading
– achieving high system capacity

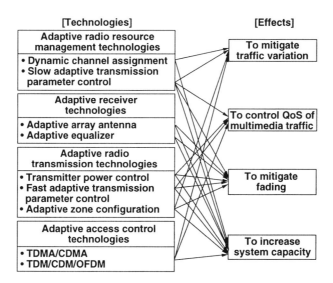

Fig. 1   Classification and effects of the intelligent and flexible radio access/transmission techniques.

Figure 1 shows classification and effects of the intelligent and flexible radio access/transmission techniques. These techniques are classified into the followings:

– radio resource management technologies, including dynamic channel assignment (DCA) [4], [5] and slow adaptive transmission parameter control [6], [7]
– adaptive receiver technologies, including adaptive array antenna [8], [9] and adaptive equalizer [10]
– adaptive radio transmission technologies, including transmission power control [11], fast adaptive transmission parameter control [12]-[15] and adaptive zone configuration [16]
– adaptive access control technologies that flexibly select access or multiplexing schemes according to the required services and capacity

The adaptive reception techniques are the most popular and practical techniques to compensate for fading. For example, when we employ a bi-directional decision feedback equalizer (BDDFE) and two-branch space diversity, we can achieve 2 Mbit/s transmission with a reasonable number of DSP computation when the maximum delay time of the delayed path (τ_{max}) is up to 10 µs [17]. However, if τ_{max} is longer than this value, which could be sometimes observed in the macrocell environments near the mountainous area or hilly terrain, the adaptive equalizer is no longer a practical solution for 2 Mbit/s transmission because it requires too large a number of DSP computation.

One solution to break through such a limitation is to employ, in addition to the adaptive receiver techniques, some adaptive transmission parameter control techniques that adaptively control its transmission parameters, such as the modulation scheme and symbol rate, according to the expected channel conditions. In addition to such effective combination of the adaptive receiver and adaptive transmission parameter control techniques, adaptive radio resource management is an indispensable technology field, especially to cope with dynamic channel and traffic variations.

Furthermore, adaptive access control techniques would also be very important to support multimedia services because the optimum access scheme greatly depends on various constraint conditions, such as the maximum user rate, its quality, maximum system bandwidth as well as the operational environments.

Therefore, this paper will discuss the adaptive transmission parameter control and the adaptive access control technologies which would be essential technologies for future multimedia communication systems.

3 ADAPTIVE TRANSMISSION PARAMETER CONTROL TECHNOLOGIES

3.1 System Design Strategy of the Adaptive Transmission Parameter Control for TDMA-based and CDMA-based Systems

Although the adaptive transmission parameter control techniques can be applied to both the TDMA-based and CDMA-based systems, their strategies are quite different. Figure 2 shows relationship between TDMA-based and CDMA-based transmission parameter control systems.

	TDMA-based system	CDMA-based system
basic strategy	Adaptive modulation parameter control	Adaptive DS/SS parameter control
constraint conditions	constant maximum signal bandwodth	constant chip rate
variable parameters	modulation level symbol rate coding rate	processing gain coding rate
support of constant bit rate transmission	fast adaptive modulation with buffer	channel activiation

Fig. 2 Relationship between TDMA-based and CDMA-based
Transmission parameter control concept.

In the case of TDMA-based systems, variable transmission parameters are the modulation level [12], symbol rate [13] and coding rate of the channel encoder [14]. These parameters are controlled according to the expected propagation path condition for the transmitter time slot considering the required BER, user rate, assigned bandwidth, and traffic conditions. In the case of CDMA-based systems, on the other hand, processing gain and coding rate of the channel encoder are the variable transmission parameters [18], [19].

One disadvantage of the adaptive transmission parameter control system is that its instantaneous bit rate is subject to the instantaneous propagation path conditions, which is a serious problem for supporting constant bit rate media. In the case of the TDMA-based system, data buffering is a solution for this problem [20]. In the case of CDMA-based systems, on the other hand, its solution is a channel activation; a shorter burst is employed in the case of smaller processing gain, and a longer burst is employed in the case of larger processing gain [18].

3.2 Classification of the Adaptive Transmission Parameter Control Systems

The adaptive transmission parameter control techniques are roughly classified into the following three types:

- **pre-assigned transmission parameter control mode (pre-assigned mode)** in which transmission parameters are preliminarily determined in each base station or supported services in the base station. During stand-by period, each terminal checks an available mode by detecting broadcasted message from the base station.
- **slow adaptive transmission parameter control mode (slow control mode)** in which transmission parameters for each call are assigned by the base station during the initial call set up process, and the assigned parameters are employed until the call is terminated, or a new channel is assigned by the handover process.
- **fast adaptive transmission parameter control mode (fast control mode)** in which transmission parameters are dynamically controlled during each call according to the fast variation of propagation path characteristics.

One of the most important issues for introduction of a new system is how to smoothly migrate from currently operated systems to a new system. From this viewpoint, pre-assigned mode is a good initial step. Therefore, it is already introduced in several dual mode cellular systems as well as the FLEX$^{(TM)}$ pager systems. Moreover, the concept was also included in some TDMA-based IMT-2000 candidate systems proposed in European and Japanese committees for standardization [21], [22].

Slow control mode is more advanced concept. A main difference between the pre-assigned mode and the slow control mode is that, the slow control mode adaptively controls transmission parameters for each terminal or base station call by call. Figure 3 shows a concept of the slow control mode (slow adaptive modulation system) for the application of TDMA-based macrocell and microcell system in comparison with the concept of conventional quaternary phase shift keying (QPSK) systems. In this figure, the modulation schemes for the slow control system are selected from full-rate 16-ary quadrature amplitude modulation (16QAM), full-rate QPSK, half-rate (1/2-rate) QPSK and quarter-rate (1/4-rate) QPSK [6].

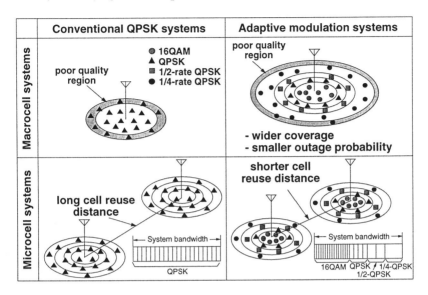

Fig. 3   Concept of the slow adaptive transmission parameter control mode for the application of TDMA-based macrocell and microcell systems.

In the case of macrocell systems, 16QAM is assigned when a terminal is located near the base station, and lower rate QPSK is assigned when a terminal is located in the fringe area. Therefore, the zone radius can be enlarged in comparison with the conventional QPSK systems because lower rate QPSK has higher receiver sensitivity than the full rate QPSK.

In the case of microcell systems, on the other hand, reduction of cell reuse distance can be expected due to higher receiver sensitivity of the lower rate QPSK. At the same time, the number of channels per assigned system bandwidth can also be increased because a higher modulation level is applicable to terminals near the base station as shown in Fig. 3. Moreover, when the slow control mode is effectively combined with the DCA technique, we can further increase system capacity [23]. At present, the slow control mode is applied to the General Packet Radio Service (GPRS) systems.

Almost the same effects can be obtained in the case of CDMA-based systems by controlling CDMA specific transmission parameters, such as the processing gain and coding rate [19].

Fast control mode is the most advanced concept. One of its distinct features is that it is alternative to the transmitter power control techniques because energy per bit to the noise spectral density (E_b/N_0) can be controlled by changing transmission parameters [24] without changing the transmitted power. Because the fast control mode is the most effective in improving transmission quality, system capacity and data throughput, we will explain its performances in more detail in the following sections.

3.3   TDMA-Based Fast Adaptive Transmission Parameter Control Systems

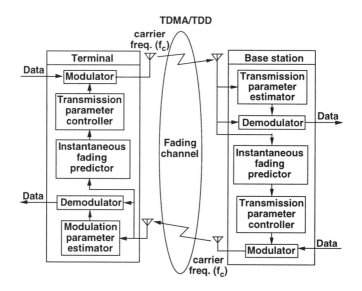

Fig. 4   Basic concept of the TDMA/TDD based fast adaptive transmission parameter control system.

Figure 4 shows the basic concept of TDMA and time division duplex (TDD) based fast adaptive transmission parameter control systems. At the instantaneous fading predictor in the receiver, instantaneous delay profile for the next transmission time slot is estimated, and the optimum modulation parameters that achieve the highest bit rate with satisfying a certain BER (BER_{th}) is selected in the modulation parameter controller according to the estimated delay profile. Because the modulation parameters are controlled slot by slot, modulation parameter estimator is prepared in the receiver to know the used modulation parameters for the received signal [13].

Ref. [25] shows laboratory experimental results of a TDMA-based adaptive transmission parameter control system. In the developed system, all the techniques necessary for slot-by-slot transmission parameter control including synchronization are installed.

Figure 5 shows frame and slot formats of the experimental system. Four channels are multiplexed on a TDMA/TDD frame with its frame length of 8 ms. Each slot consists of preamble, midamble, postamble and information symbol sections (Data in Fig. 5). Full rate (400 ksymbol/s) binary phase shift keying (BPSK) is applied to the preamble, midamble and postamble, and modulation parameter control is applied only to the information symbol sections. In preamble, midamble and postamble, eight pilot symbols (P in Fig. 5) are embedded for the pilot symbol-aided fading compensation technique. In the midamble, two types of control words are embedded; a channel estimation word (CE) that consists of a 4-stage modified pseudo-random noise (PN) sequence to measure delay profile for each reception time slot, and a modulation parameter estimation word (W) that consists of 16-symbol Walsh function to estimate modulation parameters used for demodulation of each reception time slot.

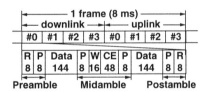

Fig. 5   Frame and slot formats of the implemented system [25].

In the receiver, delay profile for each reception time slot is measured by taking correlation between the received CE and the transmitted CE, and delay profile for the next transmission time slot is estimated by extrapolating the measured delay profile sequence. Using the estimated delay profile, C/N_0 and delay spread for the next transmission time slot is estimated, and a combination of transmission parameters that achieves the highest bit rate with satisfying BER of less than 10^{-4} is selected.

Figure 6 shows a transmission parameter selection chart for the implemented system, where coding rate (r) is controlled by a punctured code technique and its original encoder is a convolutional encoder with its coding rate of 1/2 and its constraint length (K) of 7. When C/N_0 is high and delay spread is small, transmission parameters that achieve higher bit rate transmission are selected. On the other hand, when C/N_0 is low or delay

106

spread is large, transmission parameters that achieve lower bit rate are
selected.

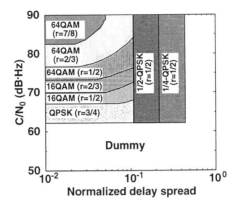

Fig. 6 Transmission parameter selection chart for the implemented system [25].

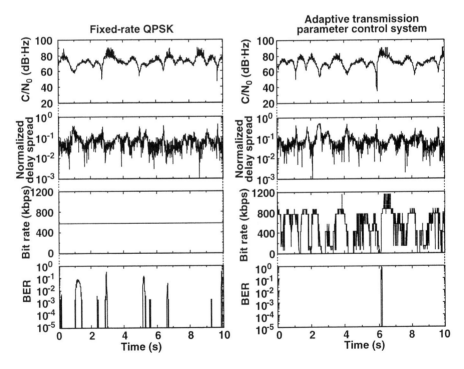

Fig. 7 Time variations of C/N_0, normalized delay spread, bit rate and BER under frequency
selective fading conditions (average $C/N_0 = 75$ dB·Hz and $f_d = 1$ Hz) [25].

Figure 7 shows time variation of C/N_0, delay spread normalized by a symbol duration for full rate transmission (normalized delay spread), instantaneous bit rate and instantaneous BER for both full-rate QPSK and the adaptive transmission parameter control modes of the implemented system under frequency selective fading conditions, where the average C/N_0 is 75 dB·Hz, normalized delay spread is 0.1, and maximum Doppler frequency is 1 Hz. In the case of the adaptive transmission parameter control system, transmission parameters for higher bit rate is selected when C/N_0 is high and delay spread is small, whereas transmission parameters with lower bit rate is selected when C/N_0 is low or delay spread is large. As a result, BER and burst error length for the adaptive transmission parameter control system are drastically lowered compared to the fixed rate QPSK system.

3.4 CDMA-Based Fast Adaptive Transmission Parameter Control Systems

Figure 8 shows base station and mobile station configurations of the CDMA-based fast adaptive transmission parameter control system [26]. In the mobile terminal, a pilot channel is prepared for the received CINR monitor as well as for the coherent Rake diversity combining at the base station. To keep orthogonality between the pilot and traffic channels, different spreading codes that consists of Walsh code and pseudo noise (PN) code concatenated codes are multiplied to both channels.

In the base station receiver, according to the measured CINR using the pilot channel, an optimum combination of the processing gain, coding rate and channel activity is selected at the transmission parameter selector. Table 1 shows relationship between the received carrier to interference plus noise power ratio (CINR) and transmission parameters. Moreover, a power control command (PCC) is selected at the PCC selection stage. The transmission parameter and PCC commands multiplexed with the traffic data are transmitted to the mobile station.

At the mobile station, after the transmission parameter and PCC commands are detected, the traffic channel is modulated according to the received transmission parameter message, and transmitted at the power level requested by the PCC.

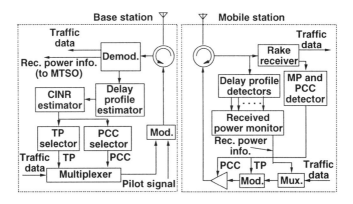

Fig. 8 Configuration of the CDMA-base fast adaptive transmission parameter control system.

Table 1   Relationship between CINR and selectable transmission parameters for a CDMA-based adaptive transmission parameter control system [26].

No.	CINR (dB)	Processing gain	Coding rate	Channel activity
1	CINR<-19.7	256	1/2	1
2	-19.7<CINR<-18.4	256	2/3	96/128
3	-18.4<CINR<-16.7	128	1/2	64/128
4	-16.7<CINR<-15.4	128	2/3	48/128
5	-15.4<CINR<-13.7	64	1/2	32/128
6	-13.7<CINR<-12.4	64	2/3	24/128
7	-12.4<CINR<-10.7	32	1/2	16/128
8	-10.7<CINR<-9.4	32	2/3	12/128
9	-9.4<CINR<-7.7	16	1/2	8/128
10	-7.7<CINR<-6.4	16	2/3	6/128
11	-6.4<CINR<-4.7	8	1/2	4/128
12	-4.7<CINR	8	2/3	3/128

In the CDMA-based adaptive transmission parameter control system, short duration peak of interference to neighboring cells (intercell interference) can be reduced because target CINR for terminals in the fringe area can be reduced. At the same time, reduction of CINR for terminals in the fringe area can be compensated for using the transmission parameter control because the adaptive transmission parameter control scheme has equivalent effect to the transmission power control. To further utilize this equivalent transmission parameter control effect, the CDMA-based adaptive transmission parameter control system in Ref. [26] also employs a soft power control technique. In this scheme, upper and lower CINR thresholds are prepared; when the received CINR is higher than the upper CINR threshold, the transmitter power is reduced to achieve CINR of equal to the upper threshold value; when the received CINR is lower than the lower CINR threshold, the transmitter power is increased to achieve the received

CINR of equal to the lower threshold value; otherwise, the transmitter power is not controlled.

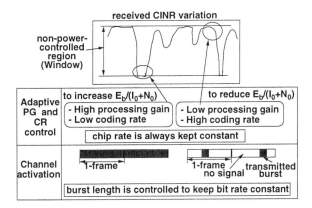

Fig. 9   Basic concept of the CDMA-based adaptive transmission parameter control combined with soft power control and channel activation.

Figure 9 shows the basic concept of the CDMA-based adaptive transmission parameter control technique combined with soft power control and channel activation techniques. In this system, processing gain and coding rate of the channel encoder are controlled while keeping the chip rate constant in order to keep the $E_b/(N_0+I_0)$ constant.

When such an adaptive control is employed in the case of a fixed-size data file transmission under good channel conditions, average bit rate can be increased thereby increasing throughput performance of the data file transmission. On the other hand, in the case of real time and constant bit rate services, a channel activation technique in which shorter burst is employed in the case of higher information bit rate, and longer burst is employed in the case of lower information bit rate.

Figure 10 shows the outage probabilities vs. the number of users in each cell for both the conventional and proposed systems, where upper and lower CINR threshold values for soft power control are -5 dB and -18 dB, and the outage probability is defined as the probability that the BER $< 10^{-3}$ cannot be satisfied [25]. When the acceptable outage probability is defined as 1%, the maximum number of users in each cell for the conventional system is 25, and that for the adaptive transmission parameter control system is 45. Therefore, the adaptive transmission parameter control system can achieve 80% higher capacity than the conventional system.

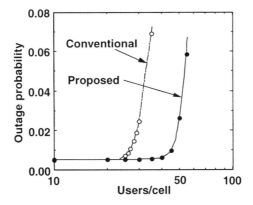

Fig. 10 Outage probability vs. the number of users in each cell for the conventional
and the adaptive transmission parameter control systems [26].

4   FLEXIBLE ACCESS CONTROL TECHNIQUES

4.1   Comparison of TDMA and CDMA features

Although CDMA is an effective access scheme for wireless multimedia
communication systems, it is not an all-round technology from the
viewpoints of service extension to various operational environments as well
as further increase in system capacity, especially under irregular traffic
conditions.

Table 2 Feature comparison between TDMA and CDMA.

Requirements	TDMA	CDMA
Flexible radio resource management	• DCA • adaptive TP control	• no frequency planning required • adaptive TP control
Max. bit rate (single code or carrier)	• 2-4 times a system bandwidth (by using QAM)	• less than 1/10 of a system bandwidth (to resolve multipath)
Extensiton of Max. bit rate	• multi-carrier	• multi-code
Inter-system co-ordination for private systems	• DCA	• difficult
Multi-layer cell	TDM or FDM-based inter-layer orthogonality required	

Table 2 summarizes comparison of TDMA and CDMA features from these viewpoints. To support variety of services regardless of its operational environments and dynamic variation of traffic, CDMA is very suitable because it requires no frequency planning. However, it has several disadvantages, i.e., maximum bit rate is limited to less than approximately 1/10 of the system bandwidth to resolve multipath, inter-system coordination is quite difficult when it is applied to the indoor private systems, and so on.

Another important issue is the multi-layer cell applicability because it is very effective in increasing system capacity and in coping with extraordinary hot spot traffic [27]. In the case of TDMA systems, inter-layer orthogonality is easily achieved by allocating different frequency band or different time slots to each layer. In the case of CDMA systems, on the other hand, allocation of different frequency band is impossible if the allocated frequency band is fully employed by the macrocell layer. Therefore, we will propose a TDM-based flexible access control technique.

4.2 Time Division Multiplexing-Based Access Control System

Because CDMA will be a strong candidate access scheme for the third generation wireless systems, this section will discuss how to include a flexible access control function in the CDMA-based wireless systems.

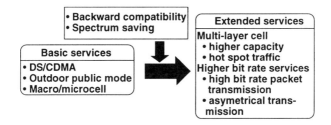

Fig. 11 A migration from basic services to the extended services.

Figure 11 shows a migration scenario from basic services to the extended services. Among many technologies, multi-layer cell and high bit rate transmission techniques are considered to be the most effective means to satisfy the requirements. In this case, backward compatibility and spectrum saving are very strong requirements. Therefore, we will propose a TDM-based access control scheme using a punctured code technique.

112

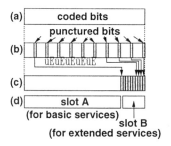

Fig. 12 Frame formats for a TDM-based access control scheme.

Figure 12 shows a frame format of the proposed system. Figure 12 (a) shows a frame format of convolutionally encoded data sequence for a CDMA system. In the proposed system, let us assume that the transmitted data for a CDMA system are convolutionally encoded as shown in Fig. 12 (a). Next, a part of this encoded bit sequence is regularly deleted using a punctured code technique as shown in Fig. 12 (b), and locates the deleted bits at the end of each frame as shown in Fig. 12 (c). In the following, the former and latter slots will be called slot A and slot B as shown in Fig. 12 (d). We can support basic services when we employ the frame format shown in Fig. 12 (c). On the other hand, when extended services will be introduced, the slot A is employed to support basic services and the slot B is employed for the extended services.

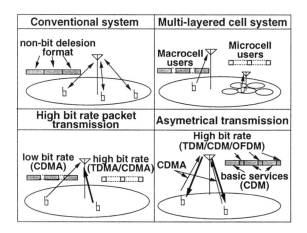

Fig. 13 Application examples of the proposed scheme.

Figure 13 shows some examples of the application of the proposed scheme. In the case of multi-layer cell application, slot A is employed for the macrocell terminals and slot B is employed for the microcell terminals. When we apply it to the high-bit-rate packet transmission system, we can employ not only CDMA but also TDMA as the access scheme for slot B. Especially, when we employ TDMA and adaptive transmission parameter control techniques for packet transmission using slot B, we can achieve very high throughput performances, say, several Mbit/s to 10 Mbit/s. In the case of applying the proposed technique to downlink of the asymmetrical transmission systems, we can increase downlink bit rate for supporting various broadcasting services. In this case, we can also employ orthogonal frequency division multiplexing (OFDM) techniques in addition to the TDM and CDM techniques.

4.3 System Capacity Evaluation of a Multi-Layer Cell System using the Proposed Scheme

For the evaluation of the proposed TDM-based adaptive access control system in the case of multi-layer cell application, we have conducted computer simulation, where coding rate of the convolutional encoder for the conventional system is 1/3 and that for the extended system (punctured code) is assumed 1/2. When we employ rate-1/2 punctured code, because slot length for slot A is 2/3 of the conventional system, slot length for slot B becomes half of the slot A. To support the same user rate for both slot A and slot B, processing gain for slot A and that for slot B are selected as 256 and 128, respectively.

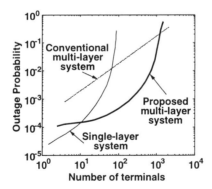

Fig. 14 System capacity of the conventional single-layer, conventional multi-layer and the proposed TDD-based multi-layer systems.

Figure 14 shows system capacity comparison between the following three systems:

conventional single layer cell configuration: processing gain (PG)=256 and coding rate (r)=1/3

conventional multi-layer cell configuration: PG=256 and r=1/3 for both macro and micro cells, where both cells employ a frame format of Fig. 12 (c), and the same frequency band is allocated to both layers.

proposed multi-layer cell system: PG=256 and r=1/2 are employed for macrocells (slot A) and PG=128 and r=1/2 are employed for microcells (slot B), where 49 microcells are overlaid to a macrocell. In any case, the number of macrocell user is assumed 2% of the total users.

When we define system capacity as the number of acceptable users for the outage probability of 1%, we can find that the acceptable users for the conventional system, that for the conventional multi-layer cell configuration and that for the proposed multi-layer cell configuration are 60, 60 and 600, respectively. This means that the conventional multi-layer cell configuration is helpless to increase system capacity due to its inter-layer interference, whereas the proposed multi-layer cell configuration can achieve approximately 10 times higher system capacity than the conventional single layer cell configuration. These results show that the proposed multi-layer cell configuration is very effective in increasing system capacity due to its high inter-layer orthogonality.

5 CONCLUSION

We have discussed intelligent and flexible radio transmission/access technologies for wireless multimedia communication systems. Especially, we have focused on the adaptive transmission parameter control and adaptive access control technologies because these two are quite important research fields and these technologies exactly meet requirements for service flexibility, adoption of phased approach, and easy introduction of the market oriented new services with satisfying backward compatibility. These technologies, we believe, will act important roles in the soon coming software defined radio era.

REFERENCES

[1] Recommendation ITU-R M. 1034, "Requirements for the radio interface(s) for future public land mobile telecommunication systems (FPLMTS)," 1994.

[2] S. Sampei, "Application of digital wireless technologies to global wireless communications," Chapter 1, Prentice-Hall, 1997.

[3] N. Morinaga, M. Nakagawa and R. Kohno, "New concepts and technologies for achieving high reliable and high-capacity multimedia wireless communication systems," IEEE Commun. Mag., Vol. 35, No. 1, pp. 34-40, Jan. 1997.

[4] T. Kanai, "Autonomous reuse partition in cellular systems," 42nd IEEE Veh. Tech. Conf. (Denver, CO), pp. 782-785, May 1992.

[5] Y. Furuya and Y. Akaiwa, "Channel segregation, a distributed adaptive channel allocation scheme for mobile communication systems," IEICE Trans. Commun., Vol. E74, No. 6, pp. 1531-1537, June 1991.

[6] S. Sampei, S. Komaki and N. Morinaga, "Adaptive modulation/TDMA scheme for personal multi-media communication systems," GLOBECOM'94 (San Francisco, CA), pp. 989-993, Nov. 1994.

[7] S. Sampei, N. Morinaga and Y. Kamio, "Adaptive modulation/TDMA with a BDDFE for 2 Mbit/s multimedia wireless communication systems," 45th IEEE Veh. Tech. Conf. (Chicago, IL), pp. 311-315, July 1995.

[8] T. Ohgane, N. Matsuzawa, T. Shimura, M. Mizuno and H. Sasaoka, "BER performance of CMA adaptive array for high-speed GMSK mobile communication –a description of measurements in central Tokyo," IEEE Trans. Veh. Technol., Vol. 42, No. 4, pp. 484-490, Nov. 1993.

[9] Y. Ogawa, K. Yokohata and K. Itoh, "Spatial-domain path diversity using an adaptive array for mobile communications," 4th IEEE ICUPC (Tokyo, Japan), pp. 600-604, Nov. 1995.

[10] R. D'Avella, L. Moreno and M. Agostino, "An adaptive MLSE receiver for TDMA digital mobile radio," IEEE J. Sel. Areas Commun., Vol. 7, No. 1, pp. 122-129, Jan. 1989.

[11] A. J. Viterbi, "CDMA, principle of spread spectrum communications," Addison-Wesley Publishing Company, 1995.

[12] W. T. Webb and L. Hanzo, "Quadrature amplitude modulation -Principle and application for fixed and wireless communications," Pentech Press, 1994.

[13] T. Ue, S. Sampei and N. Morinaga, "Symbol rate and modulation level controlled adaptive modulation/TDMA/TDD for personal communication systems," 45th IEEE Veh. Tech. Conf. (Chicago, IL), pp. 306-310, July 1995.

[14] S. M. Alamouti and S. Kallel, "Adaptive trellis-coded multiple-phase-shift keying for Rayleigh fading channels," IEEE Trans. Commun., Vol. 42, No. 6, pp. 2305-2314, June 1994.

[15] S.-G. Chua and A. Goldsmith, "Adaptive coded modulation for fading channels," IEEE ICC'97 (Montreal, Canada), pp. 1488-1492, June 1997.

[16] D. J. Y. Lee and C. Xu, "Capacity and trunking efficiency of smart antenna," VTC'97 (Phoenix, AZ), pp. 612-616, May 1997.

[17] T. Nagayasu, S. Sampei and Y. Kamio, "Complexity reduction and performance improvement of a decision feedback equalizer for 16QAM in land mobile communications," IEEE Trans. Veh. Technol., Vol. 44, No. 3, pp. 570-578, Aug. 1995

[18] S. Abeta, S. Sampei and N. Morinaga, "Adaptive coding rate and processing gain control with channel activation for multimedia DS/CDMA systems," IEICE Trans. Commun., Vol. E80-B, No. 4, pp. 581-588, April 1997.

[19] E. Berruto, M. Gudmundson, R. Menolascino, W. Mohr and M. Oizarriso, "Research activities on UMTS radio interface, network architectures, and planning," IEEE Commun. Mag., pp. 82-95, Feb. 1998.

[20] Y. Kamio, S. Sampei and N. Morinaga, "Performance of modulation-level-controlled adaptive-modulation under limited transmission delay time for land mobile communications," 45th IEEE Veh. Tech. Conf. (Chicago, IL), pp. 221-225, July 1995.

[21] A. Urie, M. Streeton, C. Mourot, "An advanced TDMA mobile access system for UMTS," IEEE Personal Communications," Vol. 2, No. 1, pp. 38-47, Feb. 1995.

[22] ARIB FPLMTS Study Committee, "Report on FPLMTS Radio Transmission Technology Special Group (Round 2 activity report), Draft v.E1.1, Jan. 1997.

[23] T. Ikeda, S. Sampei and N. Morinaga, "TDMA-based adaptive modulation with dynamic channel assignment (AMDCA) for high capacity multi-media microcellular systems," VTC'97 (Phoenix, AZ), pp. 1479-1483, May 1997.

[24] S. Sampei, "Application of digital wireless technologies to global wireless communications," Chapter 17, Prentice-Hall, 1997.

[25] S. Sampei, N. Morinaga and K. Hamaguchi, "Experimental results of a multi-mode adaptive modulation/TDMA/TDD system for high quality and high bit rate wireless multimedia communication systems," VTC'98 (Otawa, Canada), May 1998 (to be presented).

[26] M. Hashimoto, S. Sampei and N. Morinaga, "Forward and reverse link capacity enhancement of DS/CDMA cellular system using channel activation and soft power control techniques," PIMRC'97 (Helsinki, Finland), pp. 246-250, Sept. 1997.

[27] Recommendation ITU-R M. 1035, "Framework for radio interfaces and radio subsystem functionality for FPLMTS", 1994.

Chapter 7

Current Topics in Wireless & Mobile ATM Networks: QoS Control, IP Support and Legacy Integration

D. RAYCHAUDHURI
NEC USA, C&C Research Laboratories

Abstract: This paper presents a selection of current topics related to emerging wireless and mobile ATM network technologies. A brief summary of wireless ATM system architecture and related radio access layer and "M" UNI/NNI protocols is given in the introductory section. As technical approaches for these radio access and mobile network protocols are now entering the convergence phase, we identify several systems-oriented issues that need to be addressed next for efficient and flexible use of the mobile/wireless ATM networking platform. Specific topics discussed include: (1) Quality-of-service (QoS) control in wireless ATM systems; (2) Interworking with legacy wireless services such as GSM and IS-95/136; and (3) IP support alternatives such as IP-over-mobile ATM, IP switching (Ipsofacto), and IP routing over WATM link layer.

1 INTRODUCTION

Wireless ATM, first proposed in [1,2], is a specific technical approach for implementing next-generation wireless networks capable of delivering broadband services to mobile multimedia devices. During the past few years, mobile and wireless ATM technologies have steadily migrated from research stage to standardization and early commercialization. A number of R&D organizations have demonstrated proof-of-concept prototypes [3-6], and early products are expected to reach the market in the next 1-2 years. Several standards activities, including the ATM Forum's Wireless ATM (WATM) working group [7], ETSI's Broadband Radio Access Networks (BRAN) [8] and Japan's Mobile Multimedia Access Communication (MMAC) [9] are currently in progress, and should produce initial specifications for mobile network and radio air interface protocols during 1998 and 1999.

The above mentioned standards, together with the 5 Ghz U-NII [10] and Hiperlan spectrum allocations [11] in the U.S. and Europe should lead to a rapid increase in commercial activity in the broadband wireless area. We visualize the emergence of public and private microcellular wireless networks which provide a variety of broadband mobile services including high-speed Internet access and audio/video delivery. The availability of such a broadband wireless networks by the end of this decade should stimulate the evolution of a new generation of high-performance mobile computing devices in both vertical and horizontal markets worldwide. It is noted here that although wireless ATM is typically associated with mobile multimedia services, selected components (mobile network infrastructure, radio access link, etc.) of the technology can be applied to a variety of other application scenarios including PCS/cellular infrastructure, wireless Internet access, residential wireless local loop and microwave infrastructure links.

A review of wireless ATM system architecture and technology status as of Sept 1997 was presented at the last PIMRC conference [12]. During the past year, the field has matured steadily with a gradual convergence of technical approaches for core WATM subsystems such as radio/modem, MAC/DLC, and mobile ATM handoff and location management protocols (see for example the collection of papers in [13]). As these core network functions become well understood and/or standardardized, a new set of technical issues at a higher level of abstraction need to be addressed in order to complete a system solution based on wireless ATM technology.

One key system-level issue is that of providing quality-of-service (QoS) for wireless ATM services transported on shared-access radio links with handoff of connections during user movement. Robust QoS control mechanisms (possibly involving "soft" bandwidth control and network-initiated renegotiation) are required to deal with wireless channel impairments and dynamic rerouting during handoff. A second issue of particular importance is that of supporting legacy radio services (such as cellular voice or data) using the same mobile ATM network infrastructure used for future broadband applications. A suitably designed network should be able to support a mix of legacy and broadband wireless services within a single framework, thus facilitating the necessary migration from today's cellular and wireless data services towards the so-called "IMT-2000" third-generation wireless scenario being defined by the ITU. A third problem of current significance is that of efficiently supporting Internet protocol (IP) services in a wireless/mobile ATM network. As in conventional ATM, there are various technical options for IP support, ranging from existing IP-over-ATM solutions which utilize ATM signalling to alternative methods such as IP switching or routing directly over ATM/WATM hardware.

In the following sections of this paper, we present a brief review of wireless and mobile ATM network architecture, and then discuss open issues and potential technical approaches for each of the above mentioned topics, i.e. QoS control, legacy service integration and IP support.

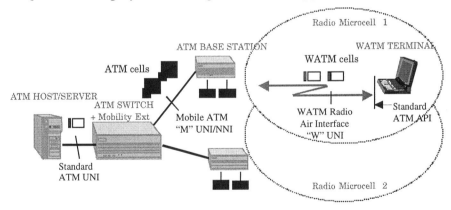

Fig. 1. Wireless ATM System Components and Interfaces

2   WIRELESS ATM ARCHITECTURE & PROTOCOLS

The major hardware/software components which constitute basic wireless ATM are shown in Fig. 1. From the figure, it can be seen that a wireless ATM system typically consists of three major components: (1) ATM switches with standard UNI/NNI capabilities together with additional mobility support software; (2) ATM "base stations" or "access points" also with mobility-enhanced UNI/NNI software and radio interface capabilities; and (3) wireless ATM terminal with a WATM radio network interface card (NIC) and mobility & radio enhanced UNI software. Thus, there are two new hardware components, ATM base station (which can be viewed as a small mobility-enhanced switch with both radio and fiber ports) and WATM NIC, to be developed for a wireless ATM system. New software components include the mobile ATM protocol extensions for switches and base stations, as well as the WATM UNI driver needed to support mobility and radio features on the user side.

The system shown involves two new protocol interfaces: (a) the 'W" UNI between mobile/wireless user terminal and ATM base station; and (b) the "M" UNI/NNI interface between mobility-capable ATM network devices including switches and base stations. The "W" UNI interface contains support for mapping of user terminal ATM connections to the shared-medium radio access link, as well as for terminal mobility (handoff and location management). The "M" UNI/NNI interface between mobility-enhanced switches and base stations contains signalling and routing protocol extensions necessary for handoff and location management.

For completeness, we briefly summarize the major functions of the radio access layer and mobile ATM protocols which form the basis of the wireless ATM network. The reader is referred to our earlier papers such as [14-16] for further technical details on the basic system.

2.1   WATM Radio Access

WATM radio access is supported by PHY, MAC, DLC and wireless control protocols which are integrated into the standard ATM UNI protocol stack, as shown in Fig. 2.

The WATM PHY consists of two parts, the PMD (physical media dependent) sublayer which specifies the modem's basic transmission frequency/format, and the RTC (radio transport convergence) sublayer which specifies the transport format and framing used by the MAC layer above. The MAC layer is responsible for assignment of RTC-level resources to multiple mobile terminals which generate ATM virtual circuits (both signalling and data) and associated wireless control messages. A key issue is that of mapping ATM services such as UBR, CBR, VBR to the radio channel so as to maintain required QoS properties for each virtual circuit. The data link control (DLC) layer provides the interface from WATM radio to the ATM data plane and is responsible for reliable, in-sequence delivery of radio cells to the ATM layer.

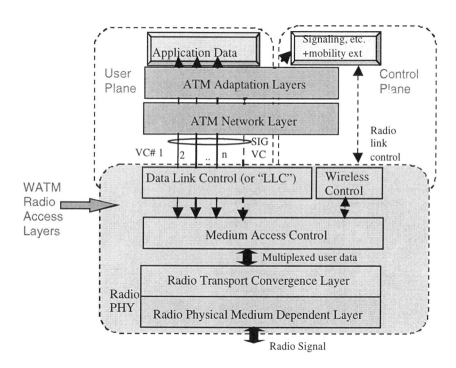

Fig. 2. WATM Protocol Stack with Radio Access Layers

The wireless control protocol interfaces the WATM radio to the ATM control plane (signalling), and is responsible for functions such as terminal registration, authentication and MAC/DLC control. Fig. 3 provides an overview of WATM MAC layer protocol functions and syntax for a representative dynamic TDMA/TDD approach. Further details can be found in [17-19].

2.2   Mobile ATM

Mobile ATM protocols ("M" UNI/NNI) are intended to provide integrated support for mobile users within the core ATM network infrastructure. As mentioned earlier, the approach is to augment existing ATM protocols (such as UNI, ILMI and PNNI) to provide additional ("+M") functions required to manage mobile users. The basic functions of mobile ATM are mobile user registration (via extensions to ILMI), location management (via extensions to connection signalling) and dynamic handoff (via extensions to signalling and routing). Mobile user registration is the process by which a mobile terminal is assigned an ATM address to initiate services on the network. While this is similar to initial address assignment in conventional ATM, mobility requires additional authentication procedures and more dynamic address management for roaming users. Location management functions provide the ability to find mobile users during connection setup. In an integrated mobile ATM approach, this is achieved without partitioning the address space between fixed and mobile users, thus avoiding arbitrary distinctions between mobile and non-mobile services. Typically, a "home switch" assigns a permanent address to each mobile user and then maintains pointers to the current temporary address of the mobile through the use of a suitable location update protocol. Location management may be integrated with the connection setup procedure to provide transparent ATM service to mobile users. Finally, once a connection has been established, the mobile ATM network also provides handoff support by dynamically rerouting connections from one point of radio attachment to another. This involves rerouting a portion of the ATM connection through a crossover switch (COS) during the course of the call, while maintaining QoS and other service parameters to the extent possible. A conceptual summary of mobile ATM network operations is given in Fig. 4. Further details can be found in [16, 20-23].

The WATM radio access and mobile ATM protocols outlined above are currently going through a technical convergence process in the technical literature and in standards committees such as ATM Forum's WATM WG and ETSI BRAN. Although several details have yet to be worked out, the outline of system design for WATM has emerged in the technical community generally consisting of the following:

-        Low power, microcell radio access, typically in the 5 Ghz U-NII or Hiperlan bands

- ~25 Mbps modem, either single carrier QPSK/GMSK or multicarrier OFDM
- TDMA/TDD based RTC format, with 1-2 ms frames and ATM cell size data slots
- Dynamic TDMA MAC protocol with UBR/VBR/CBR allocation for ATM traffic and contention access for control
- Selective reject ARQ based DLC with per-cell sequence numbering
- Location management with single address space for fixed and mobile users, integrated with connection control signalling
- Handoff control with based on extensions to UNI signalling and PNNI, with flexible COS discovery and rerouting procedures

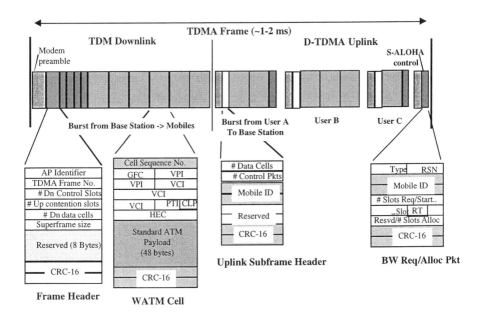

Fig. 3. Overview of WATM MAC layer protocol for TDMA/TDD example

Implementations of wireless ATM systems based on the design features listed above include.

124

- NEC's WATMnet™ (prototype demonstrated mid-95, trial product due late-98) [6]
- ORL's Radio ATM System (prototype demonstrated mid-96) [3]
- ACTS "MAGIC WAND" system (prototype demonstration due mid-98) [24]

It is expected that most of the details for the above technology components in wireless/mobile ATM will converge during the next 1-2 years. As the basic networking framework matures, it is appropriate to start consideration of a number of systems-oriented issues related to efficient or flexible use of the wireless/mobile ATM platform. Specific topics that need to be addressed include (but are not limited to):

- Quality-of-service control and traffic management
- Interworking with legacy wireless services, such as cellular voice and data
- IP support over wireless/mobile ATM networks
- Software architecture for scalable/robust mobile multimedia applications
- Audio/video support at a WATM terminal
- Network management system (NMS) extensions for mobile/wireless
- Application to UMTS or IMT-2000 scenario

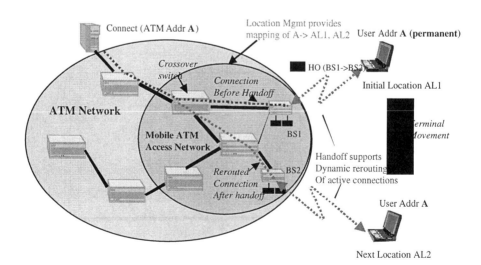

Fig. 4. Outline of mobile ATM network functionality

In the following sections, we provide a brief discussion of technical issues and potential solutions to three of the above topics, namely QoS control, legacy service integration and IP support.

3   QOS CONTROL IN WIRELESS ATM

Quality-of-Service (QoS) support in wireless ATM involves several considerations beyond those addressed in earlier work on conventional ATM networks. In particular, QoS control in a WATM scenario must deal with the following additional issues:

- potentially limited bandwidth on the shared-medium radio access link
- varying radio access link and core network traffic conditions during handoff/path rerouting associated with terminal mobility
- increased heterogeneity in terms of display, throughput and computational limitations of mobile terminals (laptops, PDA's, telephones, etc.)

The above factors imply the need for a robust and scalable QoS framework which permits some degree of variation in delivered service quality during the course of a connection [25]. Such a "soft QoS" control framework has been proposed in [26] in context of non-mobile ATM and IP networks, as a practical means for achieving high statistical multiplexing gains with non-stationary multimedia and video traffic. This approach is based on the concept of a "VBR+" service class which permits dynamic renegotiation of usage parameter control (UPC) values during the course of a connection. The renegotiation may be initiated either by end-user devices (server, terminal) or by network entities (switch, base station) using relatively simple ATM signalling extensions such as MODIFY_REQ, MODIFY_PROC and MODIFY_ACK/REJ [27]. The resulting end-to-end QoS framework for a WATM network is schematically illustrated in Fig. 5.

It is observed that the QoS model in Fig. 5 requires some degree of application scalability in order to operate robustly under varying conditions. This can be achieved if applications are designed to exhibit "soft" user satisfaction vs. bit-rate profiles typical of the "S-curves" shown in Fig. 5. Many real-time media sources (including compressed audio and video) have

126

been shown to exhibit such soft degradation [28], so that the principle can be used in most practical cases. Once the S-curve for an application component has been found, it is possible to design a QoS controller at network entities which takes into account specified parameters such as a "minimum satisfaction index" to allocate bandwidth between active connections, and to block new connection requests when congestion is anticipated.

Use of the VBR+ QoS model outlined above provides mechanisms to deal with each of the wireless or mobility specific issues identified earlier. The soft QoS approach addresses the problem of limited radio access bandwidth by providing scaling mechanisms to reduce application demands during congestion periods. The ability to start a virtual circuit at nominal QoS level and then renegotiate it downwards when additional users enter the radio microcell is an important feature for WATM. The same capability is helpful for dealing with handoff since migration of a higher bandwidth connection into a congested microcell might otherwise result in a sudden loss of service – using this approach, bandwidth can dynamically be reallocated from one active VC to another, as needed to maintain overall user satisfaction. Renegotiation capability is also useful during establishment of the new route after handoff since resource limitations may be experienced at one or more switching elements along the path. In the absence of soft QoS and VBR+ renegotiation, it would be difficult to provide statistical guarantees for services on mobile terminals without the use of overly conservative call admission control policies.

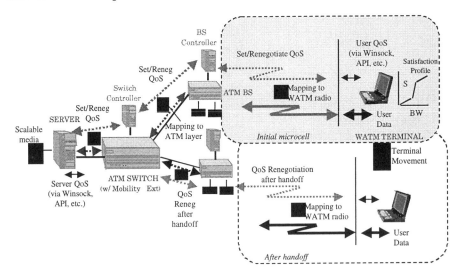

Fig. 5. End-to-end QoS Model for Wireless ATM System

Clearly, the last issue of mobile terminal heterogeneity can also be addressed once a dynamic QoS renegotiation capability is established at the network level. Application scalability with relatively large dynamic range is necessary for supporting a common set of multimedia applications on a wide range of devices. For the type of two-way QoS renegotiation model under consideration here, this requires careful definition of the application

Programming interface and the distributed software model above it. In [29], the authors outline an object-oriented software approach for adaptive QoS control using the concept of QoS "contracts" between client, server and network. The use of such contracts between service entities facilitates application scaling at during session establishment and permits adaptation to changes in both computational and networking capabilities during the course of the session.

4   LEGACY SERVICE INTEGRATION

In Sec 2, we briefly outlined the functions provided by "mobile ATM" protocols used by the core network in a wireless ATM system. As mentioned earlier, a network running the "M" UNI/NNI protocols can be viewed as a generic mobile service infrastructure which provides integrated support for mobility-related functions such location management and handoff control [16,30]. While these mobility support features are obviously applicable to new wireless ATM services, they also provide a basis for integrating legacy mobile services such as cellular/PCS and wireless data within an ATM infrastructure. The ability to integrate a variety of wireless services into a single network architecture contrasts with today's situation in the cellular industry where infrastructure networks must generally be customized to the specific radio air interface being used. For example, GSM radio access is associated with a GSM core network designed specifically for that wireless service. While such a vertically integrated approach might be acceptable for a single service telephony scenario, it does not readily provide a much needed migration path towards data (e.g. CDPD, GPRS) or integrated services (e.g., IMT-2000).

In mobile ATM, non-ATM radio access is supported generically by terminating specific radio protocols at the base station, which acts as an interworking gateway between the legacy mobility control protocol and "M"

128

UNI/NNI. [Observe that this is similar in spirit to current use of ATM edge switches as a generic network gateway for various fixed network protocols such as frame relay, IP and telephony.] This concept of a mobile ATM network supporting multiple radio access technologies is illustrated in Fig. 6 below. The mobile ATM base station is designed to accept plug-in radio modules corresponding to the wireless services being offered. The base station also incorporates processor modules which support interworking functions specific to the radio protocol(s) being terminated. Note that because the "M" UNI/NNI protocols incorporate handoff and location management functions, conventional mobile network components such as MSC, HLR and VLR are no longer needed within the mobile ATM cloud, potentially providing both cost and performance benefits over existing systems.

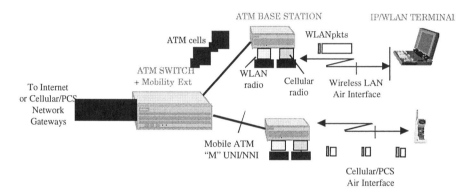

Fig. 6. Mobile ATM Network Supporting Multiple Radio Access Technologies

The details of interworking with important cellular/PCS protocols such as GSM, IS-95, IS-136, PHS, etc. are still under preliminary consideration [31]. Our preferred approach is to utilize handoff and location management capabilities of mobile ATM as long as users remain within the service area. This can be implemented using the concept of a "proxy M-UNI" process corresponding to active mobile terminal with radio attachment to a given base station. The proxy UNI process can be thought of as virtual WATM endpoint of the ATM network, thus making the mobility service model identical to that used for end-to-end WATM services. As the mobile moves, the proxy M-UNI process migrates from one base station to another while issuing appropriate handoff signalling and location update messages, etc. This model for cellular PCS (e.g. GSM) interworking via an M-UNI to radio protocol gateway at the base station is illustrated in Fig. 7 below. Of course, other options which use a subset of mobile ATM features (e.g. handoff control, but not location management, or ATM transport only without either handoff or location management) may also be used in conjunction with an overlay of servers executing legacy network protocols.

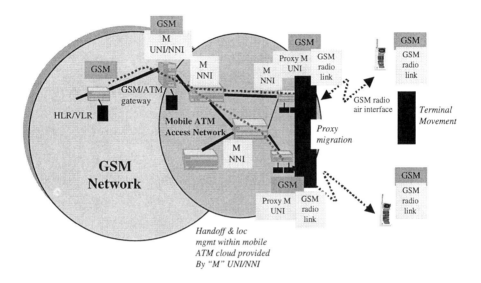

Fig 7. Use of proxy "M" UNI for supporting migration of non-WATM terminals

5   IP OVER MOBILE ATM

Efficient support of IP is clearly one of the most important system-level issues to be addressed in conjunction with initial deployment of wireless ATM networks. There are several levels at which IP service can be handled in wireless ATM, ranging from an ATM-centric solution ("IP-over-WATM") to an IP-centric solution (e.g. IP routed network with WATM link layer).

In the straightforward "IP-over-WATM" approach [16], mobility of the IP terminal within the mobile ATM cloud is handled transparently by the "M" UNI/NNI protocols in the same way as for cellular/PCS discussed earlier, as shown in Fig. 8. In this scenario, the mobile terminal has both an IP address and a (permanent) ATM address, and uses standard address resolution

130

protocols (ATM ARP or NHRP) to determine the mapping from IP to ATM. IP services within the mobile ATM cloud use ATM virtual circuits on an end-to-end basis; connections entering the Internet must pass through an IP gateway which terminates the ATM VC's and forwards packets in an IP routed mode. In both cases, the mobile ATM network provides full handoff and location management support as the terminal migrates within the service area. The mobile ATM segment appears as a mobility-enhanced link layer to the IP gateway, and thus does not require mobile IP to support local terminal movements. Of course, as described in [32], Mobile IP may still be used to handle global mobility, while mobile ATM is used for local mobility. Note that mobile ATM provides the important advantage of fast handoff when compared with mobile IP, thus reducing variations in the quality-of-service experienced by media flows or TCP connections at the mobile terminal. It is observed here, that as for legacy PCS/cellular services discussed in Sec. 4, the IP user may access the mobile ATM network using either a WATM radio or any other wireless data technology (such as IEEE802.11) and still receive the same network service (in the latter case, base stations must provide the proxy M-UNI gateway function outlined in Sec. 4).

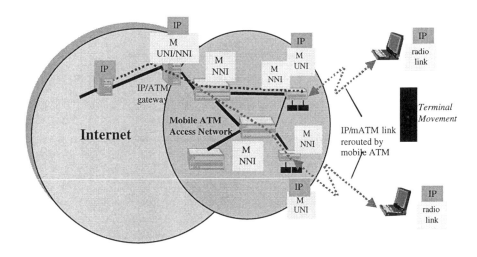

Fig 8. Principle of "IP-over-mobile ATM" approach

A second solution for IP support is based on "IP Switching" in which IP flows are mapped on to ATM hardware without the use of ATM signalling [33]. One possible IP switching method known as "Ipsofacto" (IP Switching Over Fast ATM Cell Transport) provides a direct mapping from IP routing protocols to ATM virtual circuits on a pure hop-by-hop basis [34]. In this scheme, the first packet of each IP flow is forwarded in a routed mode from one switch or base station controller to another. This packet then implicitly sets up a switched ATM path for subsequent IP packets on that flow, thus leveraging the throughput and QoS advantages of ATM hardware. To avoid signalling between network entities, Ipsofacto uses the concept of randomly selecting "unused VC's" at each routed hop and then locally binding input and output VC's to establish the desired switched path. Because Ipsofacto supports IP on a strict hop-by-hop basis, new features such as IP multicasting, RSVP and mobile IP can also be incorporated in a straightforward manner.

The Ipsofacto protocol can be readily applied to the wireless ATM scenario [35], by replacing ATM protocol stacks at each network entity (mobile terminals, base stations and switches) with lighter weight IP/Ipsofacto stacks. In the simplest case, mobility support can be provided by mobile IP running over Ipsofacto, as outlined in Fig 10. Of course, this mode does not allow for dynamic handoff of flows, but this might be acceptable for at least a subset of mobile Internet access applications. Optimizations for mobility related rerouting of switched paths in Ipsofacto are also possible. For example, if RSVP is used in conjunction with Ipsofacto, IP flows can be handed off dynamically from one base station to another using mobility extensions to RSVP (e.g. "M" RSVP introduced in [36]).

One additional element that needs to be considered in the WATM case is the dynamic mapping of VC's on the multiple access (unicast uplink and broadcast downlink) WATM radio link. Specifically, since the entire VC space must be shared by all mobiles using a given WATM link, it is necessary to add a control protocol on the wireless link for association of VC's to mobiles. For IP multicast services, the ability to specify a multicast receiver group is also required. Such capabilities may be added via extensions to IGMP ("+W") which allow mobile hosts and base stations to exchange VC establishment queries and reports. The reader is referred to [34] for further details.

Another alternative to IP support is to use the WATM radio access protocol only as a wireless link layer in conjunction with full IP routing support at the base station, as shown in Fig. 10. This model requires the WATM radio link to provide a packet interface in addition to the cell interface, a topic that is currently under consideration within ETSI BRAN. Since the WATM radio layer provides a superset of link layer services needed for IP (packet segmentation, MAC allocation and error control), it is relatively

132

straightforward to use the same radio in a packet mode if the right interfaces are specified in advance. The additional QoS capabilities offered by the WATM link should prove useful for future IP services (assuming that a suitable generic interface can be defined to serve both ATM and IP needs).

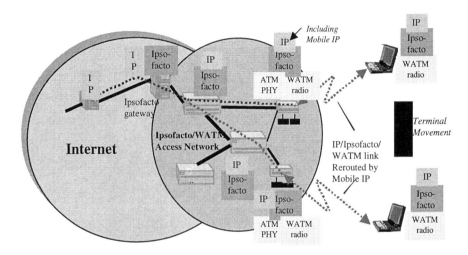

Fig. 9. Use of Ipsofacto (IP switching) in Wireless ATM

6   CONCLUDING REMARKS

This paper has presented an updated view of wireless ATM technology as it matures from R&D stage to standardization and early commercialization. A review of the basic features of the mobile ATM and radio access protocols which constitute a wireless ATM system was given. In view of the anticipated technical convergence of the core mobile/wireless ATM protocols, several systems-level issues which need to be considered next were identified. A brief discussion of technical issues and potential solutions was provided for the following selected topics: (a) QoS control; (b) legacy service integration, and (c) IP support. We are currently engaged in detailed design and prototype/trial product implementations related to each of the above issues, and expect to provide more complete results in future publications.

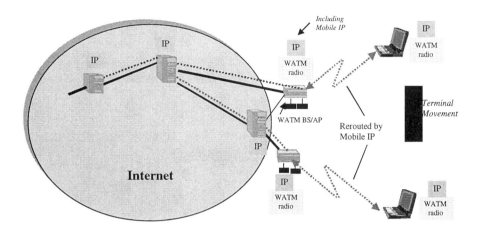

Fig. 10. WATM Radio as a Link Layer for IP Networks

Acknowledgements

The author would like to acknowledge valuable contributions from Dr. Arup Acharya and Dr. Daniel Reininger to the sections on IP support and QoS Control respectively.

REFERENCES

[1] D. Raychaudhuri and N. Wilson, "Multimedia Personal Communication Networks: System Design Issues", 3rd WINLAB Workshop on 3rd Gen. Wireless Information Networks, April 1992, pp. 259-288. (also in: "Wireless Communications", Eds. J.M. Holtzman & D.J. Goodman, Kluwer Academic Pub., 1993, pp. 289-304).

[2] D. Raychaudhuri and N. Wilson, "ATM Based Transport Architecture for Multiservices Wireless Personal Communication Network", IEEE J. Selected Areas in Comm., Oct. 1994, pp. 1401-1414.

[3] J. Porter & A. Hopper, "An Overview of the ORL Wireless ATM System", IEEE ATM Workshop, Wash. D.C., Sept. 30-Oct 1, 1995.

[4] K. Y. Eng et. al, "BAHAMA: A Broadband Ad-Hoc Wireless ATM Local Area Network", Proc. ICC' 95. pp. 1216-1123.

[5] M Umehira, et. al, "An ATM Wireless Access System for Tetherless Multimedia Services", ICUPC' 95, Tokyo, Nov. 1995.

[6] D. Raychaudhuri, et al, "WATMnet: A Prototype Wireless ATM System for Multimedia Personal Communication", IEEE J. Selected Areas in Comm. , Jan 97, pp. 83-95.

[7] "Charter, Scope and Work Plan for Proposed Wireless ATM Working Group", ATM Forum, Anchorage, April 1996, ATM Forum/96-0530/PLEN.

[8] ETSI-BRAN, "BRAN: Requirements and Architectures for BRAN", WG1 TD1 Technical Draft Report, 1997.

[9] "Multimedia Mobile Access Communication", MMAC systems promotion council activity brochure, Association of Radio Industries and Businesses (ARIB), 1997, Tokyo, Japan.

[10] U.S. Federal Comm. Commission, "Operation of Unlicensed NII Devices in the 5 Ghz Range", ET Docket 96-102, Jan 1997.

[11] ETSI-RES10, "High Performance Radio Local Area Network (HIPERLAN)", Draft Standard, Sophia Antipolis, France, 1995.

[12] D. Raychaudhuri, "Wireless ATM: Present Status and Future Directions", PIMRC'97 invited lecture, Sept 1997, Helsinki, Finland.

[13] Special Issue on Wireless ATM, IEEE JSAC, Jan 1997.

[14] D. Raychaudhuri, "Wireless ATM Networks: Architecture, System Design & Prototyping", IEEE Personal Comm. Mag., Aug. 1996, pp. 42-49.

[15] D. Raychaudhuri, "Wireless ATM: An Enabling Technology for Multimedia Personal Communication", ACM/Baltzer J. Wireless Networks, 1996, Vol. 2, pp. 163-171.

[16] A. Acharya, B. Rajagopalan and D. Raychaudhuri, "Mobility Management in Wireless ATM Networks", IEEE Commun. Mag., Nov. 1997, pp. 100-109.

[17] P. Narasimhan, et al, "Design and Performance of Radio Access Protocols in WATMnet, a Prototype Wireless ATM Network", Proc. Winlab. Workshop, April 1997.

[18] C.A. Johnston, et al, " Architecture and Implementation of Radio Access Protocols Wireless ATM Networks", Proc. ICC'98, June 1998, Atlanta, GA.

[19] B. Walke, D. Petrass, and D. Plassmann, "Wireless ATM: Air Interface and Networks Protocols of the Mobile Broadband System", IEEE Personal Comm. Magazine, Aug. 1996, pp. 50-56.

[20] A. Acharya, J. Li and D. Raychaudhuri, "Primitives for Location Management and Handoff in Mobile ATM Networks", ATM Forum/96-1121/WATM, Aug 1996.

[21] G. Bautz and M. Johnsson, "Proposal for Location Management in WATM", ATM Forum 96-1516, Dec 1996.

[22] B. Rajagopalan, A. Acharya and J. Li, "Signalling and Connection Rerouting for Handoff Control Management" ATM Forum 97-0338,April 1997.

[23] H. Mitts, et al, "Microcellular Handover for WATM Release 1.0: Proposal for Scope & Terms of Reference", ATM Forum 97-0226, April 1997.

[24] "The MAGIC WAND Demonstrator", http: //www.tik.ee.ethz.ch/~wand

[25] A. Campbell, C. Aurrecoechea and L. Hauw, "A Review of QoS Architectures:, ACM Multimedia Systems Journal, 1996.

[26] D. Reinginer, D. Raychaudhuri & J. Hui, "Dynamic Bandwidth Allocation for VBR Video in ATM Networks", IEEE J. Selected Areas in Commun., August 1996, pp. 1076-1086.

[27] D. Reininger, B. Rajagopalan, and S. Gopalakrishnan, "Dynamic Bandwidth Renegotiation for Wireless ATM", ATM Forum 98-0363, April 1998, Berlin.

[28] D. Reinginer and R. Izmailov, "Soft Quality-of-Service with VBR+ Video", Proc. 8th Intl. Workshop on Packet Video, Aberdeen, Scotland, Sept. 1997.

[29] M. Ott, G. Michelitsch, D. Reininger and G. Welling, "QoS Aware Browsing in Distributed Multimedia Systems", Proc. IDMS'97

[30] D. Raychaudhuri and Y. Furuya, "ATM-based Wireless Personal Communication System with Migration to Broadband Services", Americas Telecom '96 Forum, Rio De Janeiro, Brazil, May 1996.

[31] M. Cheng, S. Rajagopalan, F-F., Chang, G. Pollini and M. Barton, "PCS Mobility Support Over Fixed ATM Networks", IEEE Comm Mag, Nov. 1997, pp. 82-92.

[32] B. Rajagopalan, "An Architecture for Wide-Area Mobile Internet Access", Proc. PIMRC'97, Sept 1997, Helsinki, Finland.

[33] P. Newman, T. Lyon and G. Minshall, "Flow Labelled IP: A Connectionless Approach to ATM", Infocom 96, pp. 1251-60.

[34] A. Acharya, R. Dighe and F. Ansari, "IP Switching over Fast ATM Cell Transport (IPSOFACTO)", IEEE Broadband '97 Workshop, Tucson, AZ, Jan 1997.

[35] A. Acharya, J. Li, F. Ansari and D. Raychaudhuri, "Mobility Support for IP over Wireless ATM", IEEE Comm. Mag., April 1988, pp.84-88.

[36] A. Talukdar, B.R. Badrinath and A. Acharya, "MRSVP, A Resource Reservation Protocol for an Integrated Services Network with Mobile Hosts" submitted to Mobicom'98.

Chapter 8

PROVIDING INTERNET SERVICES TO MOBILE PHONES

A case study with Email

THOMAS Y.C. WOO, KRISHAN K. SABNANI, and SCOTT C. MILLER
Bell Laboratories, Lucent Technologies
{ woo,kks,scm } @research.bell-labs.com

Abstract   Mobile phones are quickly becoming one of the most ubiquitous wireless consumer devices. Separately, Internet services are growing by leaps and bounds. Thus, an interesting area of research is to see if and how the two can be married together to provide wireless ubiquitous access to the ever-growing Internet services.

In this paper, we highlight the challenges and issues in providing Internet services to mobile phones. As an example, we describe and examine a research prototype called Wireless Data Server, which provides, among other services, wireless email service to mobile phone users.

1. INTRODUCTION

Digital mobile telephony presents an interesting opportunity for wireless data. Most second generation cellular/PCS air interface protocols such as IS-95 [13], IS-136 [14] and GSM [10], include data capabilities in addition to voice. A mobile handset, thus, can potentially provide both wireless telephony and data services in a single device. This is particularly appealing to service providers as mobile handsets are projected to become the most popular wireless consumer device in the coming years.

Unlike voice services however, wireless data services, with the exception of a few primitive ones such as fax and short message service (SMS) [5], have not been standardized, and are very much areas of ongoing research. Indeed, a lot of discussion has centered around finding the "killer application" for wireless data.

Our belief is simple: what will jumpstart the demand for wireless data should be no different from what is currently propelling the unprecedented growth of wireline data, that is, the Internet. Therefore, an interesting research challenge is how to effectively provide Internet services to mobile phones.

To properly address this, a number of inter-related questions must be answered:

1. What are the constraints imposed by a wireless environment?

2. What is the right set of Internet services that can be effectively provided to mobile telephony users under the constraints of a wireless environment?

3. What kind of network infrastructure should be used for providing these selected Internet services?

In this paper, we examine these questions and provide some of our answers. For concreteness, we reference in the discussion the Wireless Data Server (WDS), a prototype application layer server we have built at Bell Laboratories for providing advanced wireless data services to mobile users [12, 17].

Because of length limitation, we also focus our description to a particular Internet service, namely, electronic mail (email), whenever appropriate. The reason for interest in email is simple. With the growth of global Internet, electronic mail is rapidly becoming the second most common means of communication, right behind telephony. Thus, a service package that provides both forms of communication in a single device, i.e., the mobile handset, is attractive for users, and can represent an interesting entry-level horizontal wireless data applications to service providers.

The balance of this paper is organized as follows. In Section 2., we describe the different kinds of constraints imposed by a wireless operating environment and specifically mobile telephony. In Section 3., we examine the nature of services that can be effectively provided to mobile telephone users. In Section 4., we present the design, architecture, and implementation of the Wireless Data Server, with special emphasis on its email subsystem. We conclude in Section 5..

2. CONSTRAINTS OF MOBILE TELEPHONY

Mobile telephony imposes two main kinds of constraints: those that are related to the wireless operating environment and those that result from the limitation of the end devices.

Wireless is generally a harsh environment for data services, which typically requires reliable exchanges and predictable quality of service. More specifically, wireless access is marred by a number of problems: (1) High-error rate — Wireless channels exhibit order of magnitude higher error rate, 10^{-2} not being uncommon, than wireline channels. (2) Low and bursty bandwidth — Bandwidth of wireless channels is typically lower and burstier than their wireline counterparts. For existing mobile telephony, the raw bandwidth ranges from 9.6kbps to about 20kbps. (3) Asymmetry — The actual bandwidth available on the forward and reverse wireless channels can also be asymmetric. Even when the raw channel rate is identical in the two directions, potential media access contention in the reverse channel often lowers its effective bandwidth. (4) Availability — Wireless channels are not as available as wireline ones. Strictly speaking, this is only partially caused by the wireless nature. User mobility and the use of portable end device are the main factors. User mobility leads to handoff and out-of-range conditions, while portable end devices can be disconnected from the network for various reasons (e.g., power-off, low battery).

As a platform for wireless data services, mobile handsets also present a number of limitations. (1) Input/output — Because of their form factor, mobile phones can support only limited input/output options. Most have only keypads and small text-only screens. This significantly affects the mode of user interactions. In particular, a well-designed application should minimize the number of inputs, and either produce highly efficient output or allow alternate output devices. (2) Processing power and memory — The processing power and memory of a mobile end device is necessarily limited when compared to the network. Thus, local computation should be kept simple, or if possible, offloaded completely to the network. The lack of persistent storage is particularly constraining for many data applications. (3) Battery capacity — Advance in battery technology is still painfully slow. Despite the many power-saving features in the electronics and the air interface protocols, the standby and talk times of a mobile phone are still measured in hours. To optimize power consumption, a data application should try to minimize transmission at the expense of reception, as it takes a lot more power to send than to receive. (4) Programmability — Most data applications follow a client-server model. In case of mobile telephony, the client must reside in the handset. However, there is no standardized programming environment for mobile handsets, thus making integration of such a client difficult. A number of ideas such as HDML [15] and SIM Toolkit [6, 7] have been proposed to address this problem. They are, however, not yet widely implemented.

Most of the wireless channel constraints can be shielded from the applications with an adaptive middleware layer that is optimized for wireless channels.

Device limitations, on the other hand, can not, and in some cases arguably should not, be overcome in a totally transparent manner. We will present in later sections a number of approaches such as protocol conversion and service restriction to address them.

3.   WHAT KIND OF SERVICES?

Ideally, wireless data services should be no different, if not more versatile, from wireline data services. Tetherlessness and mobility should only enhance the services that can be provided. In practice however, the constraints described in Section 2. have narrowed the kinds of wireless data services that can be effectively and economically provided in the near future.

From the market perspective, the kind of wireless data services that should be offered depends very much on the composition of the mobile users. A low-end user may use voice exclusively, while a power user may use data services as heavily as voice. Consequently, the mobile handset market is similarly segmented. Most mobile phones can not support data applications other than plain SMS, while a few high-end phones (e.g., Nokia 9000) are a hybrid between phones and personal digital assistants (PDAs). Therefore, to decide the service offering, one must first identify the intended user segment and end device.

In our research, our initial focus is on horizontal consumer data applications with fairly standard mobile handsets. We adopt a communication-centric view of mobile data usage. In other words, a mobile handset is predominantly used for communication, rather than computation purposes. Thus, instead of adapting computation-oriented data applications, we focus on identifying and implementing a highly integrated suite of communication-oriented data services, which can also complement the existing voice service available on a handset.

The types of services that can be provided are also dependent on the particular types of data channels used. In this paper, we describe only those that can be supported over the SMS channels. To that end, it is helpful to first examine the key characteristics of the SMS channels: (1) Universal — SMS is lowest common denominator of wireless data channels. It is available in all major cellular/PCS air interface protocols. Thus, an SMS-based service could potentially be portable across different systems. (2) Low bandwidth — The maximum size of an SMS message is a few hundred bytes. They are typically carried on control channels, which means they must contend with higher priority signaling messages. Thus, the effective bandwidth of an SMS channel is fairly low, and is unsuited for data intensive application. (3) High latency, bursty — Because of contention and queuing, SMS delivery tends to be bursty. As a result, it offers a highly variable quality of service. (4) Store-and-forward — SMS is based on a store-and-forward model. An application does not have explicit control on when an SMS is delivered. As a result, applications of real-time or strong interactive nature can not be supported over SMS. On the other hand, asynchronous nature of SMS makes it ideal for carrying notifications.

In the following, we describe different classes of services that we are targeting in our research. As we will see, the characteristics of these services match very well with that of SMS.

Messaging. One of the most basic wireless data applications is messaging. It is simple to understand and use, and it complements the voice functionalities very well.

Wireless messaging can be one-way or two-way. One-way wireless messaging allows a handset to receive a message from the network. Two-way wireless messaging allows two-way exchange of messages between a handset and the network.

SMS can be viewed as a very primitive wireless messaging application. It offers a closed form of communication among mobile users, and is fairly tedious to use. Our view is that SMS provides greater value as a wireless transport than as an end-user application.

A simple yet powerful way to enhance SMS has been proposed in [17]. The key idea is that messages (or *active messages* as they are called) are *forms* that can be pre-stored or sent to a handset and customized on demand at the time of use. Sending or replying to a message is reduced to no more than filling out a form. This significantly simplifies the input problem. Active messages can be programmed by the end user and dynamically downloaded to the handset.

Another useful form of messaging is to allow mobile handsets "access" to wireline mailboxes. Because of various limitation, the kinds of "access" allowed from a mobile handset will be different from a desktop. In fact, a different model of interaction is needed. We defer the discussion to Section 4..

Information Services. Information services are data services which deliver information to users. Some examples of information services are news, weather information, traffic condition.

Information services can be provided on a push basis or a pull basis. In the former, information is delivered on some pre-specified schedule (based on time or trigger). In the latter, a user must explicitly request the information. Clearly, a hybrid of the two is possible, in the form of a wireless analog of Pointcast.

The usefulness of an information service is highly dependent on the degree of personalization. A great deal of research effort is currently focused on personalization technologies.

Information services can also be enhanced by the location tracking abilities of mobile telephony. By leveraging the cell location information, an information service provider can further personalize the content it offers.

Personal Information Management (PIM). Applications such as address book, calendar, notepad, and task list are collectively referred to as PIMs. Increasingly, mobile handsets support a subset of PIMs.

Standalone PIMs are not as useful. Their utility is maximized if they can synchronize with their network-based masters.

More generally, this suggests the need for a generic over-the-air data synchronization service.

World Wide Web. Web browsing is the primary driver for wireline data applications. A Web browser can also serve as an universal client for all other applications. Naturally, wireless Web browsing should be an ideal candidate for wireless data applications. The richness of Web-based information, however, does not match well with the constraints of existing mobile telephony. In particular, SMS is a poor transport for rich data. Browsing also implies a synchronous model of interaction, which can not be supported effectively over SMS.

Without browsing, useful access to WWW can still be provided. Specifically, WWW is a rapidly growing repository of information, and can simply serve as a source for information services. To this end, content is first retrieved from the WWW, then filtered to extract the information requested, and then packaged into messages that are delivered to the user.

The main challenges here are implementing a general set of filtering mechanism, and specifying the filters. In the simplest case, filters can be pre-defined by the service providers for selected WWW pages. Ideally though, an end user would like to define and install his or her own filters.

4.   WIRELESS DATA SERVER

As part of our research, we have built a prototype application server, which we call Wireless Data Server (WDS), for providing the types of services we mentioned earlier. WDS follows a client-server model. For its operation, it requires a "thin" client to be integrated in the mobile handset. The thin client is added via a firmware upgrade, and its code size is only about 8KB.

To better understand WDS, it is useful to first examine some design principles and decisions behind it:

- All services offered by WDS follow a messaging model. Each request is sent as a self-contained message independent from others, and all responses are filed as messages in various inboxes. A messaging model automatically entails a asynchronous mode of interaction.

- The basic metaphor used in the design of WDS is as follows: All the complexity resides inside the network. A wireless end device serves simply as a remote control for invoking the services inside the network. It does so by sending simple commands uplink.

  Input and output are completely decoupled. In other words, an uplink command can invoke processing whose output is delivered to a completely different output channel.

- WDS is designed to be intelligent about bandwidth usage on the wireless link. Specifically, it tries to reduce the amount of communication with the end device, and sends only what is needed by the user. This is accomplished by offering service granularities that match with the typical user demands.

- WDS provides a high degree of personalization. It includes a component called *user agent*. Each end device logically has a user agent working inside the network on its behalf. A user agent interfaces between the end device on one side and the rest of the network on the other. It can be used to provide personalized or value-added functions.

  For convenience, most of WDS personalization can be done via a WWW-based interface.

- WDS is designed to be modular and extensible. This is particularly important considering the constantly evolving nature of wireless data technology.

  As we will describe below, a number of WDS components can be "dropped in" on demand.

WDS provides a number of different services and includes a distributed collection of components. It is not possible to describe them all here. Instead, we describe in the following the email service and the associated components of WDS. Another description of WDS focusing on its other services and components can be found in [12].

4.1 WDS EMAIL SERVICE

In one aspect, WDS email service is a restricted version of what can be done with desktop email. In another aspect, WDS email provides features (e.g., text-to-speech) that are specifically of interest to mobile phone user, and are not available in desktop email,

A service provider deploying WDS will assign a mailbox address to a subscriber for his or her email service, in addition to a phone number for his or her voice service. WDS maintains a mailbox in the network for each WDS email subscriber.

WDS email service comprises of two main sub-services: *smart notification* and *selective retrieval*. Smart notification refers to the delivery of a notification to the user upon the arrival of a new email at the user's WDS mailbox.

The "smartness" comes from the fact that the criteria for sending notification, the content of the notification, and the delivery format of the notification can all be personalized by the user. Typically, the criteria for sending notification are defined by conditional rules whose conditions are expressed using standard attributes (e.g., sender, subject, time) of email and whose actions specify various delivery options (e.g., SMS, voice, fax, email).

Typically, the notification contains a description (e.g., sender, subject, time, size, and attachments) of the newly arrived email detailed enough for the user to decide on possible actions on the email.

Once notified, a user can decide to retrieve the mail. WDS allows a user to retrieve an email in whole or by parts, and have the result delivered in any format convertible from the original received format. This is what is referred to as selective retrieval.

Selective retrieval is most useful in handling large email and/or email with multimedia attachments. It allows a user to receive an email part without the need to process and/or download the whole email, which may consume nontrivial resources. We defer a more detailed operational description of smart notification and selective retrieval to Section 4.3.

In the above description, a user's WDS mailbox is distinct from other mailboxes the user may have. The user is responsible for directing email to his WDS mailbox. WDS email service can also be provided for corporate mailboxes, thus eliminating the multiple-mailbox problem. One way to accomplish this is to set up some type of *virtual private network* (VPN) arrangement. Alternatively, an Intranet version of WDS can be deployed inside the corporation, we omit the details here.

Although our discussion is centered around email, similar functions can be offered for a unified mailbox, thus providing a form of wireless unified messaging.

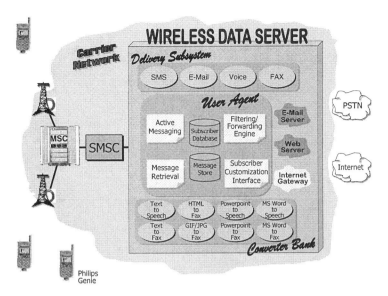

Figure 8.1 WDS Architecture

4.2 ARCHITECTURE

WDS consists of a distributed collection of modules, as shown in Figure 8.1. In terms of connectivity, WDS is connected to the Short Messaging Service Center (SMSC) using industrial standard protocols (e.g., SMPP [1]), to the telephone network for voice and fax services, and to the Internet for email (SMTP [11]) and WWW access (HTTP [2, 9]).

The three main components related to WDS email service are the *user agent*, the *converter bank*, and the *delivery subsystem*. Among the components inside the user agent, the key ones are the *filtering/forwarding engine* which handles smart notification and the *message retrieval module* which is responsible for selective retrieval. The *subscriber customization interface* provides the backend logic for subscriber personalization. The *active messaging module* is not relevant for email.

The converter bank houses individual format converters that may be called upon by the user agent. All converters support a uniform interface: each takes an input object of a specified type and output an object of a different type. A new converter can be easily added to the system by installing inside the converter bank.

By the same token, the delivery subsystem houses interface modules for the different delivery options. WDS supports all the important existing interfaces: SMS, voice, fax and email. A new interface, such as to paging systems, CDPD [3], or GPRS [8], can be quickly implemented and added as desired.

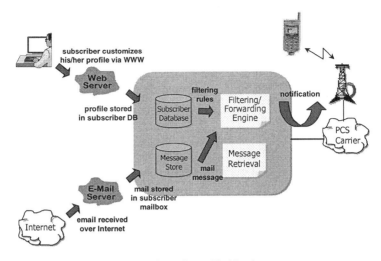

Figure 8.2   Smart Notification

Apart from the above key components, WDS also includes a email server for sending and receiving email to and from the Internet, and a WWW server providing a Web-based interface to subscribers. The Internet gateway as shown in the figure does not provide functions related to email.

It should be clear that WDS has an architecture designed for easy extensibility. By following reasonably well-defined interfacing procedures, WDS can be extended to support a new input method (e.g., phone, WWW), a new end device (e.g., PDA), a new delivery option (e.g., paging), or a new format converter (e.g., Postscript-to-text).

4.3   OPERATION SCENARIOS

In this subsection, we describe some operation scenarios of WDS.

Smart Notification.   Figure 8.2 shows the operation of smart notification. For simplicity, it illustrates only notification via SMS. The flow for other forms of notification is similar.

Through a Web-based interface, a subscriber can define his or her own set of filtering rules, the associated notification format and notification delivery format. These subscriber data are stored in a subscriber database.

Each arriving email is stored in a message store, and a trigger is sent to the filtering/forwarding engine to initiate its processing. To process an email, the rule set of the recipient is extracted from the subscriber database. The rules are computed using the attributes of the email to determine possible matches. A notification is generated when a match is found.

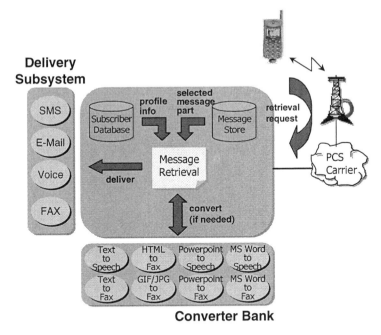

Figure 8.3 Selective Retrieval

Each notification includes, among other things, a unique message reference id that uniquely identifies the original email in the message store. This message id is used in the selective retrieval step to be discussed next.

Selective Retrieval. Figure 8.3 shows how selective retrieval is processed in WDS.

A mobile handset initiates a retrieval request by originating an uplink request. The request contains, among other things, the message reference id for the message to be retrieved, the part of the message desired (in the form of a part number and an offset), and the desired delivery format (with the associated delivery parameters, e.g., phone number).

The retrieval request is handled by the message retrieval module, whose processing steps are straightforward: (1) It first extracts the desired message (or message part) from the message store. (2) It then determines if a conversion is needed by comparing the format type of the message (or message part) with the desired delivery format. (3) It calls the appropriate converter if a conversion is needed, or signals an error if no appropriate converter is available. (4) It delivers the result to the appropriate interfacing module in the delivery subsystem.

The input and output are completely decoupled by including a delivery format in the retrieval request. It should be clear that the retrieval mechanism is independent of how the retrieval request was originated. In other word, a voice-based retrieval interface can be easily added to WDS by adding an interfacing module that interacts with a user using voice and generates the corresponding retrieval request.

4.4   PROTOTYPE IMPLEMENTATION

The server side of WDS runs on standard off-the-shelf hardware: a SUN Ultra 2300 running Solaris 2.5.1 and a 300MHz Pentium II PC running Windows NT 4.0. The SUN server implements all of the processing logic of WDS, while the NT server is responsible for the voice and fax functions. In a production version of WDS, the NT server can be replaced with rack-mounted PC-based modem banks.

Most of WDS is implemented in Java using standard APIs provided in Java JDK 1.1.5. All the processing code is multi-threaded, and makes use of a custom-designed message-passing mechanism for interprocess communications.

Apart from the application logic, WDS uses off-the-shelf software from different vendors: `sendmail` as the SMTP mail server, SUN's `imapd` for IMAP4 [4], and Apache as the WWW server. It also uses an Oracle relational database (via JDBC) for its database needs. On the NT server, the telephony and fax functions are programmed in Visual Basic using libraries from Visual Voice Pro.

The client side of WDS has been implemented on a Philips IS-136 phone and the Philips GSM Genie phone. We have added about 8KB of code to each phone. The key functions of the client side code are for menu interactions, and parsing and sending WDS control messages.

5.   CONCLUSION

Existing wireless data services and the infrastructure providing them tend to be ad-hoc. In this paper, we have identified a number of low-bandwidth, entry-level valued-added wireless data services that can be effectively implemented over existing mobile telephony. We have also presented a novel paradigm for approaching wireless data services. Specifically, we view wireless end devices as low-powered remote controls, while all the complexity and processing reside within the network.

We support our new paradigm by proposing and implementing a systematic architecture for providing wireless data services, called WDS. A WDS prototype has been completed, and is being readied for production trials. To this end, we are researching issues concerning reliability, scalability, and manageability.

With higher speed packet data channel (e.g., GPRS [8]) and more capable end device, the nature of wireless data services is bound to change. For example, wireless Web browsing may become the dominant application because of its universality.

There are also various efforts addressing the wireless transport protocol and client-side issues. Some of the notable ones respectively are the Wireless Application Protocol (WAP) [16] and SIM Toolkit [6, 7].

References

[1] Aldiscon. *Short Message Peer to Peer (SMPP) Interface Specification*, January 1996.

[2] T. Berners-Lee, R. Fielding, and H. Frystyk. *Hypertext Transfer Protocol — HTTP/1.0*. RFC 1945, May 1996.

[3] CDPD Forum. *Cellular Digital Packet Data System Specification — Release 1.1*, January 19 1995.

[4] M. Crispin. *Internet Message Access Protocol — Version 4 Rev 1*. RFC 2060, December 1996.

[5] European Telecommunications Standards Institute. *GSM 03.40 Digital Cellular Telecommunications System (Phase 2); Technical Realization of the Short Message Service (SMS) Point-to-Point (PP)*, June 1996.

[6] European Telecommunications Standards Institute. *GSM 11.11 version 5.5.1 Digital Cellular Telecommunications System (Phase 2+); Specification of the Subscriber Identity Module — Mobile Equipment (SIM-ME) Interface*, October 1997.

[7] European Telecommunications Standards Institute. *GSM 11.14 version 5.4.0 Digital Cellular Telecommunications System (Phase 2+); Specification of the SIM Application Toolkit for the Subscriber Identity Module — Mobile Equipment (SIM-ME) Interface*, October 1997.

[8] European Telecommunications Standards Institute. *GSM 02.60 version 5.2.0 Digital Cellular Telecommunications System (Phase 2+); General Packet Radio Service (GPRS) — Service Description; Stage 1*, January 1998.

[9] R. Fielding, J. Gettys, J. Mogul, H. Frystyk, and T. Berners-Lee. *Hypertext Transfer Protocol — HTTP/1.1*. RFC 2068, January 1997.

[10] M. Mouly and M.B. Pautet. *The GSM System for Mobile Communications*. 1992.

[11] J. Postel. *Simple Mail Transfer Protocol*. RFC 821, August 1 1982.

[12] K.K. Sabnani, T.Y.C. Woo, and T.F. La Porta. A system for wireless data services. In Savo G. Glisic and Pentti A. Leppänen, editors, *Wireless Communications: TDMA versus CDMA*, pages 205–229. Kluwer Academic Publishers, 1997.

[13] Telecommunications Industry Association. *TIA/EIA IS-95 Mobile Station — Base Station Compatability — Standard for Dual-Mode Wideband Spread Spectrum Cellular System*.

[14] Telecommunications Industry Association. *TIA/EIA IS-136 800 MHz TDMA Cellular — Radio Interface — Mobile Station — Base Station Compatability — Digital Control Channel*, December 1994.

[15] Unwired Planet Incorporated. *HDML 2.0 Language Reference*, July 1997.

[16] WAP Forum. *Wireless Application Protocol Architecture Specification Version 30-Apr-1998*, April 30 1998. Available from http://www.wapforum.com.

[17] T.Y.C. Woo, T.F. La Porta, and K.K. Sabnani. Pigeon: A wireless two-way messaging system. *IEEE Journal on Selected Areas in Communications*, 15(8):1391–1405, October 1997.

Chapter 9

Design, Implementation, and Performance of a Cluster Mobile Switching Center

R. RAMJEE, K. MURAKAMI, R. W. BUSKENS, Y-J. LIN, and
T. F. LA PORTA
Bell Labs, Lucent Technologies

Abstract   In this paper, we present the design and implementation of a software system
that performs the control functions of current Mobile Switching Centers
(MSCs) and Visitor Location Registers (VLRs), and provides support for third
generation services. The novelty of the system design is in its software
architecture. To allow flexible deployment of the system as the migration from
second generation to third generation systems occurs, the software is modular.
To provide scalability in terms of capacity, distributed processing is used. This
type of system design is becoming increasingly more desirable and common as
commodity processor prices drop and large capacity switches are demanded.
One serious concern with these approaches is whether the required
performance of the target switching systems can be met. We show that a
deployment of this system using three Sun Ultra workstations achieves a call
throughput of 300,000 calls/hour with a latency of below 300 milliseconds.
This meets the latency requirements of third generation systems.

1   INTRODUCTION

Requirements and protocols for third generation wireless communication systems are currently being defined [1]. Due to the large investment in second generation wireless systems, third generation systems will likely evolve from second generation systems and interwork with them.

Third generation wireless systems will support multimedia services and provide access to ISDN and Intelligent Network value-added services such as time-of-day call forwarding. To do this, these systems will require more complex processing in network elements to provide multimedia and new value-added services, and use new air interfaces, which support higher bandwidth network access. Figure 1 shows the network architecture for second generation cellular telecommunication systems, which will be the basis of third generation systems.

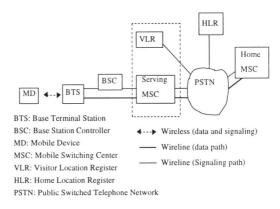

Figure 1. Cellular Telecommunication Network Architecture

A Base Terminal Station (BTS) terminates the air interface protocol with the Mobile Device (MD). It forwards user traffic to the Mobile Switching Center (MSC), and control traffic, i.e., signaling messages, to the Base Station Controller (BSC). A BSC provides the intelligence for multiple BTSs, which includes performing power control, managing soft handoffs, and controlling paging to locate mobile devices.

A Home Location Register (HLR) contains a permanent service profile for each user and mobile device in the network, and tracks the approximate location of the mobile devices. The home MSC[30] serves as a bridge for

---

30. In some systems, these procedures may be started by a *gateway* MSC. We use the term home MSC interchangeably.

incoming calls to a mobile device. Calls for a mobile device are routed to its home MSC which, through interaction with the HLR, determines the serving MSC at which the mobile device is located, and then sets up a connection to the mobile device through the serving MSC via the Public Switched Telephone Network (PSTN).

The serving MSC performs tasks related to both transport of user information and signaling. Tasks related to user information transport include switching, voice coding and frame selection. The various signaling tasks include *mobility management*, i.e., registration, paging, assignment of temporary routing numbers, and handoffs; *connection control*, i.e., routing; and *call control*, i.e., providing access to processing logic for value-added services. The VLR functions as a database for location and service profile information. In current systems, these two elements are typically co-located as shown by the dashed box in Figure 1.

In this paper, we present the design and implementation of a software system that performs the integrated control functions of current Mobile Switching Centers (MSCs) and Visitor Location Registers (VLRs), and provides support for third generation services. We call the system a cluster MSC (cMSC) because in general, it is made up of a cluster of processors. To allow flexible deployment of the system as migration from second generation to third generation systems occurs, the *software is modular*. To provide scalability in terms of capacity, *distributed processing* is used. To lower cost, commodity software and hardware platforms are used. Our implementation uses workstations running the UNIX operating system. The Common Object Request Broker Architecture (CORBA) [2] is used as the communications middleware. Software techniques are used to ensure system reliability [3] but are outside the scope of this paper.

To determine if such a modular system built from commodity computing platforms meets the performance requirements for third generation systems, we perform a detailed experimental evaluation of the cMSC. Our results show that a three processor Sun Ultra Sparc cMSC has a call handling throughput of over 300K calls/hour with an average latency of under 300 milliseconds.

Many current research projects in control software for broadband and third generation wireless networks also stress flexibility, either through layered software or object-oriented design [4][5][6]. Distributed processing techniques are also widely used in these systems to achieve scalability and allow for increased flexibility to incorporate algorithms which may improve overall system performance. To reduce the time required to develop distributed systems, several efforts propose using CORBA as a communications middleware [5][6][7].

Except for [7], the research efforts referenced above do not report on actual system implementations, but rather architectures and concepts. The system described in [7], called *xbind*, has been implemented using CORBA, with performance measurements being reported in [8]. The focus of the *xbind* experiment was to improve performance of connection establishment by building a CORBA-based control system distributed over a wide area network and defining new network algorithms for connection establishment. In the cMSC, the focus is to build a single network element that is compatible with second generation wireless standards and is evolvable to support advanced services.

As can been seen by the amount of research work in this area, modular, distributed architectures for call processing are becoming increasingly important. The two main concerns with this type of system are performance and reliability. This paper addresses the performance concerns. We present the first extensive set of experimental results for a large call processing system (the core call processing consists of over 65,000 lines of C++ code, not including the protocol stacks) built using these technologies. Our results show that the performance requirements of second and third generation systems can be met.

Many of the concepts of the cMSC originate from earlier work on the Wireless Distributed Call Processing Architecture (W-DCPA) [9][10]. W-DCPA was designed and prototyped to support third generation services such as high bit-rate voice and multimedia communication whereas the cMSC is designed to evolve from a second generation system to a third generation system. This led to innovations regarding software structure and protocol interworking managers.

The remainder of the paper is organized as follows. In Section 2, we present the cMSC system design and implementation. In Sections 3-6 we present the test environment and results of performance experiments. In Section 7 we conclude.

2 SOFTWARE DESIGN AND IMPLEMENTATION

In this section we present the software architecture and cMSC implementation.

2.1 Architecture

In order to reduce the cost of building the cMSC and increase portability, widely available commercial software and hardware platforms are used instead of custom-built counterparts. In the current implementation, the cMSC executes on both HP-UX and Solaris operating systems using CORBA as communications middleware. The software is written in C++.

Software objects are defined to perform specific tasks and manage particular resources. These objects interact to provide end-to-end services. Each object has a well-defined interface through which others may access its services. As long as its interface is kept unchanged, a single object may be modified to change its behavior or upgrade its functionality without affecting other existing objects. This makes the system *scalable in the functional dimension* and aids in the evolution from second to third generation systems. Objects that perform strongly related functions are grouped together into a *server*.

The objects within each server are implemented as C++ objects; servers are implemented as CORBA objects, each with its own interface defined in the CORBA Interface Definition Language (IDL). This is the only interface to the server; interfaces to the individual internal objects of a server are not accessible to objects outside the server.

Servers each run as a single UNIX process. The servers may be replicated and distributed across processors to allow the system to be *scalable in the capacity dimension.*

Using CORBA for communications middleware eases the implementation of the distributed application. CORBA also allows a deployment of processes (servers) on different processors without a priori knowledge by the clients of the server location. This is important as the cMSC may be realized with many processor configurations depending on capacity and reliability requirements and failure conditions. Another major benefit of using CORBA is the separation of interfaces from implementation by using IDLs. This allows new algorithms to be easily incorporated into the servers. The main concern when using any communications middleware is that performance degradation may occur. The performance of the cMSC in terms of call handling capacity and call establishment latency is presented in Sections 4 through 6.

Figure 2 shows the complete architecture of the cMSC. Each box represents a CORBA server within the cMSC. The interconnection of servers may be over a high speed LAN or over a backplane depending on whether or not the servers are co-located on the same processor.

158

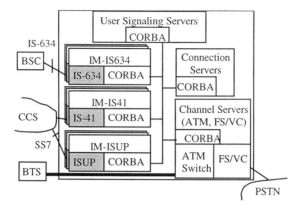

Figure 2. cMSC Software Architecture

The cMSC has two classes of servers: *Interworking managers* (IMs) that act as gateways and provide interfaces to external network elements, and *core servers* that perform call processing functions and communicate with each other using CORBA. IMs terminate standard protocols with the external elements and use CORBA to communicate with the core servers.

The IMs allow the call processing in the core servers to be designed independently of the external signaling protocols. As new protocols are standardized, the cMSC could deploy new IMs to support the new protocols without modifying the core servers. This is important as networks evolve from second to third generation systems and as new protocols emerge.

The current cMSC has three interworking managers. An IM-IS634 terminates the IS-634A protocol with the BSC across the standard IS634 A1-interface [11]. The IM-ISUP terminates the SS7 protocol suite for call/connection control with the common channel signaling network across the Network Node Interface (NNI) [12]. The IM-IS41 terminates the IS-41 [13] protocol stack, which is used for mobility management. Other IMs for B-ISUP [14] and MAP [15] can be envisioned.

There are four core servers in the cMSC: the frame selector/voice coder channel server, ATM channel server, user signaling server, and connection server. The servers each provide a specific service to the other servers, and may in turn, request services from the other servers.

Frame selector/voice coder (FS/VC) and *ATM channel servers* manage the resources of their respective elements. The FS/VC channel servers manage the assignments of frame selectors and voice coders. The ATM channel server manages VCI space and bandwidth in the ATM switch.

The *user signaling server* performs mobility management functions of a VLR and provides access to IN-type services consistent with second and third generation systems. It also maintains call and connection state from the user's perspective. Mobility management functions include coordinating mobile device registration with the home network, managing paging, and assigning temporary routing numbers which are used by other network elements to route incoming calls to a mobile device. To provide access to IN services, the user signaling server maintains a temporary copy of a user's service profile, which is obtained during registration procedures. It also checks service triggers to determine if value-added services should be activated.

The *connection server* determines a route between the BTS serving the mobile device and a circuit to the next hop switch at the edge of the PSTN. This includes choosing a frame selector/voice coder. The connection server interacts with the FS/VC channel server, the BSC (through the IM-IS634), and the PSTN (through the IM-ISUP) to reserve resources for this portion of the connection.

2.2   Procedures

To external elements, the cMSC behaves as a standard integrated MSC and VLR. We now describe the procedures for three key call processing scenarios: mobile registration, call origination, and call termination. These procedures play an important role in the performance experiments in Sections 4 through 6. The procedures are shown using message flows in Figures 3 through 5. In these figures, the servers inside the cMSC are shown grouped together in an oval. Each server is shown as a box. Note that for simplicity, only a single instance of each server is shown. External entities are shown as boxes outside the cMSC.

The names of the messages that flow within the cMSC correspond to the CORBA IDL methods invoked to generate the message. Messages that flow between the cMSC and the external elements shown in italics correspond to messages that are part of the standard indicated in the figures.

The procedure for a successful mobile registration due to power up is shown in Figure 3. During this procedure, a mobile device registers with the user signaling server in the cMSC (via the UPDATE_LOCATION message), which in turn, registers the device with its remote HLR (NOTIFY_POWER_UP). Any service profile information required for the cMSC to serve the mobile device is downloaded to the user signaling server, which creates a USER object and populates it with the profile information. Similar procedures exist for power down registration and for registration and deregistration due to movement.

The procedure for a mobile call origination is shown in Figure 4. In these procedures, a mobile device requests a call be established to a dialed number.

The connection server performs digit analysis to determine on which trunk group the outgoing connection should be routed. The trunk group is mapped to a possible frame selector and voice coder. Then, a frame selector and vocoder is assigned at the FS/VC channel server (SETUP_FS), traffic channels are assigned at the base-station (SETUP_BS), and voice trunks to the next hop switch at the edge of the PSTN are assigned (SETUP_SEG).

The procedure for a mobile call termination is shown in Figure 5. Call termination, unlike call origination, takes place in two phases. In the first phase, the cMSC serving the mobile device is located. This is accomplished by the originating switch querying the serving cMSC through the HLR for a routing number called a Temporary Local Directory Number (TLDN) (REQUEST_TLDN).

In the second phase, the connection to the mobile device is established. The originating switch uses the TLDN to route a connection request (IAM) to the serving cMSC. A connection is requested from the connection server (SETUP_SEG_ORDER), which in turn, offers the connection to the user (OFFER_CONN). The user signaling server pages the mobile device to determine its current serving base-station (PAGE), and informs the connection server. The connection server coordinates the assignment of FS/VC resources at the channel server (SETUP_FS) and the traffic channels at the BSC (SETUP_BS). At this point, an end-to-end connection is established, and the connection server instructs the FS/VC channel server to generate an in-band ringback tone (START_ALERT). Finally, when the mobile device answers the call, the ringback tone is stopped (STOP_ALERT).

3   EXPERIMENTAL TESTBED

The testbed consists of several Sun Sparc 5 (85 MHz clock, 32 MB memory), Sparc 20 (125 MHz clock, 64 MB memory) and Ultra 2[31] (200 MHz clock, 128 MB memory) workstations connected to a FORE ASX-200 ATM switch. The cMSC control traffic is carried over dedicated ATM links. The cMSC servers use Iona's multi-threaded CORBA implementation, Orbix2.2MT.

---

31. Note that only a single UltraSparc I processor in each Ultra 2 workstation was used in our experiments.

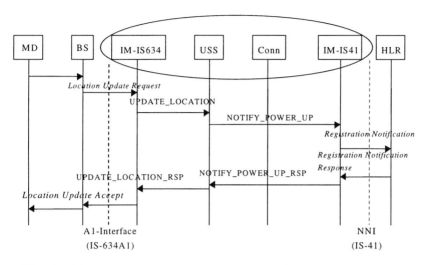

Figure 3. Power-Up Registration

MD: mobile device
BS: base terminal station and base station controller
IM-IS634: interworking manager - IS634A
USS: user signaling server

Conn: connection server
IM-IS41: interworking manager - IS-41
HLR: home location register
NNI: network node interface

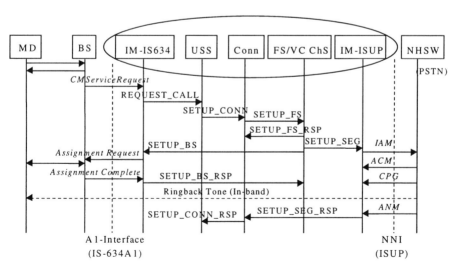

MD: mobile device
B3: base terminal station and base station controller
IM-IS634: interworking manager - IS634A
USS: user signaling server

Conn: connection server
IM-ISUP: interworking manager - ISUP
NHSW: next hop switch
NNI: network node interface

Figure 4. Mobile Call Origination

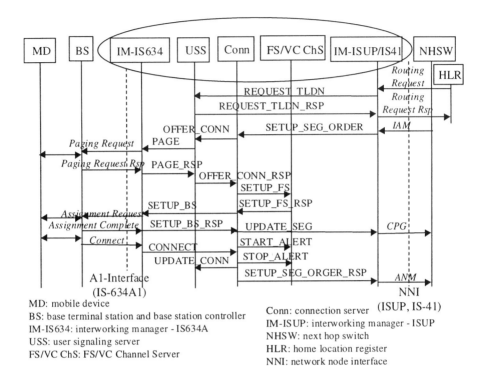

Figure 5. Mobile Call Termination

Simulators were built for the base-station, network switches in the PSTN, and Home Location Register (HLR) to generate control traffic to the cMSC. In our experiments, handoff registration and call setup requests are generated based on a Poisson distribution, and calls are released after an exponentially distributed holding time

We focus on two performance measures for the cMSC: call throughput and call setup latency. Call throughput is measured in number of calls (one call includes both the call setup and call teardown) processed per hour by the cMSC, and call setup latency is the average latency experienced at the MSC for setting up a call. From [1], an average latency of 300 milliseconds is acceptable for the call processing latency of a single MSC.

We measure call setup latency for call origination as the time between the base station requesting a connection and a traffic channel on the air interface being requested for assignment. In Figure 4, this is the time between the base station simulator initiating a CMService Request message and receiving an Assignment Request message. For incoming calls, also known as call termination, call setup latency is measured as the time difference between the PSTN requesting a connection from the MSC and the MSC indicating to the PSTN that the mobile device is ringing. In Figure 5, this is the time between the PSTN simulator initiating the transmission of the IAM and receiving the CPG message.

In Section 4, we examine the impact of several factors related to the implementation platform, such as processor speed and number of processors, on the performance of the cMSC. In Section 5, we study the impact of mobility, using parameters such as call origination percent and mobility-to-call ratios, on cMSC performance. In Section 6, we discuss the performance overhead of CORBA. The parameters used in these tests are listed in Table 1. The *Range* column indicates the range of values covered in our experiments.

Table 1. Parameters that affect MSC Performance

Parameter	Meaning	Range
N	Number of registered users	20,000 - 60,000
λ_o	Mobile call origination rate	0 - 5 calls/hour/user
λ_t	Mobile call termination rate	0 - 5 calls/hour/user
η	Call origination percent - $\lambda_o/(\lambda_o+\lambda_t)$	0 - 100
ζ	Registration rate (idle handoffs)	0 - 20 regs/hour/user
σ	Mobility-to-call ratio - $\zeta/(\lambda_o+\lambda_t)$	0 - 4
P	Processor type	Sparc 5, 20, Ultra 2
M	Number of processors	1 - 3

4   IMPACT OF PLATFORM ON PERFORMANCE

In this section, we highlight the basic relationship between two performance measures, call setup latency and call throughput, for a typical set of values for the parameters in Table 1. Subsequently, we examine the effect on performance as the parameters P and M are varied. The baseline set of values for the parameters, unless otherwise specified, is N=20,000, η=64, and σ=1. For each set of measurements, steady-state was reached requiring between 80,000 and 300,000 calls to be established and released.

164

4.1 Basic Results

To determine the baseline performance of the system, the baseline parameters were used to test the cMSC software executing on a single Ultra 2 workstation. The mobile call origination and termination rates, λ_o, and λ_t, were varied while maintaining a constant mobility-to-call ratio, σ, and a constant call origination percent, η.

Figure 6 shows the average call setup latency in milliseconds for both mobile call origination and mobile call termination, versus call throughput. First, note that the call setup latency for call termination is higher than the latency of call origination. This is because call termination procedures involve extra signaling and control for mobility management. Second, the graph has a knee at 114,000 calls/hour indicating that this is the maximum load the cMSC can handle in this configuration. The corresponding average call setup latency of 300 milliseconds for call termination and 135 milliseconds for call origination meets the requirements of third generation systems [1].

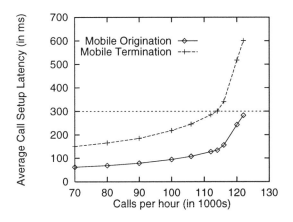

Figure 6. Call Setup Latency vs. Throughput

Figure 7 shows the cumulative probability distribution function of call setup latencies for the given cMSC configuration at 114,000 calls per hour. From the figure, the 95th percentile call termination latency is 595 milliseconds and 95th percentile call origination latency is 283 milliseconds. These values are within the target range for third generation systems.

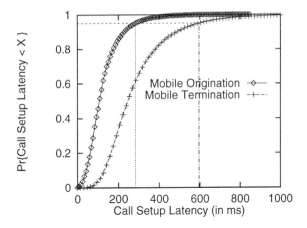

Figure 7. Cumulative Distribution Function of Call Setup Latency

The performance of the cMSC is independent of the number of active calls in the system. As the number of active calls in the system increases, the number of objects maintained by the cMSC servers also increases. In order to eliminate the effect on performance of the increased number of objects, we maintain pointers to objects in hash tables which results in relatively constant access times regardless of the number of objects stored.

4.2 Scalability with Processor Type (P)

One goal of using commodity processing platforms is to take advantage of the increasing speed of processors to improve the performance of the cMSC. The cMSC was tested with the baseline parameters using the three different types of workstations in the testbed. To determine the relative performance of these workstations, a simple benchmark that executed a "for loop" iteration was run on each workstation. The performance ratio of Sparc 5:Sparc 20:Ultra 2 was found to be 0.20:0.36:1.0.

For an average call termination setup latency for the cMSC to be within 300 milliseconds, the maximum average call throughputs for the three workstations were measured as 20K, 43K and 114K calls per hour respectively. This corresponds to a ratio of 0.18:0.38:1.0 for the three workstations. Note that this agrees closely with performance ratio computed using our simple benchmark. Thus, we conclude that the performance of the cMSC scales almost linearly with the speed of different processor types.

4.3   Scalability with Number of Processors (M)

The main motivation for using a distributed architecture for the cMSC is to provide freedom in deploying the system so that call handling capacity may be scaled by increasing the number of processors in the cluster. The cMSC was tested using the baseline parameters with configurations of 1-3 Ultra 2 workstations. The communication between the workstations is through an ATM switch. To keep the comparison fair, we maintain a constant average number of users per workstation of 20,000. Thus, N=40,000 for the two-workstation case and N=60,000 for the three-workstation case.

To obtain the maximum performance for each configuration, the server processes were distributed among the workstations in an optimal configuration subject to the following system constraints. First, only a single instance of each external interface exists, i.e., there can only be a single IM-ISUP, IM-IS41, and IM-IS634. Second, there is only a single FS/VC channel server. The other servers may be replicated as necessary to achieve optimal performance.

The bottleneck in the cMSC is the communication overhead. Therefore, to optimally load balance the system, the number of messages being transmitted and received by each processor should be made to be as close to equal as possible.

For the two processor case, the best division of the servers from nine possible scenarios is for IM-IS634, CONN, and USS to reside on Processor 1, and IM-IS41, IM-ISUP, FS/VC, CONN, and USS to reside on Processor 2. In the three processor case, the best division of the servers from 40 possible scenarios is for IM-S41, IM-UP, FS/VC, and USS to reside on Processor 1, CONN to reside on Processor 2, and IM-IS634, and USS to reside on Processor 3.

Figure 8 shows the average call setup latency for mobile call termination versus call throughput for the three systems. Given the constraint that the average call setup latency for a given cMSC is within 300 milliseconds, the call throughputs for the three systems are 114K, 227K, and 312K calls per hour respectively. This is close to linear speed-up but due to the configuration constraints, the systems cannot be perfectly load balanced.

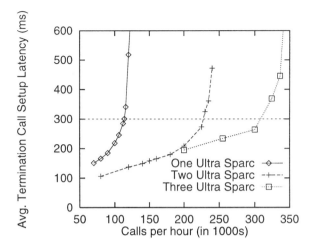

Figure 8. Call Setup Latency vs. Call Throughput

5   IMPACT OF MOBILITY ON PERFORMANCE

In this section, we examine the impact of mobility-related parameters on cMSC performance. Note that while actual call throughput and latency values are dependent on the hardware and software platforms chosen for the experiments, the general impact of mobility on performance is inherent to managing user mobility and is independent of the cMSC implementation. Our experiments were run using the single Ultra 2 configuration.

5.1   Impact of Mobility-to-Call Ratio (σ)

As the mobility of user increases, the call handling capacity of the cMSC will decrease as a larger percentage of CPU time will be devoted to processing registration messages due to movement. The system was tested with the baseline parameter values except for the mobility-to-call ratio, σ, which was varied from 0 to 4. A value of 0 for σ corresponds to the wireless local loop situation where there is no user mobility, a value between one and two corresponds to regions with low mobility, and values of three and higher correspond to regions with high mobility. Call load was kept constant at 100,000 calls/hour.

Figure 9 shows the average call setup latency for both call origination and call termination for different values of σ. As shown in the figure, the call setup latency is lower than the requirements for third generation systems for a wide range of mobility/call ratios, indicating that a call throughput of up to 100,000 calls per hour is a suitable operating point for the cMSC in this configuration.

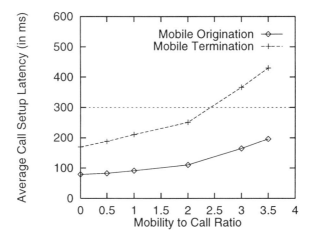

Figure 9. Call Setup Latency vs. Mobility to Call Ratio

5.2   Impact of Call Origination Percent (η)

Since call termination procedures require mobility management procedures in addition to call/connection control procedures, we expect the call handling capacity of the cMSC to decrease as the percentage of mobile-terminated calls increases. Experiments were run using the baseline parameter values except for the call origination percent, η, which was varied from 0 to 1. The call load was kept constant at 100,000 calls/hour.

Figure 10 shows the average call setup latency vs. call origination percent. As shown in the figure, the latency for call termination more than doubles as we move from 95% call origination to 5% call origination. Even so, the average call establishment latencies are within the third generation performance targets.

6   IMPACT OF CORBA

The largest performance bottleneck in the system is the overhead of using CORBA. To quantify the overhead of using Orbix, we measured the processing time, using a performance profiling tool, QUANTIFY, of various functions of Orbix 2.2 MT. Table 1 summarizes the processing overhead of various tasks for sending and receiving a CORBA message.

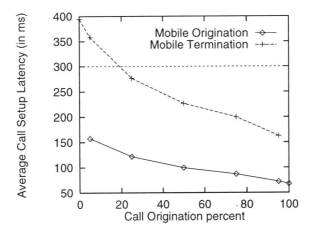

Figure 10. Call Setup Latency vs. Call Origination Percent

Table 2. CORBA Overhead

Task	Sender	Receiver
write system call	67%	NA
read system call	NA	30%
marshalling/un-marshalling	21%	22%
memory allocation/locking	12%	40%

From these results, we conclude the overhead at the sender and receiver of using CORBA as a general purpose middleware layer is about 12% and 40%, since the write, read and data marshaling functions are necessary in our system. These overhead values agree with earlier results [16] and may drop as future implementations of CORBA eliminate some of the inefficiencies of memory management. Two other possibilities for improving the performance of CORBA in our system are to use a single threaded version of CORBA or use server co-location.

6.1 Single Threaded CORBA

The current cMSC implementation uses multi-threaded Orbix. Multi-threaded implementations introduce overhead due to the need for locking of shared data structures. This, in part, is the reason for the high overhead in the locking and memory allocation functions in our system. Using a single threaded version of CORBA may reduce this overhead.

In addition, the multi-threaded implementation of Orbix has an architectural bottleneck in that Orbix2.2MT uses one thread for each TCP connection. Each of these threads enqueues incoming requests at a common queue which is guarded by a semaphore. One master thread is then responsible for dispatching the queued requests to the appropriate method within the cMSC server. This results in a bottleneck at the message queue semaphore and the need for some unnecessary locking. Since the current implementation of cMSC servers does not use multi-threading, the use of a single threaded CORBA implementation can improve the cMSC performance. However, this may be a limitation as additional features of third generation wireless networks are added to the cMSC servers.

6.2   Server Co-Location

When using server co-location, two or more servers executing on the same processor are merged into a single process. In this way, each signaling message between any given pair of processes is optimized from a CORBA inter-process message to a function call. Thus, the use of server co-location can result in dramatic improvements in the performance of the cMSC. The use of IDLs and CORBA facilitates object co-location.

A disadvantage of such an approach for the cMSC arises when cMSC reliability and scalability must be ensured. In [3], we present a software-based approach as a solution for making the cMSC reliable. In this approach, failing hosts and servers results in new cMSC servers being instantiated, with checkpointed state information, on new processors in real-time. In this case, there may be a need for separation of co-located objects at run-time. This requires storing executable files for all possible configurations of co-located servers, and managing the deployment of these processes at run-time. While this may render server co-location too complex for large multiprocessor configurations, it may be possible to use co-location in the case of smaller deployments of the cMSC to improve performance.

7   CONCLUSIONS

In this paper, we have examined the design, implementation and performance of a novel Cluster Mobile Switching Center. By using distributed call processing, the system scales in terms of call handling capacity. The modular design of the cMSC lends itself well for feature additions and load balancing in a network of processors. The cMSC servers were built on top of a CORBA-based middleware, thus easing the development of the distributed servers.

Our performance experiments indicate that a cMSC configuration of three Ultra 2s can support over 300,000 calls per hour with an average call setup latency of 300 milliseconds which meets the requirements of third generation systems. Thus, we conclude that the software design, CORBA-based platform, and commodity processors are suitable for call processing applications in a wireless environment. Given that the performance-to-price ratio of processors will continue to improve dramatically, we expect solutions based on these principles to have long life cycles. These results are very encouraging considering the growing importance of modular design and distributed implementations being researched for future call processing systems.

REFERENCES

[1] *IEEE Personal Communications Magazine*, Special Issue on IMT2000: Standards Efforts of the ITU, Vol. 4, No. 4, Aug., 1997.

[2] Object Management Group (OMG), The Common Request Broker: Architecture and Specification, Rev. 1.2, Dec., 1993.

[3] K. Murakami, R. Buskens, R. Ramjee, Y-J Lin, T. La Porta, "Design, Implementation, and Evaluation of Highly Available Distributed Call Processing Systems," *FTCS'98*.

[4] K. Basu, D. Robertson, M. Rau, F. Homayoun, "Wireless Network Architecture or Migration to ATM and B-ISDN," *Proc. of International Switching Symposium (ISSŌ97)*, Toronto, 1997.

[5] G. Enoroth, M. Johnsson, "ATM Transport in Cellular Networks," *Proc. of International Switching Symposium (ISSŌ97)*, Toronto, 1997.

[6] H. Blair, S.J. Caughey, H. Green, S.K. Shrivastava, "Structuring Call Control Software Using Distributed Objects," *International Workshop on Trends in Distributed Computing*, Aachen, Germany, 1996.

[7] A.A. Lazar, K.-S. Lim, F. Marconcini, "Realizing a Foundation for Programmability of ATM Networks with the Binding Architecture," *IEEE Journal on Selected Areas of Communications*, Vol. 14, no. 7, 1996.

[8] M.C. Chan, "Architecting Control Infrastructure of Broadband Networks," Ph.D. Thesis, Columbia University, 1997.

[9] La Porta, T.F. Veeraraghavan, M., P. Treventi, R. Ramjee, "Distributed Call Processing for Personal Communications Services," *IEEE Communications Magazine*, Vol. 33, No. 6, June 1995.

[10] T. F. La Porta, M. Veeraraghavan, R.W. Buskens, "Comparison of Signaling Loads for PCS Systems," *IEEE/ACM Transactions on Networking*, Vol. 4, No. 6, Dec. 1996.

[11] IS-634 revision A, 2nd Ballot Version, October 1997.

[12] ITU Recommendations Q.700-Q.795, "Specifications of Signaling System No. 7," 1989.

[13] TIA/EIA IS-41 (Revision D): "Cellular Radio-Telecommunications Intersystem Operations," 1997.

[14] ITU Recommendations Q.2761, "General Functions of Messages and Signaling of the B-ISDN User Part of SS7," 1995.

[15] ETSI, "European Telecommunications Standards Institute GSM Recommendations."

[16] A. Gokhale, D.C. Schmidt, "Measuring the performance of Communication Middleware on High-Speed Networks," *Sigcomm'96*.

Chapter 10

Challenges of Higher Speed Wireless Internet Access

Ahmad R S Bahai
ALGOREX Inc.

Abstract

High speed wireless data services have been recently highlighted in many recent standard activities. The prerequisite for the growth of wireless data services is an efficient, low cost, and user friendly access method combined with proper application software. Third generation wireless systems promise more efficient wireless data services than those of current cellular systems. Recent proposals for next generation wireless communication technology have prompted numerous research efforts comparing different technologies for next generation systems. Variations of wide-band CDMA and TD-CDMA have been the strongest candidates for the physical layer of next generation cellular systems. Third generation technology promises data rates of up to 2 Mbps in small coverage areas for low mobility users, whereas, for some applications higher data rate in larger area is very desirable. That brings about another challenge regarding flexibility of third generation proposals to accommodate new services such as higher speed packet data, particularly in downlink, for wireless Internet access. Standardization activities of wireless LAN and some other high speed data services add another dimension to ubiquitous wireless data services. In this Chapter, we discuss technical challenges and requirements of these proposals and briefly study complexity of network and terminals for multi-mode, multi-band systems. Moreover, we discuss the proposed concept of integrating asymmetric high-speed wireless packet services with current cellular systems for third generation wireless data services.

1   INTRODUCTION

While wire-line data traffic has surpassed voice traffic in some countries, wireless load is predominantly voice. Wireless data market has experienced slow growth due to several reasons including slow data rate, expensive services, and lack of application software. Many of these bottlenecks are being addressed in several new wireless standards.

Proposals for third generation wireless systems and other wireless data services have instigated many interesting technical debates ranging from compatibility and flexibility issues to signal processing challenges. Attempts to converge main proposals have been partially fruitful. However, some concerns about system complexity, introduction of new services, and co-existence with current technology have been raised. New enhancements for higher data rates and new services of successful second generation mobile systems such as GSM, on one hand, and amount of investments on infrastructure, on the other hand, can potentially delay introduction of third generation systems in several big markets.

A quick review of third generation proposal objectives set the ground for a comparative study of different systems and their potentials. Priority of different objectives varies for different system and service providers. In general, *flexibility* is a prime goal for new systems. Variable bit rate or bandwidth on demand for circuit switched **and** packet switched services can boost spectral efficiency of wireless systems. Another aspect of flexibility is seamless operation of mobile terminals in unregulated and regulated environments. Operation in a mixed cell environment such as hierarchical cell structure and overlay architecture is another requirement for mobile terminal and infrastructure. New architecture should be expandable to new services, some of which are not currently well defined. Compatibility with some current services is a highly debated feature of third generation systems. Obviously, for current successful systems compatibility has an important impact on the speed of transition to new technology. *Integrated solution* for a variety of current and future services is another demanding aspect of third generation systems. Indoor high speed services, outdoor mobile services, wide area low mobility high speed data services and satellite services are proposed to be integrated. The integration can impose a big burden on handheld terminals and system backbone design. Increased *capacity* measured in Erlang (or Kbps) $/MHz/km^2$ is a driving force for earlier introduction of third generation systems. While the urgency of higher capacity varies from one geographic area to another, demand for higher capacity in the future is inevitable. Currently, providing enough capacity for mere voice channels in hot spots of second generation digital systems is becoming a complicated task. *Toll quality voice* is another goal for new systems. Second generation digital mobile systems' voice quality does not show a big improvement, if any, over their predecessor analog systems.

Improved voice quality needs more efficient source coding and a better bandwidth efficient system architecture. Reducing *transmit power* to achieve longer battery life and minimize potential health hazards is another attractive feature. Recent activities in broadband wireline modems such as ADSL and cable modem create an expectation of high speed Internet access through wireless services. Using voice channels for data services is neither economical nor efficient. Emergence of other wireless data services such as wireless LAN and wireless ATM in conjunction with data service capability of enhanced second generation and third generation systems can provide a seamless and efficient solution for wireless data access.

In this Chapter, we offer an overview of technical challenges of introduction and utilization of third generation and some other wireless data systems. For this purpose, we briefly review the main objectives of third generation systems, then elaborate on challenges of physical layer design for current main proposals. Future demands for high data rate packet and circuit switched services beyond what third generation proposals can offer and some candidates for these services are discussed in Section III. The impact of phased introduction of new wireless services including high data rate packet services is discussed in Section IV. In Section V we look at the challenges of terminal design for new systems.

2   PHYSICAL LAYER DESIGN

First, we start with an overview of different proposals for third generation cellular systems. The main third generation proposals considered in this study are the WCDMA system approved by ARIB in Japan and ETSI in Europe, and WCDMA proposed by CDG group. We briefly refer to other technologies such as TD/CDMA adopted by ETSI as part of UTRA system, and high speed TDMA adopted in US. Considerable effort is made to converge different WCDMA proposals for third generation yet some differences in physical layer makes system design a challenging task. While the main proposals are variations of Wide band CDMA, they follow two different approaches in system design [1]. The CDG proposal is an evolutionary enhancement to the IS-95 [2] A/B systems. The design is based on minimizing maximum power of interferers. By using fairly long spreading pseudo-noise code covering about 512 symbols the interference from other users is treated as uncorrelated Gaussian interferers. A known fact in information theory [3] states that among additive interferences of a power constrained communication system, a Gaussian one reduces the upper bound of achievable capacity to

$$C = 1/2 \log(1 + E/\sigma^2)$$

where E is the input power and σ^2 is noise variance.

Hence, a powerful error correcting code in the uplink is used to improve the performance of an interference limited system. Without any pilot signal (CDMA-2000 has a coherent uplink) in the uplink, and weak correlation between users, multi-user detection or interference cancellation is not particularly attractive. The CDMA architecture requires a robust and efficient power control as overall interference power should be kept to minimum and usually no attempt is made to identify and cancel the effect of each interferer. Any excessive delay or deficiency in power control impacts the performance, particularly when the user is further away from base station than the interferers. Another class of WCDMA proposals supported by ARIB [4] and ETSI [5] uses a set of quasi-orthogonal short codes and multi-user detection, and interference cancellation is considered to improve system performance significantly. In this design, users are not necessarily uncorrelated. In multi-user detection, correlation between users is taken into account to provide a close to optimum detector [6]. This approach can withstand the near-far problem of typical CDMA systems as strong interferers can be detected and removed reliably. Smaller code size is helpful in reducing the self-noise effect, but may adversely affect the sensitivity of the system to inter-symbol interference. A Rake receiver can constructively combine scattered replicas of the received signal but, in the presence of long delay spread, comparable to a major fraction of a symbol duration, adjacent symbols may interfere with the detection. More precisely, the received signal is presented as:

$$r(t) = \sum_{k=1}^{K} A_k \sum_{i=-M}^{M} b_k(i) S_k(t - iT - \tau_k) + n(t)$$

where K is the number of users, A_k and τ_k are the received amplitude and delay of kth user, $b_k(i)$ is the ith symbol of kth user, $n(t)$ is additive noise of power σ^2 and $s_k(t)$ is the received signal waveform of the k-th user

$$S_k(t) = \int s_k(t - \tau) h_k(\tau) d\tau$$

For a simple two-ray channel model of the form

$$h_k(t) = \delta(t) + \alpha_k(t - T)$$

and a two-user case the received signal is

$$\hat{y}(i) = \begin{bmatrix} A_1 & \rho_{12}A_2 \\ \rho_{21}A_1 & A_2 \end{bmatrix} \begin{bmatrix} b_1(i) + \alpha_k b_1(i-1) \\ b_2(i) \end{bmatrix} + n(i)$$

for $i = -M, \cdots, M$ and ρ_{12} is the correlation between two users signatures:

$$\rho_{12} = \int_0^T s_1(t)s_2(t)dt$$

As shown, the interference is caused by a combination of frequency selective fading and multi-user environment. The optimum receiver has a two-dimensional MLSE structure, which can be approximated by a more realizable format.

2.1 HIGHER DATA RATE CHALLENGES

Second generation systems have limited data capability. TDMA systems such as GSM transmit at 9.6 kb/s and CDMA IS-95 at 14.4 kb/s. With new enhancements to these systems higher data rates in both packet switched and circuit switched modes are feasible. Enhancement of GSM system (EDGE) aims to provide data rates of up to 384 kb/s by using higher level modulation schemes and efficient FEC and ARQ techniques. GPRS (General Packet Radio Services) promises data rates of up to 115 kb/s with network connections to TCP/IP or X.25. Third generation proposals offer variable data rates of up to 2 Mbps. In current proposals, high data rate services are restricted to low mobility users within small coverage areas such as indoors. In addition, higher data rate in a multi-code system comes at the expense of occupying valuable traffic channels, which drives up the usage cost. Some upcoming applications are severely restricted by constraints of data rates in third generation systems. Affordable wireless access to Internet services with high data rates of up to 10 Mbps is an appealing feature not yet offered by third generation systems. Recently, some supplementary high data rate services have been proposed for high-speed wireless access [7,8]. In this section, we briefly review some of the offered technologies and their merit for co-existence with third generation systems. Physical layer design for high speed packet switched data such as Internet access requires fast synchronization as the overhead for synchronization or any adaptation in the system should be kept to minimum. In addition, the modulation scheme should withstand severe delay spread relative to symbol duration, commonly encountered in outdoor environments. It is preferable to reduce system sensitivity to timing jitter as the receiver clock is not constantly powered on and may experience some jitter in standby mode.

For data rates higher than 1 Mbps, single carrier transmission with no spreading requires a very fast and complicated equalizer to reduce the effect of intersymbol interference. Since channel characteristics may change from one packet to another and delay spread may extend to more than several symbols in an outdoor environment, a fairly complicated equalizer should be retrained for each packet. One remedy to this problem is spreading of the signal by using a short code to relax equalization requirements. This concept is close to design philosophy of TD-CDMA proposal.

Direct sequence spread spectrum techniques combine multipath components constructively and resolve high delay spread impairment effectively. However, they do not offer high processing gain for high data rates and require fast synchronization in each frame before despreading and Rake combining. In addition, power control for different users is important to maintain system capacity. Some proposals include a pilot signal to monitor channel condition and speed up synchronization of the system. Pilot insertion requires a periodic monitoring mechanism, which adds to system complexity and power consumption.

Another promising option for high-speed packet services is Coded Orthogonal Frequency Division Multiplexing (COFDM)[7]. This technique partitions a highly frequency-selective wide band channel to a group of non-selective narrow band channels. Therefore, it is robust against large delay spreads by preserving orthogonality in the frequency domain. The multi-carrier technique is robust against timing jitter as it translates to a phase rotation after Fourier transform in the receiver. Usually simple channel estimation is used in detection and does not require significant training overhead. Other attractive features of an OFDM system are simple fallback mode. Depending on severity of delay spread and data rate, the number of sub-carriers can change without any major impact on overall receiver structure. FFT, as a major receiver block, lends itself to fallback mode very efficiently. A COFDM system is robust against impulsive noise and narrow band interference. However, frequency offset and phase noise may result in inter-channel interference and degrade the performance considerably. Another major issue about OFDM is high peak to average ratio of the transmitted signal, which results in clipping at the transmitter power amplifier and bandwidth regrowth. In another study, our analysis shows that in wireless system with raw bit error rates of 0.001 this effect can be controlled efficiently.

Design of an OFDM system for high-speed packet switched application requires a careful selection of system parameters. The number of carriers and any pre/postfix should be optimized properly. While a larger number of carriers is desirable to combat large delay spread in outdoor areas, it makes the system more sensitive to frequency offset and phase noise. The effect of pulse shaping can reduce inter-channel interference and improve receiver robustness against frequency offset or oscillator phase noise. Pulse shaping and the possibility of interference between OFDM blocks requires cyclic extension of the frame. The extension length depends on the sensitivity of the system to inter-block interference. A proper two dimensional coding and interleaving can withstand time and frequency selective fading by correlating symbols with independent fades spaced more than the coherence time and bandwidth of the channel. In packet mode, time variation within one packet is insignificant; therefore, frequency diversity of different bins is taken into account for optimum code design [10,12]. Some standards for WLAN systems have adopted OFDM for air interface. For example, IEEE 802.11 is proposing OFDM technique for 5 GHz Wireless LAN. Other WLAN standards such as Multi-Media Access Communication (MMAC) in Japan and BRAN in Europe are converging to OFDM technique. WLAN can be a critical element of a wireless data network as most of ultra high data accesses are needed in an indoor environment. Its extension to wide area service are now being considered.

Some other promising high speed wireless systems such as LMDS and MMDS are being introduced for local access. Analysis of those systems are not in the scope of this Chapter.

3   SYSTEM INFRASTRUCTURE AND NETWORK

Supporting existing and future physical layer architectures and multiple access schemes adds to complexity of network and infrastructure design. Therefore, flexible network architecture to accommodate a variety of current and new services is essential. That ranges from traditional circuit-switched services of second generation systems to more advance packet switched services such as SMS and GPRS. Recent developments of UMTS network architecture address the concept of radio independent architecture to cope with several radio access techniques. By providing a Generic Radio Access Network (GRAN), which separates the development of access (radio-dependent) and core (radio-independent) parts and generalizing mobility management and hand-over control functions, network structure can work with different radio interfaces with minimum change, mainly in mobile terminal (MT) and base station (BS) structure. Soft hand-over in WCDMA systems requires multicasting and synchronization of macro-diversity through different base stations. In the case of using an ATM or IP network between base stations and mobility management servers, some additional functional modules are required to coordinate base stations and provide multicasted data within a short time span. Providing wireless Internet access service requires an infrastructure for efficient packet switched access supported by higher layer protocol functions. Compared to speech and real time services, IP services can experience longer delay yet require low bit error rate. Low BER requirement can be partly mitigated by using ARQ retransmission protocol in the TCP/IP network.

From a network view, Service Adaptation Layer mainly handles Internet services and mobility management is controlled by mobility adaptation layer [9]. Separating speech, real time, and packet data at the service layer makes system design more efficient in handing delay and error protection. However, using the same bandwidth for higher data services results in eliminating many voice channels and has an adverse impact on system capacity. In particular, in WCDMA the voice activity factor is important for system capacity and using voice channels for data services may become costly. Using a different frequency band dedicated to high speed packet services while sharing some of network infrastructure of third generation systems is another path toward a rapid, low cost high speed system. In most of current data services, data flow is likely to be asymmetric with much higher data rate in the downlink and lower data rate in the uplink, such as web browsing applications. This can be implemented by a highly asymmetric link supported by network backbone. Use of broadcasting structure such as DAB for downlink and multi-slot GSM (HSCSD) or other cellular channels for uplink with network coordination is one example of such an asymmetric link as in Figure 1.

Time variation and losses of a wireless link in addition to asymmerty between the link capabilities on the data and acknowledgement path may seriously limit the performance of a wireless TCP/IP network by triggering congestion control mechanism when it is not needed. One solution for support of TCP over wireless is to hide the losses over wireless link from TCP by using link layer ARQ techniques [13]. Mobility of terminals and handoff also result in fluctuation in availability of network resources. Some proposals such as scalable data streams attempt to address Quality of Service provisioning in a wireless network [14].

In addition, an efficient wireless network design using some of proposed features for third generation such as dynamic channel allocation (DCA), hierarchical cell structure (HCS), and intelligent MAC layer can improve the capacity for a high speed packet data service.

Other attractive features of future wireless networks such as intelligent billing system and one number operation are not discussed in this work.

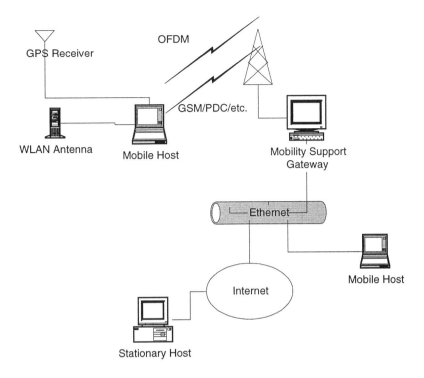

Figure 1: Wireless Data Network

4 TERMINAL DESIGN

Reaching a worldwide consensus for a third generation wireless system is an invaluable goal for standard bodies. However, recent debates about next generation wireless systems indicate that several physical layer architectures will co-exist for next few years. Enhancements to second generation systems and diversity of proposals for third generation requires a multi-mode multi-band terminal capable of handling different standard requirements. Emergence of in-door high speed packet switched data access in Wireless Local Area Networks (WLAN), satellite services, and some upcoming wireless wide area networks add to desired capability of future wireless data terminals. In addition to traditional wireless phones, many other data communication devices are and will be introduced to consumer market. PDAs and notebook computers contain built-in data communication devices and mobile phones can handle many data processing and storage tasks. Market demand has resulted in convergence of data processing and communication devices in an evolutionary fashion. However, a more optimum combination of data processing and communication tasks requires an efficient design and architecture where resources such as processing unit, memory access, and bus speed are properly shared between different tasks. For example I/O centric architecture has recently been proposed as a potential solution for future wireless terminals[15]. Adaptivity and reconfigurablility of wireless data terminals are important features for multi-mode terminals to support a variety of services.

Complexity of terminal design is partly due to different requirements for RF front-end. These include variable bandwidth, linearity requirements, power control dynamic range, and sensitivity of the system to impairments such as phase noise and frequency offset. The requirements set forth by different standards and provisions for an efficient system design from a terminal designer view vary significantly from one system to the other. Power consumption and terminal cost are important factors and are highly affected by the RF front-end.

The software radio concept is an attractive solution to avoid excessive complexity and cost of analog front-end in a multi-mode terminal. A programmable radio gives the designer the flexibility to handle different bandwidths, sampling rates, filtering and other front-end characteristics. Introduction of high speed sampling devices and programmable high speed digital multi-band RF converters have made a programmable front-end a realizable concept by transferring some of the IF analog functional modules to digital domain and placement of A/D and D/A closer to the antenna without overloading digital signal processors as shown in Figure 2.

Moreover, signal processing requirements of third generation proposals are significantly more than those of current wireless systems. Baseband modules such as interference cancellation, adaptive antenna arrays and synchronization require a stronger processing engine. Recent evolution of low power digital signal processors has been very promising. With the flexibility of software radios and processing power of signal processing engines the concept of high speed and multi-mode, multi-band terminal is more realistic than before. However, some consistency in system design in terms of reference clocking and frequency planning for chip rate, symbol rate, frame rate, and frequency planning, can make the design more efficient. This concept has been followed in TD-CDMA and WCDMA design as a 13 MHz reference oscillator can be used for both systems and is highly desirable to be extended to other proposals. Co-existence of a transceiver for high speed packet data services and one for third generation systems requires some consistency in terms of clocking and frequency plan and should be taken into account for new packet service system parameters.

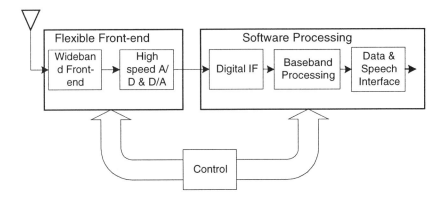

Figure 2: Basic Software Radio Architecture

Recent development of mobile IP and Wireless Application Protocol (WAP) is another dimension for integrating mobile telephony and Internet access device. WAP are designed to provide advanced Internet services for many different wireless systems including cellular, SMS, and wireless LAN.

Another defining factors in general architecture of wireless data terminal is the partitioning of tasks between terminal and network. This partitioning is important for quality of service monitoring by which terminal contributes more actively to network data collection and enables user to access different services by downloading required software over the air.

Use of location aware navigation guide such as GPS can provide additional set of features for users and more information for network operators to analyze and optimize the performance. Other features such as low power consumption, Man-Machine Interface (MMI), and security are important for wide market penetration.

5 CONCLUSION

Proposals for an IMT-2000 system are technically promising, yet demand for new services and innovative concepts for physical layers and phased introduction of third generation features may require co-existence or integration of several radio systems. Therefore, flexibility of network architecture and terminal design is fundamental to rapid and efficient introduction of new technologies and new services. On the network side, a radio independent and flexible core network is essential. New technologies in software radio and high speed, low power DSP can help to provide flexible multi-mode terminals. In spite of current packet data services provisioned in third generation proposals, higher data rate access services may require some other innovative radio access schemes. COFDM is a strong candidate for high-speed wireless Internet access. It can provide reasonable flexibility for system design and has already been utilized for applications such as radio broadcasting. It is also proposed for WLAN, Wireless ATM, and MMAC applications for higher data rates. Integration of an asymmetric high data rate packet switched service with third generation system can address requirements of higher data rates up to 10 Mbps in the same system. Since uplink data rate is typically much lower than that of downlink, it can use voice channels for uplink data transmission. Selection of the frequency band for downlink high-speed service is another critical issue in system design.

In light of many standardization activities and new proposed services based on high speed wireless communications, it is essential to engineer a flexible platform both at network side and terminal side to accommodate future services with minimum change.

REFERENCES

[1] S. Vembu, A. J. Viterbi, "Two Different Philosophies in CDMA-a Comparison": IEEE 46th Vehicular Technology Conference. Vol 2. p. 869-73, 1996

[2] TIA/EIA/IS-95 Interim Standard, Mobile Station-Base Station Compatibility Standard for Dual-Mode Wideband Spread Spectrum Cellular System, July 1993

[3] R. Gallager, Information Theory and Reliable Communication, John Wiley, 1968

[4] F. Adachi, M. Sawahashi, "Wideband multi-rate DS-CDMA for next generation mobile communications systems", Wireless Communications Conference. pp. 57-62. 1997

[5] T. Ojanpera et al., "Comparison of Multiple Access Schemes for UMTS," Proc, Vehicular Tech. Conf. 97, Phoenix, AZ, May 1997

[6] H. V. Poor, S. Verdu, "Probability of error in MMSE multiuser detection". IEEE Transactions on Information Theory, vol.43, (no.3): 858-71, May 1997

[7] J. C. Chuang, N. R. Sollenberger, " Medium Access Control for Advanced Cellular Internet Services", PIMRC, pp 673-677, Sept. 1997

[8] R. Ludwig, "Downlink Performance of General Packet Radio Services for GSM", Proc. 3rd Int'l Workshop on Mobile Multimedia Comm.

[9] G. Fleming, A. Hoiydi, J. De Vrient, G. Nikoladis, F. Piolini, M. Maraki, "A Flexible Network Architecture for UMTS" IEEE Personal Communications, pp. 8-15, April 98.

[10] M. Alard and R. Lassale, "Principles of Modulation and Coding for Digital Broadcasting for Mobile Receivers" EBU Technical Review No 224, pp. 168-190

[11] P. Lattieri, M. Srivastava, "Advances in Wireless Terminals", IEEE Personal Comm. Mag. pp: 6-18; Feb. 1999.

[12] A.R.S. Bahai, B. Saltzburg, " Multi-Carrier Digital Communications", To be published, Kluwer/Plenum, 1999

[13] H. Chaskar, T. V. Lakshman, U. Madhow, " On the design of interfaces for TCP/IP over Wireless," Proc. Milcom 96.

[14] M. Naghshineh, M. WillebeekLeMair, "End-to-End QoS Provisioning in Multimedia Wireless/Mobile Networks Using an Adaptive Framework" IEEE Comm. Mag. Nov. 1997, pp. 72-81

[15] P. Lattieri, M. Srivastava, "Advances in Wireless Terminals", IEE Personal Comm. Mag. Feb. 1997, pp. 6-18

Chapter 11

ENERGY EFFICIENT PROTOCOLS FOR WIRELESS SYSTEMS

PRATHIMA AGRAWAL[1], JYH-CHENG CHEN[1], and KRISHNA M. SIVALINGAM[2]

[1] *Telcordia Technologies*
[2] *Washington State University*

Abstract   This paper summarizes recent advances in energy efficient and low power design of networking protocols. We focus on the MAC (Medium Access Control) and application layers of the protocol stack. Of the five MAC protocols studied, EC-MAC has the best energy efficiency. EC-MAC also offers the best scalability with respect to the number of mobile terminals being serviced, as its energy consumption remains bounded. Power conservation techniques discussed for processing multimedia information at the application layer include transferring majority of the processing out of the terminal and into the network, tolerably deteriorating the video quality to lower data rates, and selectively discarding video frames.

1. INTRODUCTION

Energy efficiency is crucial due to the limited battery life of the mobile wireless terminal. Power conservation techniques are routinely used in the hardware design of such systems. Power saving techniques are also common in the MAC (Medium Access Control) and application layers. This is because the MAC layer is adjacent to the physical layer in the protocol stack and directly controls the operations of the wireless transceiver. The application layer directly controls the quality of the information presented to the user. However, inputs from the user of a mobile terminal also play a significant role in the operation of the application layer. Screen saving is an example of energy conservation when the user input is absent for a predetermined amount of time. Energy saving techniques in portable wireless terminals become increasingly important as wireless networks expand from voice only to multimedia service capability.

An indirect implication of terminal portability is the use of batteries as power sources in mobile terminals. Since batteries provide limited energy, a general constraint on wireless communications is the short continuous operation time. For example, studies have shown that significant consumers of power in a typical laptop are the microprocessor (CPU), liquid crystal display (LCD), hard disk, system memory (DRAM), keyboard/mouse, CDROM drive, floppy drive, I/O subsystem and the wireless network card in case of a mobile computer [Udani and Smith, 1996]. A typical example from Toshiba 410 CDT mobile computer shows that nearly 36% of energy consumed is by the display, 18% by the wireless interface, 21% by the CPU/Memory, and 18% by the hard drive.

Typically, energy conservation is considered in the hardware design of the mobile terminal such as CPU, disks, displays, etc. Significant power saving can result by incorporating low-power strategies in the design of network protocols used for data communications. This paper exams the opportunities for energy conservation at all layers of the protocol stack for wireless networks. Moreover we focus on MAC and application layers.

The rest of the paper is organized as follows. Section 2 briefly describes an organization of the protocol stack for wireless networks. Section 3 describes work done on energy efficient MAC protocols for wireless networks. Section 4 describes some opportunities for saving battery energy in network and transport layers, and work done on power saving in OS (operating system) and middleware layers. Section 5 describes the application layers from an energy efficiency perspective. Section 6 summarizes the paper.

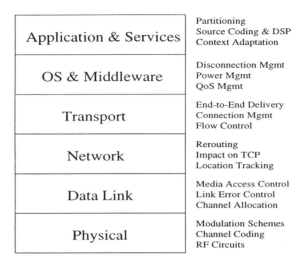

Figure 11.1 Protocol stack of a generic wireless system.

2.   PROTOCOL LAYERS

This section gives an introduction to the software used in wireless network systems. Application programs using the network do not interact directly with the network hardware. Instead, an application interacts with the protocol software. The notion of protocol layering provides a conceptual basis for understanding how a complex set of protocols work together with the hardware to provide a powerful communication system. The protocol stack for a generic mobile and wireless communication system can be divided into five layers as shown in Fig. 11.1. The application and services layer occupies the top of the stack followed by the OS and middleware, transport, network, data link, and physical layers. Unlike a fixed wired network, problems peculiar to the wireless channel and mobility issues challenge the design of the protocol stack adopted in a mobile wireless communication system. In addition, networking protocols for these systems need to be designed for energy efficiency.

The physical layer consists of Radio Frequency (RF) circuits, modulation and channel coding systems. From an energy efficiency perspective, considerable attention has been given to the design of this layer [Chandrakasan and Brodersen, 1995]. The data link layer is responsible for establishing a reliable (and secure) logical link over the unreliable wireless link. The data link layer is thus responsible for wireless link error control, security (encryption, decryption), mapping network layer packets into frames, and retransmissions. The MAC protocol is responsible for allocating the time-frequency or code space amongst mobiles sharing wireless channels in a region. The network layer is responsible for routing packets, establishing the network service type (for example, connection-less vs. connection-oriented), and transferring packets between the transport and link layers. In a mobile environment, this layer has the added responsibility of rerouting packets and mobility management. The transport layer provides a reliable end-to-end data delivery service to applications running at the end points of a network. The OS and middleware layer handles disconnection, adaptivity support, power and quality-of-service (QoS) management. This is in addition to the conventional tasks like process scheduling and file system management. Finally, the application and services layer deals with partitioning of tasks between the fixed and mobile hosts, source coding, digital signal processing, and context adaptation in a mobile environment.

In the following sections, we shall describe the role played by the key ingredients of each layer in energy efficiency.

3.   MAC LAYER

This section addresses energy efficiency in MAC protocols for wireless networks. A framework to study the energy consumption of a MAC protocol from the transceiver usage perspective is developed. This framework is then applied to compare the performance of a set of protocols that includes IEEE 802.11[IEEE, 1997], EC-MAC [Sivalingam et al., 1999], PRMA [Goodman et al., 1989], MDR-TDMA [Raychaudhuri and Wilson, 1994], and DQRUMA [Karol et al., 1995]. The performance metrics considered are transmitter and receiver usage times for packet transmission and reception. The time estimates are then combined with power ratings for a Proxim RangeLAN2 radio card to obtain an estimate of the energy consumed for MAC related activities. Detailed analysis can be found in [Chen et al., 1999].

3.1   MAC PROTOCOLS

This section briefly describes the wireless access protocols studied in this section. Fig. 11.2 shows the channel access methods for these protocols.

The IEEE 802.11 standard for wireless LANs defines multiple access using a technique based on Carrier Sense Multiple Access / Collision Avoidance (CSMA/CA). The basic access method is the Distributed Coordination Function (DCF) shown in Fig. 11.2(a). A backlogged mobile may immediately transmit packets when it detects free medium for greater than or equal to a DIFS (DCF Interframe Space) period. If the carrier is busy, the mobile defers transmission and enters the backoff

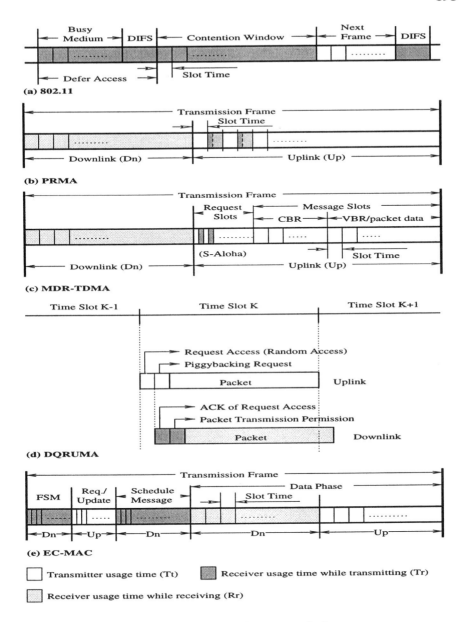

Figure 11.2 Channel access methods.

state. The time period following the unsuccessful transmission is called the contention window and consists of a pre-determined number of slots. The mobile, which has entered backoff, randomly selects a slot in the contention window, and continuously senses the medium during the time up to its selected contention slot. If it detects transmission from some other mobiles during this time period, it enters the backoff state again. If no transmission is detected, the mobile transmits the access packet and captures the medium. Extensions to the basic protocol include providing MAC-level acknowledgments (ACKs)and ready-to-send (RTS) and clear-to-send (CTS) mechanisms.

Packet reservation multiple access (PRMA) was proposed for integrating voice and data traffic. The PRMA system is closely related to reservation ALOHA since it merges characteristics of slotted ALOHA and TDMA protocols. Packets in PRMA are classified as periodic information and random information packets. Once a mobile with periodic information transmits successfully a packet in an available slot, that slot in future frames is reserved for this mobile. However, mobiles with random information need to contend for an available slot each time. The protocol is depicted in Fig. 11.2(b).

The multiservices dynamic reservation TDMA protocol (MDR-TDMA), shown in Fig. 11.2(c) supports CBR, VBR, and ABR traffic by dividing TDMA frames for different types of traffic and allocating them dynamically. The TDMA frame is subdivided into N_r request slots and N_t message slots. Each message slot provides for transmission of a packet or an ATM-like *cell*. Request slots are comparatively short and are used for initial access in slotted ALOHA contention mode. Of the N_t message slots, a maximum of $N_v < N_t$ slots in each frame can be assigned for CBR voice traffic. VBR and packet data messages are dynamically assigned one or more 48-byte slots in the TDMA interval following the last allocated voice slot in a frame. The basic channel access scheme follows a combination of circuit mode reservation of slots over multiple TDMA frames for CBR voice calls with dynamic assignment of remaining capacity for VBR or packet data traffic. In addition to first-come-first-served (FCFS) scheduling, time-of-expiry (TOE) approach has been studied to improve delay performance of real-time data traffic. Energy efficiency issues, however, are not specifically addressed in the protocol definition.

The distributed-queuing request update multiple access (DQRUMA) protocol is shown in Fig. 11.2(d). The base station employs a random access protocol and packet scheduling policy based on traffic and service requirements. Mobiles send a transmission request only when packet(s) join an empty queue. All subsequent packets that arrive at the queue can piggyback transmission requests. Two request access protocols have been studied: the ALOHA random access protocol and a generalization of the Binary Stack Algorithm. The scheduling policy considered is a round-robin packet transmission policy. Since the slots are scheduled on a finer grain in DQRUMA, the requirement that the mobile should listen during every slot places a high burden on the mobile's power resources.

The protocol design of energy-conserving medium access control (EC-MAC) protocol is driven by energy consumption, diverse traffic type support, and QoS support considerations. The protocol is defined using fixed-length frames since each mobile receiver will precisely know the time of the next beacon transmission. This enables the receiver to power off knowing precisely when the next frame will start. The frame is divided into multiple phases: frame synchronization, request/update phase, new-user phase, schedule message, and data phase. At the start of each frame, the base station transmits the frame synchronization message (FSM) on the downlink. This message contains framing and synchronization information, the uplink transmission order for reservations, and the number of slots in the new user phase. The request/update phase is composed of uplink request transmissions from the mobiles. During this phase, each registered mobile transmits new connection requests and queue status of established queues according to the transmission order. The base station then broadcasts the transmission schedule for the data phase using a schedule message. Mobiles receive the broadcast and power on the transmitters and receivers at the appropriate time. The new-user phase allows new mobiles that have entered the cell coverage area to register with the base station. The comparison analysis in next section assumes that all mobiles in the cell coverage area have already registered with the base station. Fig. 11.2(e), therefore, does not incorporate the new-user phase.

A number of other access protocols for wireless multimedia networks have been proposed in the literature. The protocols described here are chosen to represent the major categories of multiple access protocols for local area wireless networks.

3.2 ENERGY CONSUMPTION ANALYSIS

This section analyzes the energy consumption during two important protocol activities at the mobile's MAC layer: packet transmission and reception. For each of the two activities, we quantify the time spent utilizing the transmitter and receiver. For networks with power control, the energy spent for transceiver is varying in time which depends on the power management schemes. Since the MAC protocols studied here are not specified for any particular power control schemes, we assume each protocol uses fixed transmit power. Hence, energy consumed is proportional to the amount of time spent utilizing each resource. For transmitting packets (either single or periodic packets), T_r and T_t are defined as the average time spent using the receiver and transmitter, respectively. For receiving packet(s), the average receiver usage time

Name	Description
L	Time to receive/transmit a packet
l	Time to receive/transmit a reservation/contention packet
a	Time spent decoding a slot
N	Number of mobile stations ($N \geq 1$)
$E[L_t]$	Average time to receive/transmit voice talkspurts
T_s	Average transmitter/receiver usage time for a successful contention
T_f	Avg. transmitter/receiver usage time for a failure contention
p	Probability of a failed contention (other than 802.11 analysis)
P_s	Probability of a successful contention (802.11)
P_f	Probability of a failed contention (802.11)
L_A	Length of an acknowledgment packet (PRMA)
δ	Number of transmission permission issued (DQRUMA)
Δ	Queue length (DQRUMA)

Table 11.1 System parameters.

is denoted by R_r. Table 11.1 summarizes the system parameters and definitions used in the analysis.

3.2.1   Analysis of 802.11 Protocol.   During packet transmission in 802.11, the mobile needs to listen to the medium until it is free. The average time spent using the receiver is the time the receiver is turned on when some other mobile is currently transmitting, plus the time spent using the receiver when this mobile stays in backoff procedure until capturing the medium successfully. The average time the receiver is turned on when some other mobile is currently transmitting is calculated as $\frac{L}{2}$ + DIFS. By applying the *regenerative method* [Walrand, 1991], we can quantify the time spent using the receiver when the mobile is in backoff procedure. Therefore, T_r can be obtained by:

$$T_r = \left(\frac{L}{2} + \text{DIFS} \right) + \frac{P_{f1}T_{f1} + P_{f2}T_{f2} + P_s T_s}{1 - P_{f1} - P_{f2}} \qquad (11.1)$$

where the second term is obtained from the regenerative method. P_{f1}, P_{f2} and P_s can be derived by considering Poisson process and binomial distribution. T_{f1}, T_{f2}, and T_s are based on P_{f1}, P_{f2} and P_s in addition to considering the conditional probability. Derivation and detailed analysis can be found in [Chen et al., 1999].

During the backoff period, the transmitter is used only to transmit the contention packet. If the medium is captured, then the data packet is transmitted. If there is collision during contention packet transmission, the mobile then enters backoff again. Assume the mobile detects the collision after one slot time. The average transmitter usage time (T_t) can also be given by the regenerative model as follows:

$$T_t = \frac{P_{f2}\left(T_f - T_s\right) + T_s}{1 - P_{f2}} \qquad (11.2)$$

where P_{f2}, T_f, T_s are same as those in equation (11.1).

During packet reception, the receiver has to be turned on during the entire downlink transmission. It reads the header of every downlink packet, and moves to standby mode if the packet is not destined for it. If the receiver senses X slots and a is the time spent decoding each slot, the average receiver usage time is given by $aE[X] + L$. It is reasonable to assume that destinations of packets sent by the base station are uniformly distributed over all the mobiles in the cell. The expected number of slots a mobile has to receive before its intended packet is then obtained by $E[X] = N$. Therefore,

$$R_r = aN + L \qquad (11.3)$$

The analysis above is based on the transmitting and receiving of data packets. Since the 802.11 standard does not detail the handling of voice traffic, we ignore voice packets in our analysis of 802.11.

3.2.2 Analysis of PRMA, MDR-TDMA, and DQRUMA Protocols.

Both PRMA and MDR-TDMA are closely related to reservation ALOHA, where they merge characteristics of slotted ALOHA and TDMA protocols. For periodic packets, the same time slot can be reserved once the mobile contends successfully for the first packet. However, MDR-TDMA uses shorter packets than PRMA for contention and acknowledgment. DQRUMA also uses slotted ALOHA to contend for a channel. However the mobile in DQRUMA needs to listen to the transmission permission explicitly for every slot, and can piggyback the transmission request for next packet. The analysis for these three protocols are similar since they all are based on slotted-ALOHA. The regenerative model applied here is much simpler than that in 802.11. Tables 11.2 and 11.3 list the $T_t, T_r,$ and R_r for signal packet and periodic packets of all protocols. Detailed analysis can be found in [Chen et al., 1999].

Protocol	T_t	T_r
802.11	$\dfrac{P_{f2}\left(T_f - T_s\right) + T_s}{1 - P_{f2}}$	$\dfrac{L}{2} + \text{DIFS} + \dfrac{P_{f1}T_{f1} + P_{f2}T_{f2} + P_s T_s}{1 - P_{f1} - P_{f2}}$
PRMA	$\dfrac{L}{1-p}$	$\dfrac{L_A}{1-p}$
MDR-TDMA	$\dfrac{l}{1-p} + L$	$\dfrac{l}{1-p}$
DQRUMA	$\dfrac{l}{1-p} + L$	$\dfrac{l}{1-p} + \delta l$
EC-MAC	$l + L$	$2\,l \le T_r \le l\,(\eta + \psi)$

Protocol	R_r
802.11	$aN + L$
PRMA	$aN + L$
MDR-TDMA	$aN + L$
DQRUMA	$aN + L$
EC-MAC	$l + L \le R_r \le l\,\psi + L$

Table 11.2 Transceiver usage time for single packet.

Protocol	T_t	T_r
PRMA	$\dfrac{L}{1-p} + E\left[L_t\right] - L$	$\dfrac{L_A}{1-p}$
MDR-TDMA	$\dfrac{l}{1-p} + E\left[L_t\right]$	$\dfrac{l}{1-p}$
DQRUMA	$\dfrac{l}{1-p} + L + (\Delta - 1)(l + L)$	$\dfrac{l}{1-p} + \Delta \delta l$
EC-MAC	$l + E\left[L_t\right]$	$2\,l \le T_r \le l\,(\eta + \psi)$

Protocol	R_r
PRMA	$aN + E\left[L_t\right]$
MDR-TDMA	$aN + E\left[L_t\right]$
DQRUMA	$\Delta\,(aN + L)$
EC-MAC	$l + E\left[L_t\right] \le R_r \le l\,\psi + E\left[L_t\right]$

Table 11.3 Transceiver usage time for periodic packet.

3.2.3   Analysis of EC-MAC Protocol. In EC-MAC, a mobile first listens to the transmission order, and then sends out its request or update. After that, the mobile sends its packet in the data phase during its scheduled time. Therefore,

$$T_t = l + L \qquad (11.4)$$

In the FSM phase, mobile listens to downlink until it gets the transmission order. The maximum time spent using the receiver is $l\,\eta$, where η is the maximum number of downlink transmissions. Similarly, the maximum time the receiver is utilized during schedule reception is $l\,\psi$, where ψ is the maximum number of permissions in the schedule phase. The expected time receiver is turned on for sending a packet is given by:

$$2\,l \leq T_r \leq l\,(\eta + \psi) \qquad (11.5)$$

To achieve downlink packet reception, the receiver has to be turned on during the schedule message. After the mobile gets the schedule, it powers on its receiver at the appropriate time in data phase. Let ψ be the maximum number of schedule beacons as discussed above. Then,

$$l + L \leq R_r \leq l\,\psi + L \qquad (11.6)$$

The analysis above is valid for a single packet and for a data packet if data packets need to contend for an available slot each time. However, once a mobile successfully transmits a voice packet in an available slot, that slot in future frames may be reserved for this mobile until the end of the talkspurts. Table 11.3 lists the results for periodic (voice) packets.

3.3   COMPARISON OF MAC PROTOCOLS FOR ENERGY EFFICIENCY

This section provides the numerical results for the analysis presented in the previous section. Figs. 11.3 and 11.4 show the results of the comparison in terms of energy consumption for single and periodic packets, respectively. The results obtained from mathematical analysis and discrete-event simulation are presented. This comparative study demonstrates how careful reservation and scheduling of transmissions avoid collisions that are expensive in battery power. Key results observed from this study are:

1. EC-MAC's energy consumed is almost independent of traffic load and the number of mobiles. **2.** Depending on the slot allocated to a mobile in EC MAC, the energy consumed by a mobile can be different. Best case occurs when a mobile is assigned an early and the worst case occurs when it is a late slot. **3.** Best case of energy consumption in EC-MAC is lower than the energy consumed in all other compared protocols

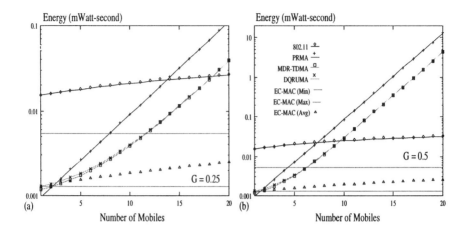

Figure 11.3 Energy spent per useful bit transmitted using Proxim's RangeLAN2 2.4 GHz radio, versus number of mobiles for transmitting a single packet. G is offered traffic load including newly generated and retransmitted packets. The figures are plotted for $G \in \{0.25, 0.5\}$. Points represent simulation and lines represent analysis.

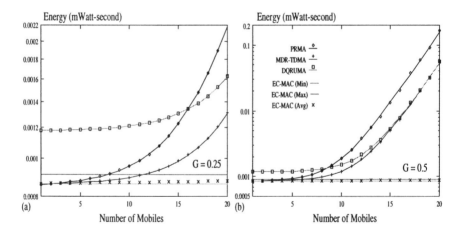

Figure 11.4 Energy spent per useful bit transmitted using Proxim's RangeLAN2 2.4 GHz radio, versus number of mobiles for transmitting periodic packets. The figures are plotted for $G \in \{0.25, 0.5\}$. Points represent simulation and lines represent analysis.

except when there is only one mobile in the system. **4.** Worst case of energy consumption in EC-MAC is lower than the energy consumed in other compared protocols when there are more than 12 mobiles with 25% traffic load (G=0.25). **5.** Energy consumed in 802.11 is worse than the EC-MAC in all cases.

This section considers mobile battery power conservation from the MAC layer perspectives. The analysis here shows that protocols that aim to reduce the number of contentions perform better from an energy consumption perspective. The receiver usage time, however, tends to be higher for protocols that require the mobile to sense the medium before attempting transmission. For messages with contiguous packets, our analysis shows that reservation is more energy conservative than piggybacking. For example, IEEE 802.11 minimizes the transmitter usage time while sending a packet, but requires the receiver to be turned on for the longest period of time for channel sensing. On the other hand, EC-MAC uses an explicit transmission order for sending reservations and a broadcast schedule to reduce energy consumption.

4. NETWORK, TRANSPORT, OS AND MIDDLEWARE LAYERS

A significant portion of battery energy consumed by mobile hosts is due to the retransmission of lost packets over the wireless link. Thus, error control mechanisms favorably impact energy conservation. Energy saving error control schemes for wireless networks are described in [Lettieri et al., 1997]. The authors describe an error control architecture for wireless networks where each packet stream has its own time-adaptive customized error control scheme. Such a scheme is based on a run-time estimated channel model and can set up parameters such as packet size, QoS requirements. Studies based on analysis and simulation under different scenarios were presented as a guideline to choose an error control scheme to achieve low energy consumption while trading off QoS, traffic types, and packet sizes. An energy efficient ARQ protocol for error control in link layer is also proposed in [Zorzi and Rao, 1997]. The proposed protocol enters a *probing mode* when the channel conditions deteriorate. The short probing packets are sent to probe the channel condition until an ACK is received. It then switches back to the *normal mode* and restarts transmission from the point it was interrupted.

The main functions of the network layer are routing packets and congestion control. In wireless and mobile networks, the network layer has added functionality of routing under mobility and mobility management including user location, update, etc. The transport layer provides a reliable end-to-end data delivery service to applications running at the end points of a network. The most commonly used transport protocol for wired networks is TCP (Transmission Control Protocol). TCP uses ARQs where error-detecting codes are used along with positive or negative ACK from the receiving end to trigger retransmissions. FEC, using error correcting codes, is commonly used while transporting real-time data. Both retransmissions and FEC waste network bandwidth as either the message has to be repeatedly transmitted or redundant bits need to be used for error correction. The transport layer will resort to a larger number of retransmissions and will frequently invoke congestion control measures confusing wireless errors and handoffs for congestion. This can significantly reduce the throughput and introduce unacceptable delays [Caceres and Iftode, 1995].

One important advantage of integrating wireless communications with computing is that it facilitates user mobility and connectivity to the network while carrying a portable computer. Mobility, directly or indirectly, impacts the design of OS, middleware, file systems and databases. It also presents a new set of challenges that result from power constraints and voluntary disconnections. To be consistent with their fixed counterparts like PCs and workstations, mobile computers need to process multimedia information. Such processing is expensive in terms of both bandwidth and battery energy. In general, majority of the techniques used in the design of today's applications to conserve bandwidth also conserve battery life.

The main function of an operating system is to manage access to physical resources like CPU, memory and disk space to the applications running on the host. To reduce power dissipation, some of the CPUs used in the design of portable devices can be operated at lower speeds by scaling down the supply voltage [Chandrakasan and Brodersen, 1995]. Due to the quadratic relationship between power and supply voltage, halving the supply voltage results in one fourth of the power consumed. To maintain the same throughput, the reduction in circuit speed can be compensated by architectural techniques like pipelining and parallelism. These techniques increase throughput resulting in an energy efficient system operating at a lower voltage but with the same throughput. The operating system has a part to play in relating scheduling and delay to speed changes.

Another technique of power management at the OS and middleware layer is predictive shutdown [Chandrakasan and Brodersen, 1995]. This method exploits the *event driven* nature of computing in that sporadic computation activity is triggered by external events and separated by periods of inactivity. A straightforward means of reducing average energy consumption is to shut down the system during periods of inactivity. However, preserving the latency and throughput of applications requires intelligent activity predictive shutdown strategies.

5.   APPLICATION LAYER

The application layer in a mobile and wireless system is responsible for partitioning of tasks between the fixed and mobile hosts, audio and video source encoding and decoding, and context adaptation in a mobile environment.

Challenged by power and bandwidth constraints applications can be selectively partitioned between the mobile and base stations [Weiser et al., 1994; Narayanaswamy et al., 1996]. Thus, most of the power intensive computations of an application are executed at the base station and the mobile host plays the role of an intelligent terminal for displaying and acquiring multimedia data [Narayanaswamy et al., 1996]. Another means of managing low power and bandwidth for applications on mobile clients is to use proxies. Proxies are middleware that automatically adapt the applications to changes in battery power and bandwidth. A simple example of proxy usage during multimedia transmissions under low battery levels or low bandwidth is to suppress video and permit only audio streams.

Multimedia processing and transmission require considerable battery power as well as network bandwidth. This is especially true for video processing and transmission. Battery power can be conserved by reducing the effective bit rate of video transmissions. For reducing video processing power at the mobile, lightweight video coding and decoding techniques [Chandrakasan and Brodersen, 1995; Gordon et al., 1996] can be used. Under severe bandwidth limited situations, video frames are carefully discarded, while maintaining tolerable video quality, before transmissions.

In [Agrawal et al., 1998], research on processing encoded video for transmission under low battery power conditions is presented. The basic idea of this work is to decrease the number of bits transmitted over the wireless link to relay a given video stream. The challenge is to accomplish this goal while preserving or minimally degrading the video quality. Decreasing the number of transmitted bits reduces the energy consumption due to reduced transmitter usage. In fact, several studies have shown that transmission accounts for more than a third of the energy consumption in video processing and exchange in a portable device.

The approach works as follows. The portable device runs a daemon in the background which periodically monitors the battery power level. If the power level is judged to be *almost full*, then the daemon does not impose any energy conservation mechanism on the video streaming applications. When the battery power level drops, the daemon requests a reduction in the number of bits transmitted over the wireless link to re-

---

DISCARD_FRAME (N, r)
    /* Reduce average bit rate by factor r over N frames */
    /*
        f_B: Fraction of B frames in the video stream;
        f_P: Fraction of P frames in the video stream;
        f_I: Fraction of I frames in the video stream;
       B_{ave}:   Average number of bits in a B frame;
       P_{ave}:   Average number of bits in a P frame;
       I_{ave}:   Average number of bits in a I frame;
        A: Average number of bits in the video frame;
    */
    $A = f_B \cdot B_{ave} + f_P \cdot P_{ave} + f_I \cdot I_{ave}$;
    $k_B = r \cdot A \cdot N / B_{ave}$; /* Number of B frames to be discarded */
    If $k_B \leq f_B \cdot N$ **then**
        Discard one out of every $f_B \cdot N/k_B$ B frames;
    Else {
        Discard all B frames;
        /* Number of P frames to be discarded */
        $k_P = \frac{r \cdot A \cdot N - f_B \cdot N \cdot B_{ave}}{P_{ave}}$
        If $k_P \leq f_P \cdot N$ **then**
            Discard one out of every $f_P \cdot N/k_P$ P frames;
        Else {
            Discard all P frames;
            /* Number of I frames to be discarded */
            $k_I = \frac{r \cdot A \cdot N - f_B \cdot N \cdot B_{ave} - f_P \cdot N \cdot P_{ave}}{I_{ave}}$;
            Discard one out of every $f_I \cdot N/k_I$ I frames;
        }
    }
    End.

Figure 11.5 Proposed strategy for discarding video frames.

duce energy consumption. The amount of reduction requested depends on the battery power level. For example, the battery power level can be considered to be in one of the following four distinct levels: *almost full, half-full, low-power,* and *nearly empty.* For each one of these levels, the daemon associates a reduction factor r by which the application must reduce its transmitted bit rate. More specifically, if the application transmits at A bps when the battery level is almost full, then the application must transmit at a rate $(1 - r)A$ bits/second when the requested reduction factor is r. Fig. 11.5 lists the algorithm for discarding video frames. Further details on how the different encoding schemes affect the choice of discarding may be found in [Agrawal et al., 1998].

6. CONCLUSIONS

The analysis of MAC protocols presented in this paper allows proper selection of protocols for the given wireless application. For example, networks supporting large number of mobile users will significantly benefit from the use of a protocol like EC-MAC. Techniques similar to those discussed in this paper may also be applied to the OS, transport and network layers of the protocol. A possible direction for energy reduction, not discussed in adequate detail in this article, is to transfer as many of the software functions as possible out of the terminal to the network [Weiser et al., 1994; Narayanaswamy et al., 1996].

References

Agrawal, P., Chen, J.-C., Kishore, S., Ramanathan, P., and Sivalingam, K. M. (1998). Battery power sensitive video processing in wireless networks. In *Proc. IEEE PIMRC '98*, Boston, MA.

Caceres, R. and Iftode, L. (1995). Improving the performance of reliable transport protocols in mobile computing environments. *IEEE Journal on Selected Areas in Communications*, 13:850–857.

Chandrakasan, A. and Brodersen, R. W. (1995). *Low Power Digital CMOS Design*. Kluwer Academic Publishers, Norwell, MA.

Chen, J.-C., Sivalingam, K. M., and Agrawal, P. (1999). Performance comparison of battery power consumption in wireless multiple access protocols. *ACM/Baltzer Wireless Networks*. To Appear.

Goodman, D. J., Valenzuela, R. A., Gayliard, K. T., and Ramamurthi, B. (1989). Packet reservation multiple access for local wireless communications. *IEEE Transactions on Communications*, 37(8):885–890.

Gordon, B. M., Tsern, E., and Meng, T. H. (1996). Design of a low power video decompression chip set for portable applications. *Journal of VLSI Signal Processing Systems*, 13:125–142.

IEEE (1997). Std 802.11: Wireless LAN medium access control (MAC) and physical layer (PHY) specifications.

Karol, M. J., Liu, Z., and Eng, K. Y. (1995). An efficient demand-assignment multiple access protocol for wireless packet (ATM) networks. *ACM/Baltzer Wireless Networks*, 1(3):267–279.

Lettieri, P., Fragouli, C., and Srivastava, M. B. (1997). Low power error control for wireless links. In *Proc. of ACM/IEEE International Con-*

ference on Mobile Computing and Networking (MobiCom), Budapest, Hungary.

Narayanaswamy, S., Seshan, S., Amir, E., Brewer, E., Brodersen, R. W., Burghart, F., Burstein, A., Chang, Y., Fox, A., Gilbert, J. M., Han, R., Katz, R. H., Long, A. C., Messerschmitt, D. G., and Rabaey, J. M. (1996). A low-power, lightweight unit to provide ubiquitous information access applications and network support for Infopad. *IEEE Personal Communications*, pages 4–17.

Raychaudhuri, D. and Wilson, N. D. (1994). ATM-based transport architecture for multi-services wireless personal communication networks. *IEEE Journal on Selected Areas in Communications*, 12(8):1401–1414.

Sivalingam, K. M., Chen, J.-C., Agrawal, P., and Srivastava, M. B. (1999). Design and analysis of low-power access protocols for wireless and mobile ATM networks. *ACM/Baltzer Mobile Networks and Applications*. To Appear.

Udani, S. and Smith, J. (1996). Power Management in Mobile Computing (A Survey). http://www.cis.upenn.edu/~udani/papers.html. University of Pennsylvania.

Walrand, J. (1991). *Communication Networks*. Aksen Associates, Inc.

Weiser, M. et al. (1994). Scheduling for reduced CPU energy. In *Proc. of First USENIX Symposium on Operating Systems Design and Implementation*, pages 4–17.

Zorzi, M. and Rao, R. R. (1997). Error control and energy consumption in communications for nomadic computing. *IEEE Transactions on Computers*, 46(3):279–289.

Chapter 12

Guaranteed Quality-of-Services Wireless Access to Broadband Networks[32]

KWANG-CHENG CHEN[33], CHIA-SHENG CHANG, RAMJEE PRASAD[34]
Institute of Communications Engineering, College of Electrical Engineering, National Taiwan University

Abstract: A wireless broadband network requires guaranteed quality-of-service (QoS) in its wireless access to satisfy end-to-end QoS. We shall categorize various existing approaches into three general categories to provide solutions for this active research and critically important subject in wireless networks. Along with empirical pros and cons from earlier research, directions for future development are highlighted.

[32] This research was supported by the National Science Council under the contract NSC-88-2217-002-002.
[33] The authors are with the Institute of Communications Engineering, College of Electrical Engineering, National Taiwan University, Taipei, Taiwan 10617, ROC. E-mail: chenkc@cc.ee.ntu.edu.tw Fax: +886 2 2368 3824
[34] The author is with the IRCTR, Delft University of Technology, Netherlands.

1   INTRODUCTION

With the tremendous success of wireless cellular voice communications and the need for broadband multimedia information systems and custom peripherals equipment, researchers all over the world are developing wireless broadband networking technologies to serve the need in the next century [42][43]. Figure 1 depicts the basic model of wireless broadband networks discussed in this paper. Broadband networks provide multimedia communication services to a set of multimedia terminals through some access points (base stations in cellular systems) that interface broadband network and wireless networking portion. However, in spite of different nature of wired backbone networks and wireless access to broadband backbone, one of the primary challenges in designing wireless broadband networks lies in how to guarantee end-to-end quality of service (QoS) for CBR, VBR, and ABR traffic. Future broadband backbone networks are typically realized by packet switching technology, while traditional QoS constraints are usually guaranteed by circuit-switching concept. The main obstacle in wireless broadband networks comes from how to guarantee QoS in wireless access due to the nature in wireless transmission and networking. It is the performance "bottleneck" while guaranteed QoS services have been successfully deployed in ATM/BISDN or other wired networks, no matter it is ATM, broadband Internet, or other types. Wireless networking has unique features to create research challenges:

- Wireless Links: Typical wireless links suffer from severe fading, shortage of bandwidth, propagation, interference, and distortion. Therefore, low bit error rate physical transmission like optical fiber is not always possible. In state-of-the-art wireless networks operating in severe fading mobile environments, bursty (and random) errors may result in significant link control problem, time-varying reduction of available bandwidth and huge delay to dissatisfy QoS.

- Mobility: Wireless networks allow stations/nodes in the networks to move around. It induces a special feature in wireless networks known as mobility management. In other words, network topology is dynamically changing as stations move in/out cell coverage or turn on/off the transmission. A more complicated scenario is multimedia terminals to transport all possible CBR, VBR, and ABR traffic to introduce more complicated handoff and routing.

- Distributed Nature: A Wireless channel is a distributed environment, and queue information of these mobile sources is not automatically available to the base station. Since all information exchange is occurred in a common channel, we must explicitly deal with the exchange of traffic information between each connection and the scheduler.

We summarize a series of efforts to provide guaranteed QoS wireless access to broadband networks such as ATM networks and others with end-to-end QoS constraints. We shall provide wireless access to serve three kinds of mobile traffic sources: available-bit-rate (ABR), constant-bit-rate (CBR), and variable-rate-rate (VBR) sources in a wireless multimedia networking environment depicted in Figure 1. End-to-end QoS requirements can be divided into two parts: within the wired broadband network and in wireless part. The latter obviously plays a dominant role due to the features of wireless networking stated in above, and is our primary concern.

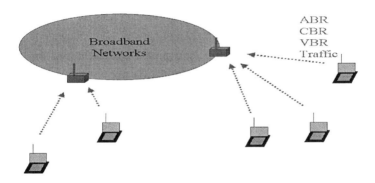

Figure 1 Wireless Access to Broadband Networks

In wireless broadband networks, the QoS requirements depend on the traffic types: ABR traffic, which has the characteristics of conventional data traffic, requires stringent packet error rate (PER) and is insensitive to delay and jitter. CBR (e.g. voice) and VBR (e.g. video) traffic require moderate packet error rate and usually very sensitive to delay and jitter.

2 MULTIPLE ACCESS FOR WIRELESS BROADBAND NETWORKS

ABR traffic, which has the characteristic of traditional data traffic, has a larger tolerance on packet delay and delay jitter. Due to their bursty characteristic, the performance metric for ABR multi-access is usually the throughput and average delay. There have been a lot of study to construct more efficient protocol for ABR (data) traffic, including CSMA (carrier sense multiple access) family protocols, CRA (collision resolution algorithm) family protocols, and a new family of multiple protocols RAP (Randomly Addressed Polling) primarily designed for wireless environments. A general form of multiple access protocols known as Multi-Layer Collision Avoidance and Resolution (MULCAR) has also been developed [1][45]. [1] provides a complete logic development regarding wireless packet networks, primarily for data/ABR traffic.

A well known, straightforward, and widely accepted approach to design wireless access protocols in wireless packet switching broadband networks typically consists of a random access scheme for initial call set-up and possibly ABR traffic transport, and certain reservation scheme to create certain logical circuit-switching connections. One of the very first proven concept of such an approach was done by well known packet reservation multiple access (PRMA). It is actually a combination of ALOHA and TDMA reservation, which might be traced back as R-ALOHA in early days of computer networking research. However, PRMA indeed provides an important step in wireless networking research as it demonstrated that packet switching by random access with reservation can grant effective connections for on-off CBR traffic, later adding data traffic [38].

In the past few years, a series of efforts have been conducted toward a new random access mechanism originally primarily designed for wireless networks and those networks with topologies that distributed transmission/carrier sensing is not reliable. From this research, a general concept has been developed for multiple access protocols, which is known as MULCAR (multi-layer collision avoidance and resolution) [45] consisting of two kinds of tree expansions based on collision avoidance (such as carrier sensing or its generalization) and collision resolution. One of the best examples to conform this concept is the group randomly addressed polling (GRAP) that implements both collision avoidance expansion and collision resolution expansion [6]. Many other protocols only implement one of the expansion techniques/trees such as auction demand multiple access [2]. It may also account why CSMA/CD is good due to its simple realization of both expansions. However, we consider that an effective random access is always the basis to design wireless access protocol to meet QoS constraints due to its fundamental role. The concept of applying RAP/GRAP/MULCAR is not only for efficiency consideration but also for reliable operation. In order to serve CBR and VBR traffic effectively, false of transmission coordination (multiple access) such as hidden terminal problem for CSMA, and reliable bandwidth allocation that is usually done in centralized way for wireless networks due to their dynamic characteristics, are primary concerns to design guaranteed QoS wireless access. RAP family protocols is initiated in a distributed way but is transformed into a centralized polling turns out capable of successfully carrying out above two issues. Such a general reliable wireless access procedure was summarized in [1][2], and that can always serve the key of guaranteed QoS wireless access.

Up to this point, we may conclude: When both ABR and CBR sources are considered, the reservation-added counterparts of the existing multiple access protocols (e.g. PRMA or R-ALOHA, R-GRAP,) are also proved very efficient solutions for the wireless access problem due to the very regular behavior of CBR traffic. All of these wireless access protocols share a common foundation, the frame-structured concept. A random access takes care of contention traffic that is primarily for ABR and control packets to set up links, and logic-channel based reservation scheme handles the CBR or on-off CBR traffic. Unlike the well-behaved CBR, the data generation rate of most VBR traffic fluctuates in a wide range with time, which is widely believed in the form of complicated dynamic system forms. Suppose our object is to provide wireless access service to CBR, VBR and ABR sources, the efficient resource allocation and QoS guarantee of VBR will be the most challenging problem for the scarce bandwidth of wireless channels.

3   CATEGORIES OF WIRELESS ACCESS PROTOCOLS TO VBR RESOURCE ALLOCATION

By looking into recently proposed QoS-guaranteed wireless access, we can be divided them into three categories based on different philosophy in design. In the following of this section, we discuss the principles and illustrate for each category.

3.1   Category 1: Access based on dynamic bandwidth sharing

Multiple access based on bandwidth sharing can be realized in several ways. A frame based TDMA is the most well known approach. The ratio of the allocated time slots to the total frame duration represents the shared fraction of the total bandwidth. Similarly, CDMA (Code Division Multiple Access) is another way to partition total bandwidth into small logic channels. By assigning these spreading code, the bandwidth can be shared accordingly. Suppose each CBR or VBR mobile node is assigned a weighting factor ϕ_i. Then Generalized Processor Sharing (GPS) processor is a work-conserving service scheme and shares the transmission bandwidth according to:

$$r_i(t) = \begin{cases} r \times \dfrac{\phi_i}{\displaystyle\sum_{j \in B(t)} \phi_j} & \text{if } i \in B(t) \\ 0 & \text{otherwise} \end{cases}$$

where $r_i(t)$ is the shared bandwidth of the *ith* VBR or CBR sources, r is the channel bandwidth, and $B(t)$ is the set of VBR of CBR with non-empty queues. However, GPS is an ideal scheduling policy assuming packets are infinitely divisible. As packets are not divisible, Parekh *et. al.* [4] presented a packet-by-packet transmission scheme, called PGPS scheduler, to approximate GPS and do the actual scheduling task. Since uses weighting factors to share the channel bandwidth, PGPS scheduler can intuitively be approximated by a TDMA sharing with the frame duration is extremely small. In terms of bandwidth sharing, TMDA, CDMA and PGPS scheduler are actually the equivalent. PGPS scheduler is a more complicated scheme to allocate bandwidth. However, PGPS is compensated by its flexibility to share bandwidth according to more generalized weighing factors and allow variable packet-size. Since the rate of VBR fluctuates with time, it is impossible to use a stationary bandwidth sharing setting to fit the bandwidth requirement of VBR connections. Therefore, for those based on bandwidth sharing, each VBR mobile node usually uses certain scheme (e.g. observing queue length variation, counting the number of arrivals in a frame duration) to estimate the instantaneous rate, and then informs the base station of this information via ABR service or by piggybacking techniques. After receiving this instantaneous rate information, base station adjusts the bandwidth sharing setting to track the bandwidth requirement of VBR connections.

3.1.1   TDMA Frame based MAC

Time Division Multiple Access (TDMA) based frame-structured access is the most common class in the literature, such as [9]- [15] [17]. In order to provide time-sensitive services (CBR and VBR) and time-insensitive services (ABR), a frame structure (the frame size if not necessarily fixed) consisting of a contention part and a reservation part is used as the basic scheduling discipline (Figure 2).

The pioneer work in wireless access with QoS constraint might be packet reservation multiple access (PRMA) proposed by D. Goodman [37]. The contention part adopts slotted ALOHA and reservation by round-robin policy serves on-off CBR (a simplified form of VBR) voice. In most recent DPRMA (Dynamic Packet Reservation Multiple Access) protocol [10], each user is responsible for providing updated information about its immediate bandwidth needs, and the base station then try to match the service rate to this instantaneous data generation rate. To accomplish this goal, thresholds are marked on the queue length and partition it into regions. As the size of the queue grows across these thresholds, the user increases its reservation request for more time slots. If the queue length decreases across these thresholds, the user requests a lower reservation rate. In short, each VBR user in DPRMA measures its instantaneous data generation rate by observing the queue length and informs the base station to adjust the number of allocated slots accordingly. For TDMA based approach, there are some factors to be determined:

- Super-frame length (that is, total of contention period and reservation period): It can be a fixed one or a variable but bounded one. Typical PRMA research uses fixed length, which has advantages of easy prediction for CBR control (also theoretical analysis) and easy network management. However, for bursty real network VBR traffic, it lacks of flexibility for bandwidth allocation and thus is hard to achieve good efficiency. [31] demonstrated that GRAP with variable super-frame length indeed provides good efficiency for multimedia traffic.

- Moving boundary of ABR and CBR/VBR: Another way to better fit bandwidth need of bursty VBR traffic is to develop a good scheme by moving boundary within a (usually fixed) super-frame.

- Admission Control: In addition to access mechanism, a successful TDMA based approach requires a matching admission control at mobile stations. Though admission control might not be specified in many research papers, it can be simple as transmission until a successful bandwidth granted from the coordinator (access point or base station) in a wireless network.

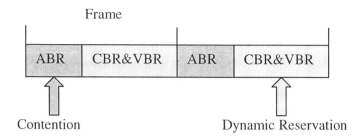

Figure 2 Frame structure: contention and dynamic reservation.

A final observation in this approach is that different contention schemes can not change a lot for throughput since CBR/VBR occupies most bandwidth. However, effective contention can improve delay performance a lot and thus QoS [28].

The most initial and successful trial of guaranteed QoS wireless access is known as polling with non-preemptive (PNP) based on the RAP family protocols. The fundamental idea is to use conceptual polling tokens to control CBR and VBR traffic, while the ABR traffic that might contain critical traffic information and other message with time constraint is exchanged based on the efficient GRAP. It is more or less considered as a combination of RAP random access and a round-robin reservation of logic bandwidth units. To design the PNP wireless access, we assume

● All packets have the same size.

● A CBR traffic source is characterized by (γ,δ), where γ is the rate of source and δ is the maximum tolerable delay. Packets generated periodically are stored in the ready-to-transmit buffer.

● A VBR traffic source is characterized by (ρ,σ,d), where ρ is the average rate, σ is the maximum burstiness, and d is the maximum tolerable delay. A VBR source is regulated by a (σ,ρ)-leaky bucket. Only packets passing through the leaky bucket can be stored in the ready-to-transmit buffer.

● An ABR traffic source has neither a jitter constraint nor a delay constraint. All ABR traffic sources share the remaining bandwidth from CBR and VBR sources in a fair and efficient way.

The associated admission control for CBR and VBR traffic is simple as a traffic can be admitted only when its maximum delay constraint can be honored for every packet that it might generate. The algorithm for PNP operation is summarized as follows:

1. For each CBR source, its polling token is generated every $1/\gamma$ in base station.

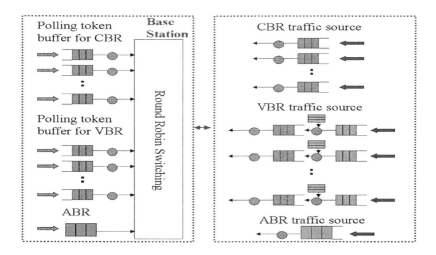

Figure 3 PNP Operation

2. For each VBR source, its first polling token is generated p second (suggested $p=d/2$) after connection is set up.
3. When a transmission ends, the base station performs:
   a) The base station scans the polling token buffer for CBR courses according to a preset priority. If a polling token is found, removes the token and polls the CBR course.
   b) If no token for CBR sources, scan VBR token buffer according to a preset priority. If any token is found, proceed the polling.
   c) When no token is found in CBR and VBR sources, start services for ABR based on GRAP.

In general, each polling token represents one packet transmission from the buffer. At the end of each VBR message, an end-of-file is transmitted and the base station removes the polling token and sets up to generate next polling token. Although the priority setting in this method is not needed, by using this simple rule, successful wireless access to guarantee QoS is achieved with rigorous proof from both simulations and mathematical bounds [3]. Of course, by playing the assignment of priority to traffic based on operating needs, we can possibly achieve somewhat better performance.

3.1.2 Code Division Multiple Access

Spread spectrum multiple access allows all mobile nodes simultaneously transmit using the same frequency band. Its typical form, Direct Sequence

Code Division Multiple Access (DS-CDMA), have been shown effective in cellular system and future wideband CDMA cellular for next generation multimedia services by using multi-code or multi-rate physical transmission for 144 kbps or 384 kbps multimedia services [46]. For broadband applications, Fischer et. al. [18] recently proposed a wireless access protocol for CDMA based wireless ATM LAN. Since the rate of VBR fluctuates with "time", unlike in a CDMA cellular telephony network, a centralized TDMA protocol is used to arbitrate access of ATM cells from different mobile nodes onto the multiple. When a mobile node receives a transmit permission from the base station, it transmit a packet into the corresponding up-link time slot according to the particular CDMA spreading code and modulation scheme. If a mobile node receives Transmit Permissions for more than one up-link time slot at the same time, it first spreads each of the symbol sequences of the different packets by the CDMA code for each channel, and then sums the resultant coded symbol sequences before transmission.

We can further generalize this concept from TDMA described earlier by considering CDMA to be a finer partition of radio resource. As long as an effective bandwidth "manager" exists, a finer partition of radio resource means more effective application of radio spectrum, and thus better QoS services. Similar story can be applied to another physical transmission technique known as orthogonal frequency division multiplexing (OFDM), which provides a way to achieve finer partition of radio resource. As long as effective "channel" (a fixed fraction of bandwidth) allocation is available, it may imply an effective realization for broadband multimedia networks, in addition to its advantage of easier equalization over frequency-selective channel [40]. Multi-carrier CDMA can provide even finer bandwidth allocation and transmission advantage along this thinking [41]. In other words, both TDMA and CDMA approaches are trying to explore the best allocation of radio resource in the time-code-frequency space.

3.1.3   Access based on Scheduling of General Processor Sharing

Instead of combination of random access and reservation, another view to tackle QoS control is just to treat the available radio resource as a bandwidth allocation for all traffic sources. Its best possible allocation like general processor sharing is naturally a solution. For this purpose of bandwidth allocation, Packet-by-Packet Generalized Processor Sharing algorithm (PGPS) [4] is well known for its flexibility, efficient sharing of resource, and is considered as a promising switching scheme to guarantee QoS in wired broadband networks and to realize the dynamic bandwidth sharing access [5][22]. Since a wireless network implies a distributed environment, the queue information of each mobile node is not automatically available to the scheduler. Since all the queue information must be transmitted to the base station in a common channel, an explicitly information exchange scheme must be specified for efficient use of the channel bandwidth. Fortunately, PGPS scheduler has been shown capable of working with only head packet information (arrival time and packet size) [5]. Therefore, mobile nodes with non-empty queue can transmit the head packet information by piggyback technique. To ensure proper scheduling, mobile mode with empty queue must be polled for head packet information periodically. With the available head packet information, the base station uses a PGPS scheduler to determine the transmission order. (Figure 4). Due to the nice feature of MULCAR while RAP is the general single-layer realization, a logical development for this approach is PGPS/RAP multiaccess [5], in which we use the PGPS scheduler to dynamically reserve bandwidth for VBR and CBR traffic in contrast to the straightforward frame-structure reservation schemes. Again, RAP random access protocol serves as the access scheme for initial call setup and the ABR service. As the rate of VBR traffic is inherently time-invariant, a feasible solution is to reserve the peak rate for each VBR stream. Although this approach may seem too conservative and not aggressive enough, PGPS scheduler's ability of dynamical bandwidth allocation enables PGPS/RAP to switch the surplus VBR bandwidth to ABR when VBR traffic is in its low-rate period. Thus in PGPS/RAP multiaccess, CBR traffic occupies a fixed amount of bandwidth while VBR and ABR share the remaining bandwidth according to the rate of VBR traffic.

To use the centralized PGPS scheduler in a wireless channel where the queues used to accommodate incoming packets are distributed, the basic idea is to generate different kinds of permits according to the traffic parameters obtained at the call setup time and then use the PGPS scheduler to arbitrate the use of the channel. A primary difference between this idea and PNP scheme is that the static priority is replaced with the more flexible weighting factor concept of PGPS scheduler. Please note that we can not apply PGPS concept directly as the non-transparent queue problem caused by distributed wireless mobile sources. We must modify the algorithm based on the initial setup head packet. Under PGPS/RAP multiaccess, we assume:

- The channel capacity is C bps

- All packets generated from a source are of the same size.

- The packets of all ABR traffic sources are of the same size but the packet of CBR or VBR traffic sources may be of different size.

- There exists a fixed parameter L_{max}, which is a size upper bound of all packets.

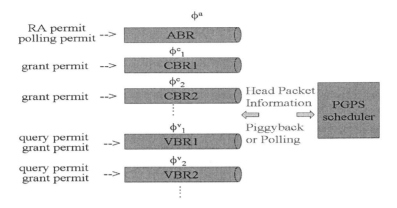

Figure 4: Access with dynamic bandwidth allocation by PGPS scheduler

A CBR traffic source is characterized by four parameters $(r^c_i, L^c_i, T^c_i, D^c_i)$, where r^c_i (bps) is the rate of this source, L^c_i is the size of packets of this source, T^c_i is the packet interarrival time, D^c_i is the maximum tolerable delay. A CBR grant permit is generated every T^c_i seconds. A VBR traffic source is characterized by five *parameters* $(r^v_i, L^v_i, T^v_i, Q^v_i, D^v_i)$, where r^v_i (bps) is the peak rate of this source, L^v_i is the size of packets of this source, T^v_i is the minimum packet interarrival time, Q^v_i is the query packet size of this source (explained later), D^v_i is the maximum tolerable delay. For VBR, a query permit is generated every NT^v_i seconds, where N is optimized such that the maximum delay is satisfied and the query operation occupies the least bandwidth. Then grant permits are generated after each query permit in accordance to its query result. Consequently, the PGPS/RAP multiaccess is:

1 Given the permit generating mechanisms stated above, set the weighting factor ϕ^c_i of the *ith* CBR source to r^c_i; set the weighting factor ϕ^v_i of the *ith* VBR source to $r^v_i(1+Q^v_i/NL^v_i)$, and set the weighting factor ϕ^a of the ABR queue to $C-\Sigma\phi^c_i-\Sigma\phi^v_i$.

2 Use a PGPS scheduler to select a permit to service
   2.1 if the permit is for the *ith* CBR user, the base station notifies the *ith* CBR to transmit a packet.
   2.2 If the permit is for the *ith* VBR, and
      2.2.1 it is a query permit, the base station queries the *ith* VBR how many packets arrive in last NT^c_i seconds.

2.2.2   it is a grant permit, the base station notifies the *ith* VBR to transmit a packet.

2.3   If the permit is for the ABR service, the base station notifies all ABR users to transmit their random addresses or polling next active address base on RAP protocol.

3   Go to 2.

If the weighting factors of PGPS are assigned according to our proposal, PGPS/RAP can be proved to guarantee worst-case delay of CBR and VBR traffic and thus provides guaranteed QoS for delay-sensitive CBR and VBR traffic in multimedia wireless networks. The QoS GPS provides is fairness and guaranteed bandwidth. Another different feature for this approach is no need for superframe-structure under the processor-sharing concept, which supports more flexibility (and thus efficiency) in network design.

As we mentioned earlier, the wireless channels usually suffer from severe fading and bursty error effect. According to the channel control and information theory, the fluctuation of bit error rate can be equivalently regarded as that the wireless channel capacity is varying with time. A PGPS scheduler is first proposed to operate with a constant transmission capacity, and the virtual time implementation of PGPS in [4] does not yield the right approximation of the GPS scheme when transmission capacity is time varying. Tseng and Chang [26] proposed that by the concept of changing virtual time, a PGPS scheduler with a time varying capacity can be viewed as a scheduler with a constant capacity under an appropriate change of time.

3.2   Category 2: Access solely based on Contention

Another approach for frame-structured access solely depends on contention. Since there exists possibly only one wireless resource for multimedia networking, it shall be effective as long as the highest priority traffic can be sent first and then lower traffic. This thinking can be solely implemented by a contention protocol as long as it can resolve contention fast enough for different priority traffic. A good example to realize this approach is the non-preemptive multiple accesses (NPMA) in the channel access control of HIPERLAN, an ETSI standard for wireless LANs. NPMA consists of different stages: prioritization phase, contention phase, and transmission phase. [29] analyzed the performance of NPMA as a special case of a multi-layer RAP family protocol. The concept surely works but suffers from potential unsatisfactory performance due to delay caused by contention to result in poor QoS or poor throughput.

As a matter of fact, the operation of this category really relies on the effectiveness of contention, especially dynamic control of its stability. MULCAR can play a critical role in this situation due to its expansion flexibility to result in excellent delay performance. MULCAR is the general form of multiple access protocols [45]. Multiple access relies on carrier sensing and collision resolution [7]. We observe that realization of both carrier sensing and resolution can be in the form of tree expansion. Such expansion can be in many ways (and thus domains). To name a few: frequency domain (such as frequency channel); time domain (such as time slot); signaling domain (such as orthogonal signaling); feature domain (such as grouping stations); probabilistic domain (such as random delay); spatial domain (such as propagation delay techniques); power domain (such as received power). By using certain of these domains for tree expansion, we can form a multiple access protocol, such as Figure 5 showing. RAP and CSMA/CD are examples to implement both carrier sensing and collision resolution. That is why they can be shown equivalent in mathematics [4], in spite of quite different operation. MULCAR therefore consists of two kinds of expansion in terms of multiple access implementation: collision avoidance/anticipation with feedback information of transmission or no transmission; and collision resolution with channel information idle, success, collision. We can use different ways to divide users into groups to reach effective channel access. In this general form, we can expect to create a powerful expansion as long as operating constraints allow. Once we have an effective contention (random access) protocol, we can expect a good performance in delay that is more critical than throughput for wireless access with QoS concerns. By looking at the CBR and VBR QoS characteristics, we can observe that delay is what we most want to control. This suggests us that control using the Markovian property of MULCAR to derive access mechanism for delay sensitive traffic is attractive. Further, since MULCAR that can pretty much represent most multiple access protocols a general methodology to control delay-sensitive traffic in a unified view is thus possible. With such a feature, the dynamic characteristics of delay sensitive traffic are represented by a Markov chain to can fully describe the dynamics of MULCAR. Its control mechanism over MULCAR is naturally an effective wireless broadband access. As the control is based on the splitting of groups that can be pre-calculated from Markov chain describing delay behavior, the algorithm of MULCAR combining feature-domain grouping and reservation can be summarized as follows [36]:

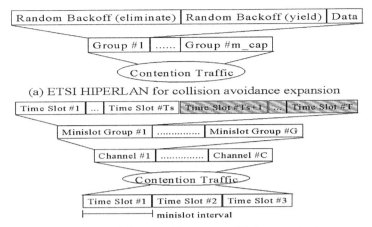

(b) IEEE 802.14 for collision avoidance & collision resolution expansion

Figure 5 Examples of MULCAR

1. Base station updates time τ since last processing real-time traffic group.
2. Based on the comparison of τ and delay-bound, we decide the operating mode that corresponds to the shape and completeness of collision avoidance expansion and collision resolution expansion.
3. Base station polls traffic in real-time reservation list.
4. Based on the tolerable dropping rate and calculated limit of delay, determine the number of groups in operation for wireless access.

3.3 Category 3: Access based on Traffic Characterization

A very different way to guaranteed QoS in wireless access is to grant access according to traffic characteristics, in stead of finding a good way to allocate bandwidth.

3.3.1 Access based on Effective Bandwidth

In multimedia wireless networks, VBR connections are statistically multiplexed and transmitted in the error-prone wireless channels. To guarantee the QoS for different kinds of traffic, we need a metric to measure the load of each connection on the multimedia wireless networks. A useful metric is the effective bandwidth of each connection [24][25]. Then, proper call admission control can be developed to guarantee the QoS on packet loss rate. Using stochastic theory, large deviation approximation, and the research result of self-similar behavior of bursty sources, there have been many studies on the effective bandwidth for a single and aggregate traffic sources. Mohammadi et. al. [25] use the concept of effective bandwidth to model both the random error and bursty error characteristics of the wireless channels and to propose a weighting function to modify the effective bandwidth relation in these non-ideal channels.

3.3.2   Access based on Deterministic Traffic Constrain Functions

In order to allocate channel resource to VBR connections more efficiently, a feasible approach is to provide more characteristic information of the VBR behavior. Due to the diverse characteristics of VBR sources, Cruz [27] proposed the concept of deterministic traffic characterization: Let $R_i(t,s)$ (bits) be the amount of packet arrival generated by the VBR source i in the interval $[t,s)$. Then VBR source i is said to be constrained by $b(\cdot)$ if there exists a nonnegative increasing function $b(\cdot)$ such that $R_i(t,t+s) \leq b(s)$ for $s \geq 0$. Since a constrain function $b(s)$ represents the upper bound of arrival in an interval with length s, it can be intuitively regarded as the integration of arrival rate with respective to time, and can convey much rate variation information. Suppose we share the link resource according to these constrain functions, the contained rate variation information should help us allocate resource to VBR more efficiently. One example is Packetized Service Curve Proportional Sharing (PSCPS) [29], where service curves are right-shifted replicas of constrain functions. There are two benefits of service curve approach: (1) Service curves are construct according to constrain functions, which contain very much information of the rate variation behavior of VBR. Therefore, resource can be allocated more precisely and efficiently than peak-rate allocation. (2) Service curve has two components: the amount of right shift is the guaranteed delay bound, which stands for the service requirement and constrain function represents the traffic characteristic. For multiple access protocols based on GPS, the packet delay bound is usually a function of the allocated rate. Hence the service requirements and traffic characteristics are coupled. Simulations show SCPS is a useful concept.

4   CONCLUDING REMARKS

Guaranteed QoS wireless access might be on the most important issues in wireless broadband networks. We categorized existing approach into three general forms and explain their different concepts to reach the goal. It is not a stand-alone design issue. Admission control and routing for handoff traffic from such research still require efforts to clearly figure out.

REFERENCES

[1] K.C. Chen, "Multiple Access for Wireless Packet Networks", presentation in the IEEE Workshop on Multiple Access, Mobility, Teletraffic for Personal Communications, also a book chapter under the same title published by Kluwer, 1997.

[2] K.C. Chen, "Medium Access Control of Wireless LANs for Mobile Computing", IEEE Network, Sep./Oct., 1994.

[3] C.S. Chang, K.C. Chen, M.Y. You, J.F. Chang, "Guaranteed Quality of Service Wireless Access to ATM Networks", IEEE Journal on Selected Areas in Communications, vol. 15, no.1, Jan. 1997, pp.106-118.

[4] A.K. Parekh, R.G. Gallager, "A Generalized Processor Sharing Approach to Flow Control in Integrated Services Networks: the Single Node Case", IEEE/ACM Transactions on Networking, Vol. 1, No. 3, pp 344-357, June 1993.

[5] C.S. Chang, K.C. Chen, "Guaranteed Quality-of-Service Wireless Medium Access by Packet-by-Packet Generalized Processor Sharing algorithm", Proc.IEEE ICC, 1998.

[6] K.C. Chen, C.H. Lee, "Group Randomly Addressed Polling for Wireless Data Networks", Proc.IEEE ICC, 1994.

[7] R.G. Gallager, "A Perspective on Multiaccess Channels", IEEE Trans. on Information Theory, vol. 31, No. 2, Mar. 1985, pp.124-142.

[8] P. Mathys, P. Flajolet, "Q-ary Collision Resolution Algorithms in Random Access Systems with Free or Blocked Channel Access", ", IEEE Trans. On Information Theory, vol. 31, No. 2, Mar. 1985, pp.217-243.

[9] J. Jiang, T-H Lai, "An Efficient Media Access Control Protocol for Delay sensitive Bursty Data in Broadband Wireless Networks," Proc. IEEE PIMRC'98, Boston, 1998.

[10] Z.J. Haas, D.A. Dyson, "The Dynamic Packet Reservation Multiple Access Scheme for Multimedia Traffic," Proc. IEEE ICC'98, Atlanta, 1998.

[11] S. Choi, K.G. Shin, "A Cellular Wireless Local Area Network with QoS Guarantees for Heterogeneous Traffic", Proc. IEEE INFOCOM'97, Kobe, 1997.

[12] T. Ors, Z. Sun, B.G. Evans, "Analysis of an Adaptive Random/Reservation MAC Protocol for ARM over Satellite," Proc. IEEE VTC'98.

[13] S. Yoon, S. Bahk, "A MAC Protocol Based on FEC for Real Time Traffic in Wireless ATM Networks," Proc. IEEE GLOBALCOM'98.

[14] S.Zhou, J. Deane, A. Alm, D. Skellern, T. Pervival, "An Efficient Media Access Control Protocol for Wireless ATM Networks," Proc. IEEE GLOBALCOM'98.

[15] S. Lee, Y.J. Song, D.H. Cho, Y.B. Dhong, J.W. Yang, " Wireless ATM MAC Layer Protocol for Near Optimal Quality of Service Support," Proc. IEEE GLOBALCOM'98.

[16] S.K. Biswas, D.Reininger, D.Raychaudhuri, "UPC Based Bandwidth Allocation for VBR Video in Wireless ATM Links," Proc. IEEE ICC'97.

[17] M. J. Karol, Z. Liu, K. Y. Eng, "Distributed-Queueing Request Update Multiple Access for Wireless Packet (ATM) Networks," Proc. IEEE INFOCOM'95.

[18] S.G. Fischer, T.A. Wysocki, H.J. Zepernick, "MAC Protocol for a CDMA based Wireless ATM LAN," Proc. IEEE ICC'97.

[19] B.H. Yae, S.C. Lee, H.S. Kim, "A Scheduling Scheme for the Heterogeneous Multimedia Services in Mobile Broadband Systems," Proc. IEEE PIMRC'97.

[20] C. Fragouli, V. Sivaraman, M.B. Srivastava, "Controlled Multimedia Wireless Link Sharing Via Enhanced Class Based Queueing with Channel-State Dependent Packet Scheduling," Proc. IEEE INFOCOM'98.

[21]   J.C. Chen, K.M. Sivalingam, P. Agrawal, R. Acharya, "On Scheduling of Multimedia Services in a Low-Power MAC for Wireless ATM Networks," Proc. IEEE PIMRC'98.

[22]   R.Kautz, Alberto Leon-Garcia, "A Distributed Self-Clocked Fair Queueing Architecture for Wireless ATM Networks," Proc. IEEE PIMRC'97.

[23]   N. R. Figueira, J. Pasquale, "Remote-Queueing Multiple Access (RQMA): Providing Quality of Service for Wireless Communications," Proc. IEEE INFOCOM'98.

[24]   B. Tse, S. Hanly, "Effective Bandwidth in Wireless Networks with Multiuser Receivers," Proc. IEEE INFOCOM'98.

[25]   A. Mohammadi, et al.,"Characterization of Effective Bandwidth as a Metric of Quality of Service for Wired and Wireless ATM Networks," Proc. IEEE ICC'97.

[26]   Y.C. Tseng, C.S. Chang, "PGPS Servers with Time-Varying Capacities," IEEE Communications Letters, vol. 2, no. 9, 1998.

[27]   R.L. Cruz, "A Calculus for Network Delay, Part I: Network Elements in Isolation," IEEE Trans. on Information Theory, vol. 37, no. 1, Jan. 1991.

[28]   D.C. Twu, "Stabilization Techniques for Multiple Access and Their Applications to Integrated Services and Mobile Networks", Ph.D Dissertation, Department of Electrical Engineering, National Tsing Hua University, 1998.

[29]   Y.K. Sun, K.C. Chen, "Multi-layer Collision Resolution Protocol for Mobile Computing", to appear in the Blatzer/ACM Wireless Networks.

[30]   H.F. Chou, K.C. Chen, "Group Randomly Addressed Polling with Reservation for Integrated Service Wireless Networks", Proc. IEEE PIMRC, 1995.

[31]   K.C. Chen, "Multimedia Wireless Local Area Networks", Records ITU TELCOM Technology Submmit, Genava, 1995.

[32]   ETSI HIPERLAN System Definition, 1995.

[33]   IEEE 802.11 Wireless LANs Standard, 1997.

[34]   IEEE 802.14 HFC Modem Draft Standard, 1998.

[35]   Y.K. Sun, K.C. Chen, "Optimization of Generalized Tree Protocols with Capture", Proc. IEEE VTC, 1998.

[36]   Y.K. Sun, K.C. Chen, "Dynamics of Generalized Tree Protocols with Application to Delay Sensitive Traffic Control", Proc. IEEE ICC, 1998.

[37]   D. Goodman, et al. "Packet Reservation Multiple Access for Local Wireless Communications", IEEE Trans. On Communications, Aug. 1989, pp.885-890.

[38]   P. Narasimhan, R.D. Yates, "A New Protocol for Integration of Voice and Data over PRMA", IEEE Journal on Selected Areas in Communications, May 1996, pp.623-631.

[39]   S. Glisic, J. Viksedt, "Effect of Wireless Link Characteristics on Packet-Level QoS in CDMA/CSMA Networks", IEEE Journal on Selected Areas in Communications, Aug. 1998, pp.875-889.

[40]   L. Cimini, N. Sollenberger, "OFDM with Diversity and Coding for High-Bit-Rate Mobile Data Applications", Proc. 3rd Int. Workshop Mobile Multimedia Communications, Princeton, 1996.

[41]   K. Fazal, G. Fettweis (ed.), Multi-Carrier Spread Spectrum, Kluwer, 1998.

[42]   L.M. Correia, R. Prasad, "An Overview of Wireless Broadband Communications", IEEE Communications Mag., Jan. 1997.

[43]   N. Morinaga, M. Nakagawa, R. Kohno, "New concepts and Technologies for Achieving Highly Reliable and High Capacity Multimedia Wireless Communications Systems", IEEE Communications Mag, Jan. 1997.

[44]   H. Zhang, "Service disciplines for guaranteed performance service in packet-switching networks", Proceedings of the IEEE, vol. 83, no. 10, Oct.1995.

[45] Y.K. Sun, K.C. Chen, D.C. Twu, "Generalized Tree Multiple Access Protocol for Wireless Communications", Proc. IEEE PIMRC, Helsinki, 1997.

[46] T. Ojanpera, R. Prasad, "An Overview of Air Interface Multiple Access for IMT-2000", IEEE Communications Mag., Sep. 1998.

[47] C.S. Chang, J. Thomas, "Effective Bandwidth in High-Speed Digital Networks", IEEE Journal on Selected Areas in Communications, Aug. 1995.

PART III

ENABLING COMPONENTS OF WIRELESS NETWORKS

Chapter 13

Integrated Circuit Technologies for Wireless Communications

BABAK DANESHRAD AND AHMED M. ELTAWIL
UCLA Electrical Engineering Dept.
Email: babak@ee.ucla.edu, ahmed@janet.ucla.edu
http://www.ee.ucla.edu/~babak

Abstract: Through advanced circuit, architectural and processing technologies, ICs have helped bring about the wireless revolution by enabling highly sophisticated, low cost and portable end user terminals. This paper provides a brief overview of present trends in the development of integrated circuit technologies for applications in the wireless communications. Two broad categories of circuits are highlighted. The first is RF integrated circuits and the second is digital baseband processing circuits. In both areas, circuit design challenges and options presented to the designer are discussed. The paper also highlights the manner in which these technologies have helped advance wireless communications.

1   INTRODUCTION

The wireless industry has enjoyed rapid growth over the past decade and a half. Advances in integrated circuit technology coupled with novel system level solutions have combined to give rise to small, low cost, low power and portable units for a host of wireless communication systems. In fact, it is the marriage of advanced system design and IC technology that has made anytime anywhere communication a reality.

The low cost paging receiver is a prime example of this marriage. The RF and analog portions of these devices consume a mere 1.5 mW - 5 mW [1]. In the cellular arena, integrated circuits have helped reduce the form factor of the AMPS phones form the bulky boxes introduced in the 80's to the sleek pocket-sized phones that are becoming a staple in our society. The move toward second generation digital cellular systems helped underscore the tremendous importance of high speed digital signal processing ICs to perform the operations for both the speech coder/decoder, as well as the physical layer. In the case of GSM and IS-136/54, it was found that the general purpose DSP architecture coupled with dedicated datapaths for Viterbi decoding can meet the processing requirements of these second generation systems. However, in the case of IS-95, the tremendous amount of processing needed to implement the rake receivers has dictated a different hardware architecture. In these systems an application specific integrated circuit (ASIC) is used to perform the chip-rate baseband signal processing at the chip rate. The role of ASICs in the realization of cellular systems is expected to grow as we move towards third generation cellular systems, which require more sophisticated baseband processing. Indoor wireless systems have also benefited from advancements in integrated circuit technology. The realization of high speed adaptive equalizer [2], beamforming [3] and FFT based ASICs [4] are key elements for realizing the physical layer of most high speed wireless indoor links. At high rates even the MAC layer functionality is typically assigned to an ASIC.

Figure 1 shows a block diagram of a typical wireless communication system. Moreover it shows the partitioning of the receiver into RF, IF, baseband analog and digital components. In this paper we will focus on two main classes of integrated circuit technologies. The first is ICs for analog and more importantly RF communications. In general the technologist as well as the circuit designer's challenge here is to first, make the transistors operate at higher carrier frequencies, and second, to integrate as many of the components required in the receiver onto the IC. The second class of circuits that we will focus on are digital baseband circuits. Where increased density and reduced power consumption are the key factors for optimization.

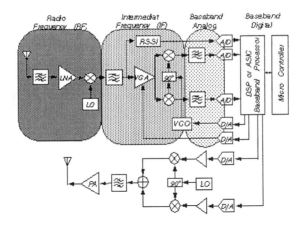

Figure 1. Block diagram of a generic wireless transceiver.

2 CMOS PROCESS TECHNOLOGY

Complementary Metal Oxide Semiconductor (CMOS) technology is by far one of the most important IC processing technologies available. The suitability and cost effectiveness of CMOS technology for the design and development of digital circuits has helped accelerate the advancement and maturity of this technology. With the exception of very high speed specialized digital circuits, CMOS is the technology of choice for all digital circuits. In fact, the rapid pace of development in this technology coupled with its cost-effectiveness arrived at partly through the economies of scale has made CMOS the technology of choice for analog circuit design as well. The tremendous advances in CMOS processing technology have shown no sign of slowing down. The current commercially available minimum size channel length is 0.25 microns (micro-meters), compared to the 2.0 micron state of the art technology that was available in 1983. The stage is set for the predicted 0.1 micron channel lengths to be available around the turn of the century. The ever shrinking transistor sizes translate into different capabilities for RF and digital circuits. In the case of RF circuits, the smaller channel lengths translate into higher operating frequencies for the transistors. This allows CMOS analog circuits to move up the transceiver chain towards the RF carrier frequency. Although CMOS process technology provides low cost and speed, circuit level innovations are still needed to meet the desired gain, linearity and noise levels in CMOS RF-IC design. Digital circuits benefit from smaller dimensions, through increased density, which implies increase functionality for the same amount of Silicon real estate. Circuit designers actually have a choice when it comes to taking advantage of the smaller dimension. They can harness the higher speeds provided by smaller transistors to clock the gates faster, or they can reduce the supply power and thus minimize power consumption while keeping the operating frequency the same.

3   TECHNOLOGY TRENDS IN RF ICS

This section attempts to provide some insights into the current trends and research in the area of low cost RF front ends. Currently, a great amount of research is underway to improve the power efficiency and lower the cost of the circuits used in the present generation of wireless products. The general approach taken towards this goal is to move towards higher levels of integration in cheap silicon based technologies, such as BiCMOS (combination of bipolar and CMOS transistors on the same substrate) and CMOS.

There are several factors that contribute to the significantly lower cost of circuits manufactured on Silicon substrates compared with other semiconductor materials, such as GaAs and SiGe. First, processes for CMOS and BiCMOS circuits provide higher yields (fewer defects per wafer). Secondly, the high volume of orders enjoyed by CMOS fabrication facilities helps reduce the cost of such circuits through economies of scale. Finally, analog circuits implemented in CMOS can be easily integrated on the same substrate with the digital baseband processing circuits for a truly optimal single chip implementation of the system.

The traditional approach towards the design of wireless transceivers has been to use GaAs based circuits for the realization of the RF components such as power amplifiers, low noise amplifiers, and switches. Silicon based analog circuits are then used for the IF to baseband sections and possibly the RF mixers. In these approaches, band selection at RF and IF are typically performed using discrete off chip components. This eliminates the difficulties associated with the realization of high-Q filters on chip. Most conventional approaches also utilize external tuned LC resonators to provide the tuning element of the VCO. The high Qs realized by the external elements are critical in ensuring good VCO phase-noise.

The above mentioned trend is still evident in the latest series of "integrated" transceivers for wireless data applications [5][6][7]. The reason for this partitioning will become clearer in the ensuing section, however, at this point it is worth noting that in general GaAs is fast, provides high gain, high selectivity (high Qs), and has lower noise than silicon based circuits, on the other hand Si based circuits are cheaper and can be used in mixed-mode circuits where digital and analog functionality is realized on the same substrate.

In an attempt to eliminate the need for the GaAs components and enjoy the cost savings associated with Silicon based circuits, researchers within industry and universities are looking for ways of realizing the same functionality now provided by GaAs based circuits on a Silicon substrate, at the expense of some degradation in the overall performance of the block. This was first initiated with Silicon bipolar circuits due to the traditionally higher bandwidths of these devices. However, in recent years advances in the development of CMOS processes and circuits have brought bandwidths, f_T, of these transistors to levels comparable with Silicon bipolar transistors. Consequently, researchers have also started to consider the use CMOS at RF.

In addition to implementing RF sub-blocks of a traditional super-heterodyne architectures, researchers are also investigating a host of alternative transceiver architectures which will help eliminate the need for high-Q, lossy, off-chip filters [8] [9].

Overall, these activities will help take us one step closer to the ultimate goal of a single chip radio, which integrates, RF, IF, analog-baseband, D/A and all the required baseband signal processing on the same substrate. Over the past several years, a number of entities have successfully demonstrated CMOS building blocks such as mixers oscillators and amplifiers [10][11][12][13] for operation in the cellular and the 900 MHz ISM bands. With the continuing trend towards the availability of smaller transistor dimensions, this trend is moving towards the PCS bands and beyond.

3.1 Technology Options for RFIC

The technology choices in Silicon are: CMOS, BiCMOS, and Bipolar, while the technologies available in GaAs are: MESFETS, HBT and PHEMT. The RF designer's choice of a circuit family is driven by the desire to simultaneously meet objectives for low power dissipation, speed, yield, component noise, linearity, gain, and efficiency. To date there is no single circuit family that will simultaneously satisfy all of these requirements at RF, IF and baseband. Baseband analog circuits that implement filters, amplifiers, and A/D and D/A's are dominated by CMOS. IF circuits such as mixers and amplifiers, and some filtering operations are realized in a mix of CMOS, BiCMOS and Bipolar technologies. Finally, the power amplifier and the LNA functions are generally realized in GaAs.

In what follows we will look at components of a transceiver and evaluate the relative merits of the different IC technologies for each application.

3.1.1 Low Noise Amplifiers

Apart from the obvious requirement for low noise, LNAs must also be linear, high gain and dissipate little DC power. In [14] a group of 6 LNAs recently reported in the literature are compared side by side. The results are summarized in Figure 2 and show that noise figures on the order of 1.5-2 dB can be achieved with Si Bipolar, and GaAs technologies. However, for the same amount of DC power dissipation, a GaAs MESFET circuit is shown to provide 1.5 dB more gain than that of a Si bipolar circuit, while GaAs HBT technology can provide 5.5 dB more gain.

The figure also shows that for the same Gain/DC- power (dB/mW) ratio as the Si bipolar LNA, a CMOS LNA will have a worse noise figure by approximately 0.8 dB. The information provided by the plot of Figure 2 can be summarized into a single figure of merit: *gain/(Pdc*NF)* which combines DC power dissipation, noise figure and gain of a given LNA. Utilizing this figure of merit, the CMOS and Si-bipolar LNA rate at approximately 0.4, while the best GaAs LNA achieves a figure of 3.0.

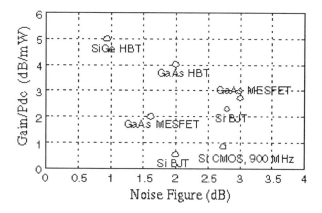

Figure 2. Gain to DC power ratio plotted versus Noise Figure for state-of-the-art 2 GHz LNAs [14]

3.1.2 Power Amplifiers

The power amplifier (PA) is probably one of the last elements to be integrated and implemented in a Si-CMOS process. Traditionally, the PA consists of discrete GaAs transistors assembled on hybrid modules. Even the papers that report on an integrated RF transceiver IC [5][6][7] do not include the power amplifier on the Silicon substrate. In general GaAs provides significantly better power added efficiencies at the high output power levels.

Figure 3. provides a summary and side-by-side comparison of some recently reported GaAs based power amplifiers.

The push towards low cost solutions has lead designers to investigate Si-bipolar and CMOS technologies in the design of the power amplifier. A recently reported PA [16] circuit implemented in CMOS and targeted for constant envelope modulation, delivers 1 W of power with a power added efficiency of 42% and operates within the 800-900 MHz range. A CMOS power amplifier [11] targeted for use in the 900 MHz band only delivers 30 mW of output power at an efficiency of 30%.

In addition to linearity, power consumption, and gain requirements of the power amplifier, its integration onto the same substrate with the rest of the transceiver elements, requires techniques that protect the other circuits from the large pulls that the PA inflicts on the power and ground lines as it sources and sinks current into and out of the load. In [15] this was accomplished via fully differential circuit design techniques.

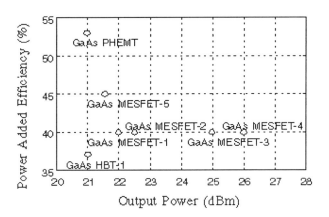

Figure 3. Power Amplifier performance [14]

3.1.3 Voltage Controlled Oscillators VCOs

VCOs are perhaps one of the most critical building blocks of RF synthesizers. In [17] an equation for the relative phase noise of a VCO is given as $\dfrac{1}{4Q^2}\left(\dfrac{w_o}{\Delta w}\right)^2 \dfrac{P_{noise}}{P_{carrier}}$, where Q is the open loop quality factor, Δw is the frequency offset, w_o is the center frequency, and P_{noise} is the spectral density of each noise source.

This equation gives rise to three rules:
1. Use high-Q passive resonators

2. Minimize the number of active (and lossy passive) devices in the oscillation path
3. Maximize the oscillation swing ($P_{carrier}$).

A completely monolithic integrated VCO suffers from low-quality monolithic inductors (typical Q-factors are less than 20 [14] while values as low as 5 [17] are not at all uncommon), lossy varactor diodes that exhibit large series resistance, and an inability to trim the center frequency to accommodate its inevitable drift due to process variations. Despite these drawbacks, substantial progress has recently been made in the development of completely monolithic silicon VCOs for wireless communications. Most of the work to date has focused on improved techniques for realizing high-Q monolithic inductors, which suffer from high resistive metal lines as well as capacitive and magnetic coupling to the substrate.

These efforts include the use of thick gold metallization to reduce the series resistance, bulk micromachining to etch away the resistive material underneath the inductor and eliminate capacitive coupling to the substrate and create a suspended inductor [19]. The realization of high-Q VCO circuits are also hampered by the difficulties associated with the realization of low resistance, wide range varactor tuning capacitance using standard IC technology [18].

3.1.4 Receiver Architectures for Integrated RF-ICs

In order to reduce (eliminate) the off chip components, such as bandpass filters that provide image rejection and channel selectivity, and thus reduce the size and power consumption of the overall transceiver, researchers have investigated various receiver architectures and approaches for monolithic implementation. These architectures range from a direct down conversion to digital IF and subsampling approaches. Some researchers are even looking towards micro-electro mechanical systems (MEMS) for the realization of RF cmoss components [20]. Although this work is a long way from commercialization, it may provide a solution for integrated monolithic transceiver architectures.

3.1.4.1 Direct Conversion and Low IF architectures

Direct conversion configurations have been intensively investigated for many years. By eliminating the need for an IF stage, these implementations not only reduce the component count associated with the receiver, but also eliminate the need for bulky off chip filters. In fact, these architectures have made their way into a few applications. The use of FSK modulation, which eliminates signal energy around DC, is particularly suited for direct conversion architectures. This has resulted in companies such as Philips and NEC to migrate from their original super-heterodyne architecture to a zero IF receiver [9]. Moreover, companies such as Alcatel have adopted the direct conversion receiver architecture into their GSM products [9].

Although direct-conversion architectures are conceptually appealing, their wide spread acceptance is hampered by serious technical challenges associated with the non-idealities encountered in the design and implementation of the required sub-blocks. These include DC offset (caused by self-mixing and I-Q mismatches), LO leakage and flicker noise of the electronics [8]

Numerous approaches have been tried to alleviate these problems. They range from various techniques for on-chip trimming of the circuits to adaptive DC cancellation using signal processing elements in the baseband digital section [18]. It is worth noting that most of the advantages of the direct-conversion architectures can still be enjoyed if a very low IF is used. In such cases the IF needs to be low enough that normal monolithic filtering techniques such as switched-capacitor filters and continuous filters can be used. In this approach, the problem of DC offset and 1/f noise are greatly reduced. However, a new problem of image rejection of the fairly close-in image frequency is introduced [18].

In [8] two common image reject architectures are outlined. These are the Hartely and the Weaver architectures respectively, and are depicted in Figures 4 and 5. In the Hartely architecture a Hilbert filter is used to introduce a -90 degree phase rotation to the negative frequency portion of the Q-rail after the LPF. This will result in the image frequencies canceling each other after the summer and passing the desired channel through. Once again the image canceling ability of this circuit is compromised by the inability to perfectly match the components in the I and Q rails. For typical matching in IC technologies, the image is rejected by about 30 to 40 dB [8].

The Weaver architecture is a two stage down conversion alternative that is a hybrid between the heterodyne and the image reject approach of the Hartely circuit. The Weaver architecture eliminates the need for high-Q bandpass filters associated with 2-stage (heterodyne) down converters and replaces the Hilbert filter of the Hartely circuit with a second quadrature mixer. Moreover, it requires a DC block circuit in front of the LPF so that any DC component created from the self-mixing of the first LO signal does not saturate any ensuing amplifiers (not shown) placed after the mixer. This architecture also suffers from the same matching requirements as the Hartely circuit and its image rejection ability is dictated by the matching achieved in the circuits.

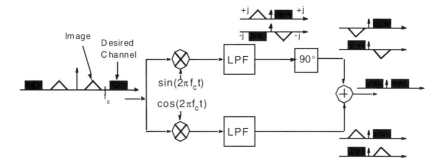

Figure 4. Hartely architecture

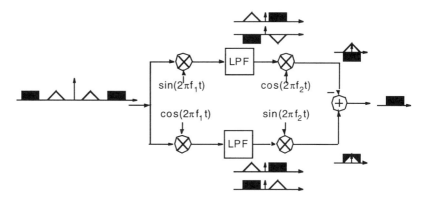

Figure 5. Weaver architecture.

4 TECHNOLOGIES FOR DIGITAL BASEBAND PROCESSING

One of the key decisions that implementers of digital communication systems must face is the choice of hardware platform for the baseband digital processing portion of the transmitter and receiver. In general, based on the nature of the data communications system and the desired data rate, three distinct hardware platforms are possible. These platforms are:

(a) Software programmable devices such as Digital Signal Processing (DSP) chips (e.g. TI TMS320Cx, Analog Devices AD21x, etc.) and microcontrollers

(b) Hardware programmable devices such as field programmable gate arrays (FPGAs)

(c) Application specific integrated circuits (ASICs) which are custom built for each application.

The optimal mapping of the functional blocks of a communication system, onto one or any combination of these three hardware platforms could seriously affect the performance, size, cost and power consumption of the overall system. Of the three platforms, only DSPs and ASICs are predominant in wireless data communication systems

4.1 Digital Signal Processors (DSPs)

The term "DSP processor" is mainly used to refer to software programmable devices that incorporate specially designed elements and architectures that are suited to the realization of repetitive, numerically intensive calculations. In particular, DSPs include dedicated circuitry to rapidly perform Multiply-ACcumulate operations (MAC), which are useful in many signal processing algorithms.

Ever since the introduction of the first single chip DSP (AMI S2811) [21] in 1970, applications of DSP have experienced an explosive growth. Today's family of low-power fixed-point DSPs deliver on the order of 40-50 MIPs [22]. This amount of processing power is sufficient to realize almost all the needed processing for the physical layer of narrowband 2nd generation cellular systems (e.g. GSM and IS-136) as well as the speech coding and decoding functionality. In [22] the MIPs requirements for various algorithms needed in the transmission and reception of an IS-136 signal are summarized. It is shown that the entire physical layer as well as the speech coding/decoding operations require on the order of 38 MIPS, just enough to fit into a state of the art DSP processor.

Although ideal for the realization of signal processing tasks, the DSP architecture is not well suited to the realization of the call processing protocol or the user interface. These tasks are traditionally realized on a general purpose micro-controllers. As a consequence, many of the existing cellular handsets have adopted a two chip digital solution consisting of a DSP chip operating side-by-side with a micro-controller. At the time of this writing, this trend has already given way to a single chip solution, which integrates both the micro-controller and the DSP. Such levels of integration will go a long way towards reducing the cost and power consumption of emerging cellular handsets.

It is interesting to note that the tremendous market potential for wireless communication systems has forced manufacturers to modify the traditional architectures for applications in wireless systems. An example is the Lucent DSP-16000 processor family which augments the architecture of the traditional processor by adding a second multiply-accumulate, a dictated hardware unit to perform the add-compare-select operations required by the Viterbi decoder operation, and a 3 input 40-bit adder [23]. Figure 6 depicts

238

the top level architecture of the DSP16000. The address space is partitioned into two parts, the X and Y address spaces, each with its own arithmetic unit (address generator) and data bus. The X-address space is reserved for program instruction, while the Y-space is reserved for data memory. This partitioning allows parallel fetching of program instructions and data and results in significant improvement in the throughput of the system.

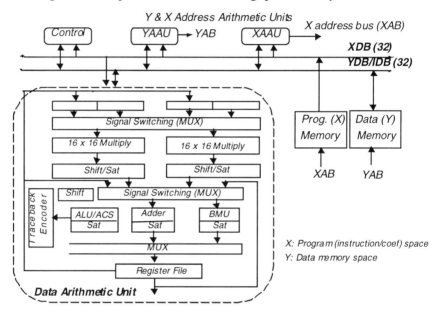

Figure 6. Architecture of the DSP16000.

The single chip DSP is at the heart of every cellular handset sold, however, a single chip is not sufficiently powerful to meet the demands of cellular base stations. In general base-stations are realized through multi-processor boards or a combination of DSP and ASIC components. A good example of the multi-processor approach is reported in [24]. The AirNet Communications VME-based cellular base-station uses a radically new multi-DSP baseband platform that overcomes the performance limitations of traditional basestation design, Figure 7. The hallmark of this platform is a reprogrammable architecture that can be reconfigured on the fly to adapt to basestation traffic load changes.

The baseband platform consists of more than 100 40-MIPS DSPs located on a total of four VME cards, running a master software program that controls the interaction between different units. Electrically arranged in a symmetrical matrix of rows and columns, the DSPs do not have preassigned functions. Instead, the baseband application software dynamically allocates tasks to each DSP. The environment is so flexible that the system can, if necessary, simultaneously support cellular telephone services operating under different cellular standards.

One of the DSPs in the system serves as a baseband controller. When a mobile telephone requests service, the basestation controller assigns one of the available channels to it and notifies the baseband control DSP of the assignment via the VME backplane. The control DSP, in turn, notifies the next available DSP that it has been allocated to handle a call on the channel assigned to the particular mobile unit. Communications between the VME bus backplane and baseband DSPs serving as call processors are handled by another baseband DSP.

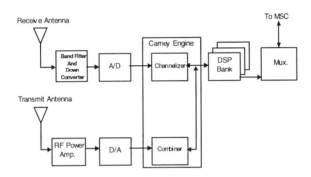

Figure 7. AirNet Multi-processor system.

4.1.1   Next Generation DSP Processors

In order to keep a foothold in the wireless data communication systems, DSP chips must evolve to meet the demands of the emerging higher speed and more complex wireless systems. Novel architectures must be developed in order to provide higher processing power and improved energy efficiency.

A prime example of this trend is the TI TMS320C62x [25] processor, which was introduced in 1996. This chip can deliver 1.6 billion instructions per second (GIPS) and an equivalent 400 MOPS on benchmark signal processing operations (MOPS: million operations per second, where an operation is defined as a multiply-add). The architecture of this chip is based on a very large instruction word (VLIW) architecture and is a significant departure from the traditional Harvard based architectures described earlier. It houses 8 parallel processing units consisting of two 40-bit ALUs, two 32-bit integer ALUs, two 16x16 multipliers, and two 32-bit adder/subtractor units. The core operates at 200 MHz and multiple instructions per cycle can be executed, one on each of the 8 processing units.

The Analog Devices ADSP-21160 processor also delivers approximately 400 MOPS of sustained throughput but uses a vastly different architecture. This part is based on a so called super-Harvard architecture (somewhat similar to that shown in Figure 6) which combines two parallel computation units (each computation unit includes a 16x16 multiplier, ALU shifter and register file) with four independent data buses, a 4 Mbit on-chip dual-port static RAM, and separate data address generator units and I/O processors. In this architecture, emphasis is placed on a balance between core performance, large internal dual ported SRAM, zero overhead DMA, I/O throughputs, and integrated multiprocessing features.

This order of magnitude increase in the processing power of these chips will enable them to meet the challenges of present and emerging computation intensive data communication systems. Some potential applications envisioned for these processors include cellular base stations, digital subscriber loop (DSL) services, 3-D graphic accelerators. The main obstacle to the use of these processors in wireless handsets and other portable terminals is their high power consumption. The current generation of the analog Devices ADSP-21160 processors dissipate approximately 2 Watts off of a 2.5V supply, while the power consumption of the TI320C6201B (the low power version of the processor) is on the order of 2.5 Watts off of a 1.8 V supply.

As the power consumption issue is addressed, it is expected that these processors will form the framework for the emerging third generation cellular systems based on DSSS W-CDMA technology. It is expected that the ultimate 3-G hardware platform will result in a marriage of a low-power version of these processors with low-power dedicated datapaths that realize the chip-rate processing.

4.1.2   Performance Measure for DSPs

As is evident from the above discussion, one can not gain any insight into the performance of a processor by simply looking at its clock frequency. In fact, the performance of a processor can be measured in many different ways. Most processor vendors use MIPS as a performance indicator. MIPS is a misleading unit because of the varying amounts of work done by instructions on different processors. A prime example of this fact is the difference between the MIPS power of the TI TMSC62x processor and the equivalent MOPS obtained in benchmarks published on the TI web site.

The best way to compare different DSPs is benchmarking them, i.e. running the same set of computations (from different applications) on different architectures and comparing the execution times. The results of such a benchmark are presented in [26].

4.2 Application Specific Integrated Circuits (ASICs)

Traditionally, application specific integrated circuits (ASICs) have found applications in high-speed data communication systems such as ATM, HDTV, wireless indoor LANs, etc. The reason why ASICs remain the popular choice despite the time consuming design cycle that they mandate is quite simple. ASICs are the most power efficient and compact solution to any application. They are capable of delivering high-speed computations at a fraction of the power and area consumed by programmable chips. The disadvantage is of course that the hardware becomes application specific and must be redesigned for different versions of the same application or for other applications.

To date, the role of ASICs in wireless communication systems has been rather limited. Almost all the signal processing demands of second generation cellular systems such as GSM and IS-136 (US TDMA) can be met with the current generation of general purpose DSP chips. The use of ASICs in wireless data communications has been limited to wireless LAN systems such as Lucent's WaveLAN and the front end, chip-rate processing needs of DSSS-CDMA based systems such as IS-95 (US CDMA).

Several major factors are redirecting the industry's attention towards ASICs for the realization of highly complex and power efficient wireless communications equipment. First the move toward third generation cellular systems capable of delivering data rates of up to 384 kbps in outdoor macro-cellular environments (an order of magnitude higher than the present second generation systems) and 2 Mbps in indoor micro-cellular environments. Second, the emergence of high speed wireless data communications whether in the form of high speed wireless LANs [27] or in the form of broadband fixed access networks [28]. A third, but somewhat more subtle factor is the increased appeal of software radios. Radios that can be programmed to transmit and receive different waveforms and thus enable multi-mode and multi-standard operation. Although ASICs are by nature not programmable, they are parameterizable. In other words, ASIC designers targeting wireless applications must develop their architectures in such a way as to provide the user with features such as variable symbol rates and carrier frequency, as well as the ability to shut-off parts of the circuit that may be unused under benign channel conditions. As an example, ASICs targeted at DSSS systems should provide sufficient flexibility to accommodate programmability of the chip-rate, spreading factor and the spreading code to be used.

Achieving this added level of flexibility requires innovations at two levels. First, advanced signal processing algorithms need to be investigated and implemented. Second, innovative circuit architectures are needed to allow variation in data rate without seriously impacting the cost and power consumption of the chip. This is a general philosophy that is taking form, and is expected to grow in the coming years. The remainder of this section we will present some signal processing algorithms and ASIC architectures for the realization of such blocks.

4.2.1 VLSI Architectures for Signal Processing Blocks

4.2.1.1 Fixed Coefficient Filters

The most intuitive means of implementing finite impulse response (FIR) filter is to use the direct form implementation presented in Figure 8a [29]. Applying the transposition theorem to this filter we get the transposed structure shown in Figure 8b. The two structures are identical in terms of I/O, however, the transposed form is ideal for high speed filtering operations since the critical path for an N tap filter is always one multiplier delay plus one adder delay. The critical path of the direct form, however, is one multiplier delay plus N-1 adder delays. The fact is that the symbol rate for most wireless communication systems is a few tens of MHz whereas a typical multiplier in today's CMOS process technologies can easily reach speeds of 80 MHz to 100 MHz. It is thus desirable to use the hybrid architecture shown in Figure 8c where each multiplier accumulator is time-shared between several taps (3 in this case) resulting in a more compact circuit for lower symbol rates.

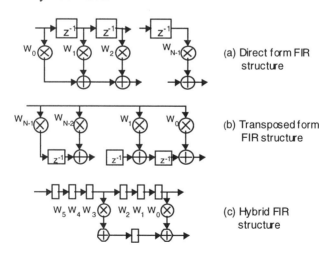

Figure 8 Alternative FIR filter structures.

The implementation of fixed coefficient FIR filters can be further simplified by moving away from the use of 2's complement number notation, and using a signed-digit number system in which each digit can take on one of three values {-1, 0, 1}. In general there are multiple signed-digit representations for the same number and a canonic signed-digit (CSD) representation can be defined for which no two nonzero digits are adjacent [30]. The added flexibility of signed-digit numbers allows us to realize the same coefficient using fewer non-zero coefficients than would be possible with a simple 2's complement representation. Using an optimization program, it is possible to design an FIR filter using CSD filters with as few as 3 or 4 non-zero digits for each coefficient. This could help significantly reduce the complexity of fixed coefficient multipliers since the number of partial products generated is directly proportional to the number of non-zero digits in the multiplier.

4.2.1.1.1 Polyphase Interpolation / Decimation Filters

Digital interpolation and decimation filters are often found in the transmitter and receiver sections of wireless data communication systems. Some examples include the transmit shaping filter (generally interpolates by 2) or the feedforward section of a fractionally spaced DFE (decimates by 2). By digital resampling of the signal, the designer can ensure that the clock frequencies at all portions of the circuit are the minimum that they need to be to properly represent the signals. This could have significant impact on the size and power consumption of the resulting ASIC.

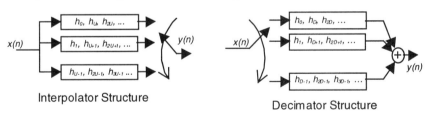

Figure 9. Polyphase filter structures for interpolation and decimation

Given the coefficients, $\{h_i\}$ for an FIR decimation or interpolation filter, the structure of choice for the realization of a decimate by D or an interpolate by D filter is the polyphase structure [29] shown in Figure 9. The attractiveness of this structure is in the fact that the filter is always operated at the lower sampling frequency.

In many cases it is desirable to resample the signal by a power of 2^N. In which case N decimate (interpolate) by 2 stages can be cascaded one after the other. Each decimator will consist of a half-band filter followed by a

244

decimator. The halfband filter could be realized using the polyphase structure to simplify its implementation. Moreover, these filters are typically very small consisting of anywhere from 7 to 15 taps depending on the specified stopband attenuation and the size of the transition band. Their implementation can be simplified by exploiting the fact that close to half of the coefficients are zero and the remainder are symmetric about the main tap due to the linear phase characteristics of the halfband filter. Finally, since these are fixed-coefficient filters, they can be realized using CSD coefficients[30].

In cases where the oversampling ratio is large (e.g. narrowband signal) an alternative approach using a cascaded integrator-comb (CIC) structure can be used to implement a multiplier-less decimator. The interested reader is referred to [31] for a brief overview of a CIC ASIC.

4.2.1.1.2 Variable rate interpolators

A variable rate interpolator takes an input data sequence $\{x(n)\}$ sampled at $1/T_s$ and produces an output sequence $y(n)$ sampled at $1/T_i$. The entire operation is performed digitally and the ratio $\alpha = T_i/T_s$ can be any value, rational or irrational, in the range from 2 to 4. Variable rate interpolation and decimation filters play a very important role in the development of highly flexible and self contained all-digital receivers. They are the critical element of all-digital timing recovery loops as well as systems with programmable symbol rates.

To better understand the operation of this block, let us define the variable μ_k, to be the time difference between the output sample $y(k)$ and the most recent input sample x_l. The job of the variable rate interpolator is to weight the adjacent input samples $(..., x_0, x_1, ...)$, based on the ratio $\mu_k T_s$, and add the weighted input samples to obtain the value of the output sample, $y(k)$. Mathematically, there are a number of interpolation schemes that can perform the desired operation. However, many of them, such as *sinc* based interpolation, require excessive computational resources for a practical hardware implementation. For real time calculation, Erup *et. al.* [32] found polynomial-based interpolation to yield satisfactory results while minimizing the hardware complexity. In this approach, the weights of the input samples are given as polynomials in the variable μ_k and can be easily implemented in hardware using the Farrow Structure [33] shown in Figure 10. In this Structure, all the filter coefficients are fixed and polynomials in μ_k are realized by nesting the multipliers as shown in Figure 10.

The signal contained in the image band will cause aliasing after resampling. However, proper choice of the coefficients in the Farrow structure can help optimize the frequency response of the interpolator for a particular application. An alternative method to determine the filter coefficients is outlined in [34].

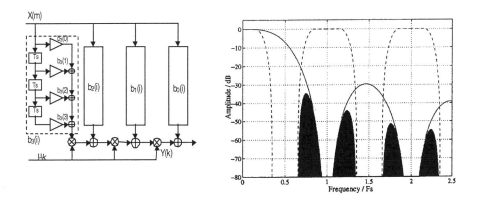

Figure 10 (a) Block diagram of the Farrow structure. (b) Frequency response of polynomial based interpolator (from [34])

4.2.1.1.3 Direct Digital Frequency Synthesizer (DDFS)

Direct digital frequency synthesizers (DDFSs) generate digital samples of sinusoids. Moreover, perfect quadrature signals can be generated due to the digital implementation of the block. Until recently DDFSs have been used in test equipment or sold as stand alone units, however, as digital-IF transceiver designs gain more popularity, they are being integrated onto ASICs along with the remainder of the baseband processing engine [35]. This allows the entire carrier recovery algorithm to be implemented in the digital domain and eliminates the need for data converters that have been traditionally used to communicate digital control signals to the VCO.

Given an input frequency word W, a direct digital frequency synthesizer will produce a sinusoid with a frequency proportional to W. The most common techniques for realizing a DDFS consist of first accumulating the frequency word W in a phase accumulator and then producing the sine and cosine of the phase accumulator value using a table lookup or a coordinate rotation (CORDIC) algorithm. These two approaches are depicted in Figure 11. The two metrics for measuring the performance of a DDFS are the minimum frequency resolution Δf and the spurious free dynamic range (SFDR). The frequency resolution can be improved by increasing the wordlength used in the accumulator, while the SFDR is affected by the wordlengths in both the accumulator as well as the sine/cosine generation block. One of the main challenges in the development of the table lookup DDFS has been to limit the size of the sine/cosine table. This has been accomplished through two steps [36]. First, by exploiting the symmetry of the sine and cosine functions it is only necessary to store ¼ of the period of a sine wave and derive the remainder of the period through manipulation of the saved portion. Second, the number of bits per entry can be reduced by dividing the sine table between a coarse ROM and a fine ROM with the final result obtained after simple post-processing of the values. Combining these two techniques can result in the reduction of the sine tables by an order of magnitude or better.

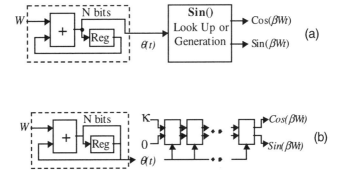

Figure 11 Two most common DDFS architectures. (a) Table lookup. (b) Coordinate Rotation (CORDIC)

In the CORDIC algorithm, Figure 11, sine & cosine of the argument are calculated using a cascade of stages each of which rotates its input complex vector by $\delta/2^k$ ($\delta=\pi/2$) if the kth bit of W is 0 and $-\delta/2^k$ if the bit is a 1. Thus each stage performs the following matrix operation:

$$\begin{bmatrix} x_{out} \\ y_{out} \end{bmatrix} = \begin{bmatrix} \cos\theta & -\sin\theta \\ \sin\theta & \cos\theta \end{bmatrix}\begin{bmatrix} x_{in} \\ y_{in} \end{bmatrix} = \cos\theta\begin{bmatrix} 1 & -\tan\theta \\ \tan\theta & 1 \end{bmatrix}\begin{bmatrix} x_{in} \\ y_{in} \end{bmatrix}$$

In [37] a simplification of the CORDIC DDFS is presented in which for small θ, $\tan(\theta)$ is simply approximated by θ. In [38] a different modification to the CORDIC architecture is proposed that will facilitate low-power operation in cases where a sustained frequency is to be generated. This is achieved by calculating the necessary angle of rotation for each sampling clock period, and dedicating a single rotation stage in a feedback configuration to contiually rotate the phasor through the desired angle.

5   CONCLUSION

This paper has provided an overview of the ever increasing role that integrated circuits play in the success of wireless communication systems and services. The trends as well as the challenges in the design of RF-ICs have been outlined. The paper also provided a brief overview of the trends in the development of digital VLSI circuits for baseband signal processing tasks.

REFERENCES

[1] Motorola MC3347Data sheet.

[2] L. Tan, et. al, "A 70 Mb/s Variable Rate 1024-QAM Cable Receiver IC with Integrated 10v ADC and FEC Decoder," *IEEE ISSCC '98, Digest of Technical papers,* vol. 41, Feb. 1998, pp. 200-201.

[3] J. Y. Lee, et. al, "A Digital Adaptive Beamforming QAM Demodulator IC for High Bit-Rate Wireless Communications," *IEEE JSSC,* vol. 33, No. 3, pp. 367-377, March 1998.

[4] N. Weste, et. al., "A 50 MHz 16-Point FFT Processor for WLAN Applications," *IEEE CICC '97,* May 1997, pp. 457-460.

[5] K. Irie, H. Matsui, T. Endo, et. al., "A 2.7V GSM RF Transceiver IC," *in IEEE ISSCC '97 Digest,* Feb. 1997, pp. 302-303.

[6] S. Heinen, K. Hadjizada, U. Matter, et. al. "A 2.7V 2.5 GHz Bipolar Chipset for Digital Wireless Communication," *in IEEE ISSCC '97 Digest,* pp. 306-307.

[7] R. G. Meyer, W. D. Mack, J. Hageraats, "A 2.5 GHz BiCMOS Transceiver for Wireless LAN," *in IEEE ISSCC '97 Digest,* Feb. 1997, pp. 310-311.

[8] B. Razavi, "Challenges in Portable RF Transceiver Design," *IEEE Circuits and Devices Magazine,* Sept. 1996, pp. 12-25.

[9] A. Abidi, "Low-Power Radio Frequency IC's for Portable Communications," *Proceedings of the IEEE,* Vol. 83, No. 4, April 1995, pp. 544-569.

[10] S. Wong, H. Bhimnathwala, S. Luo, B. Halai, S. Navid, "A 1 W 830 MHz monolithic BiCMOS Power Amplifier," *IEEE ISSCC '96 Digest,* pp. 52-53.

[11] M. Rofougaran, et. al., "A 900 MHz CMOS RF power amplifier with programmable output," *1994 Symposium on VLSI Circuits Digest of Technical papers,* pp. 133-134.

[12] N. Karanicolas, "A 2.7V 900 MHz CMOS LNA and Mixer," *in IEEE ISSCC '96 Digest,* vol. 39, pp. 50-51, Feb. 1996.

[13] J. Craninckx, M.S.J. Steyaert, "A 1.8-GHz CMOS Low-Phase Noise Voltage-Controlled Oscillator with Prescaler," *IEEE Journal of Solid-State Circuits,* vol. 30, no. 12, pp. 1474-1482, Dec. 1995.

[14] Lawrence E. Larson, "Integrated Circuit Technology Options for RFIC's - Present Status and Future Directions," *IEEE CICC '97,* May 1997, pp. 169-176.

[15] Rofougaran, et. al, "A Single-Chip 900 MHz Spread-Spectrum Wireless Transceiver in 1 µm CMOS Parts I & II," *IEEE JSSC,* vol. 33, No. 4, pp. 515-547, April, 1998.

[16] D. Su, W. McFarland, "A 2.5-V, 1-W Monolithic CMOS RF Power Amplifier," *IEEE CICC '97,* May 1997, pp. 189-192

[17] Behzad Razavi, "Challenges in the Design of Frequency Synthesizers for Wireless Applications," *IEEE CICC '97,* May 1997, pp. pp. 395-402.

[18] P.R. Gray, R.G. Meyer, "Future Directions in Silicon ICs for RF Personal Communications," *Proceedings of IEEE CICC '95,* pp. 83-89.

[19] J. Chang, A. Abidi, M. Gaitan, "Large Suspended Inductors in Silicon and Their Use in a 2-µm CMOS RF Amplifier," *IEEE Electron Device Letters,* vol. 14, No. 5, May 1993.

[20] D. J. Young, B. E. Boser, "A Micromachine-Based RF Low-Noise Voltage-Controlled Oscillator," Proc. of IEEE CICC '97, pp.431-434.

[21] Nicholson, Blasco and Reddy, "The S2811 Signal Processing Peripheral," *WESCOM Tech Papers,* Vol. 22, pp 1-12.

[22] Z. Kostic S. Seetharaman, ``Digital Signal Processors in Cellular Radio Communication," *IEEE Commun. Magazine,* vol. 35, No. 12, pp. 22-35, Dec. 1998

[23] "DSP16xxx Targets Communications Apps" *Microprocessor Report, Vol.12, No. 12, September 15, 1997.*

[24] "DSP Real-Time Operating system Simplifies Communication System Design", *Texas Instruments Communication white paper*, http://www.spectron.com/products/spox/comm.htm.

[25] "TI's New 'C6x screams at 1,600 MIPS," *Microprocessor Report*, Vol. 11, No. 2, Feb. 17, 1997.

[26] "BDTI Benchmark," http://www.bdti.com/bdtimark/results.htm

[27] K. Pahlavan, et. al., "Wideband Local Access: Wireless LAN and Wireless ATM," IEEE Commun. Mag., pp. 34-40, Nov. 1997.

[28] J. Mikkonen, et. al. "Emerging wireless broadband networks," IEEE Commun. Magazine, vol.36, no.2, pp. 112-17, Feb. 1998.

[29] J. G. Proakis, D.G. Manolakis, Introduction to Digital Signal Processing, Macmillan, 1988.

[30] H. Samueli, "An Improved Search Algorithm for the design of Multiplierless FIR Filters with Powers-of-Two Coefficients," IEEE TCAS, vol. 36, no. 7, pp. 1044-1047, July 1989.

[31] Kwentus, O. Lee, A. Willson Jr., "A 250 Msample/sec programmable cascaded integrator-comb decimation filter," VLSI Signal Processing, IX, IEEE, New York, 1996, pp.231-40.

[32] L. Erup, F.M. Gardner, R.A. Harris, "Interpolation in digital modems. II. Implementation and performance," IEEE Trans. on Commun., vol.41, No. 6, June 1993, pp. 998-1008.

[33] C.W. Farrow, "A continuously variable digital delay element," Proc. ISCAS '88, June 1988, pp. 2641-5.

[34] J. Vesma, T. Saramaki, "Interpolation filters with arbitrary frequency response for all-digital receivers," IEEE ISCAS '96, May 1996, pp. 568-71.

[35] K.H. Cho, J. Putnam, E. Berg, B. Daneshrad, H. Samueli, "30 Mbps wireless data transmission using an equalized 5-MBaud M-QAM testbed, " 1997 IEEE 6th International Conference on Universal Personal Communication. vol.1, 1997, pp.104-108.

[36] H. T. Nicholas, III, H. Samueli, "A 150-MHz Direct Digital Frequency Synthesizer in 1.25 μm CMOS with –90 dBc Spurious Performance," IEEE JSSC, vo. 25, no. 12, pp. 1959-1969, Dec. 1991.

[37] Madisetti, A. Kwentus, A.N. Willson, Jr., "A Sine/Cosine Direct Digital Frequency Synthesizer Using an Angle Rotation Algorithm," Proc. IEEE ISSCC '95. p.262-3.

[38] E. Grayver, B. Daneshrad, "Direct digital frequency synthesis using a modified CORDIC," *IEEE ISCAS,* June 1998.

Chapter 14

Amplifier Linearization
For Broadband Wireless Applications

KATHLEEN J. MUHONEN AND MOHSEN KAVEHRAD
Department of Electrical Engineering, The Pennsylvania State University

Abstract: Wireless technology is exploding. The recent and future trends in wireless technology are an increasing transmit frequency and an incredible increase in bandwidth. The lower frequency bands are extremely crowded and new services cannot be deployed there. In addition to voice, now video and data is desired through a wireless link. Video and data require much more of the spectrum than voice. Higher frequencies and broadband signals pose new design problems that have yet to be solved; but even so, companies are auctioning spectrum for these systems much faster than they can be designed. One critical item in these wireless links is the microwave or millimeter wave power amplifier. At high frequencies the cost of these amplifiers grows rapidly. Designing broadband amplifiers for digital signal transmission presents a trade off between efficient amplifiers that create distortion or unnecessarily large and inefficient amplifiers without distortion. One possible solution to have high efficiency and minimal distortion is to use a linearization technique. Of the these different techniques, which are feedforward, feedback, predistortion, RF and IF feedback, and a combination of feedback and feedforward, digital predistortion seems to be the most promising solution to the wideband applications.

1 INTRODUCTION

1.1 Amplifiers

Whether it is the power amplifier at a the base station hub, the end user in a fixed wireless loop, or the mobile handset, all need to meet the challenging specifications for each system. The broadband systems are being deployed at higher frequencies where amplifiers are harder to design for linearity. In order to meet FCC's transmission masks, the power amplifier will have to be backed off by an amount equal to the peak-to-average ratio of the signal's power to avoid distortion when the amplifier becomes nonlinear. This peak-to-average ratio can be as high as 12.5 dB for some systems.

Unfortunately, power amplifier designs often sacrifice power added efficiency, PAE, (which is a ratio of RF output power to DC power) to achieve spectral compliance that meet FCC transmission regulations. The trade-off between PAE and spectral efficiency is illustrated in Figure 1, which shows the output power and PAE versus input power characteristics of a Minicircuits Class A power amplifier. For maximum efficiency, the amplifier must be operated in gain compression where non-linearities tend to cause intermodulation distortion (IMD). The effect of distortion on the output spectrum of an amplifier is illustrated in Figure 2. This shows a $\pi/4$ Differential Quadrature Phase-Shift Keying ($\pi/4$-DQPSK) power spectra for the same Minicircuits amplifier operating in the linear regime and at 3 dB gain compression. From this figure, it is evident that distortion manifests itself by increasing the power present in the adjacent channel and first alternate, which is often referred to as spectral regrowth.

This trade off between efficiency and spectral compliance gets worse as frequency increases. Amplifier linearity is more difficult to achieve as the center frequency of operation increases. Designing a larger, inefficient amplifier is too costly and sometimes impractical due to heating problems. One possible solution that would allow the amplifier to operate efficiently without creating spectral regrowth is to implement a linearization scheme around the amplifier. Many linearization schemes have been demonstrated at lower frequencies with modest bandwidths. Investigating these techniques provides valuable information as to what linearizer, or perhaps combination of linearizers would be the best for each application.

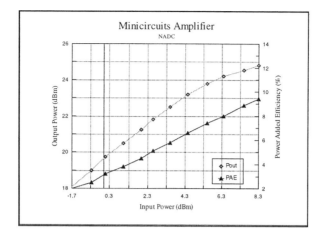

Figure 1. Minicircuits Output Power Characteristic

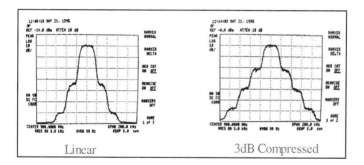

Figure 2. Spectral Regrowth when Amplifier is in Compression

1.2 Wireless Systems, Applications for Amplifier Linearization

1.2.1 Fixed Wireless Loop

Fixed wireless loop systems promise to provide services like phone, fast Internet, TV, ISDN, and video teleconferencing to small-office home-office (SOHO). These systems will be deployed using wireless technology as shown in Figure 3. Here, a central office connects to the base station via fiber, coax, or a point-to-point radio. The base station connects to the end user through a nearly line-of-sight wireless link at microwave and millimeter wave frequencies.

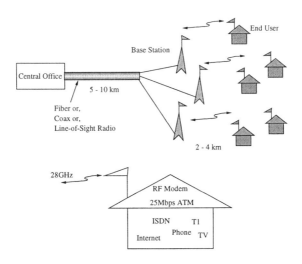

Figure 3. LMDS Example, Central Office to Base Station to End-User

For example, the Local Multi-point Distribution System at 28GHz proposes to have its downstream traffic (hub to end-user) to have 25 MHz channels and the upstream traffic (end user to hub) to have 2 MHz channels. Proposed modulation schemes are 64-QAM, which offers adequate spectrum efficiency of nearly 6 bits/Hz, or OQPSK, which offers nearly 2 bits/Hz of spectrum efficiency. The advantage to OQPSK is an amplifier can operate more efficiently with less distortion since there are a fewer number of transmit levels in this signal than in QAM.

Another proposed fixed wireless loop system is called EDGE and is part of the Global System Mobile, GSM, standard. Propagation problems and weather do not affect this signal as much at microwave frequencies, but there are other situations that require a very linear but efficient amplifier. The EDGE system proposes to use adaptive modulation. Here a linear and efficient modulation scheme like 64-QAM would be used for high data rate traffic. A less efficient scheme like QPSK would be used for lower data rate traffic such as voice communications. Without a linearizer, the amplifier would have to satisfy all specifications with either modulation scheme; thus, the amplifier is backed off a large amount in order to operate with 64-QAM and remains at this operating point for QPSK. If a linearizer was employed, the amplifier could be operated at two different levels depending on the modulation. Not only would this allow more efficient operation for 64QAM; but no matter what modulation was being used, the distortion would be reduced while maintaining that efficiency.

1.2.2 Mobile Wireless

Mobile wireless systems are faced with an even harsher environment due to the multipath and time varying nature of the channel. Mobile environments also differ from fixed wireless systems from the standpoint that they cannot take advantage of very directional antennas or some of the smart antenna technology. At no time is the location of the mobile known, thus transmission has to be omni-directional. This increases the power requirements of the amplifier.

Higher signal-to-noise ratios are needed for satisfactory BER performance; therefore, operating at a high output power with good efficiency is critical. More efficient amplifiers at the base station means smaller amplifiers with less temperature problems; for the hand held this means a longer battery life. As in fixed wireless loop, an amplifier without a linearizer will also have to be backed off in output power an amount equal to the peak-to-average ratio in order to avoid distortion.

2   LINEARIZATION TECHNIQUES

There are many different linearizers that have been demonstrated, but this chapter investigates several well-published configurations. These configurations are feedforward, feedback, and predistortion. Of the feedback linearizers, Cartesian feedback, IF/RF feedback and feedback+feedforward are discussed. Several forms of predistortion are reviewed, those being of both the analog and digital types.

2.1 Feedforward

Of all the linearization schemes, probably the most well known technique is feedforward. Not only is the technology for it mature, but its general configuration is the easiest to implement, even at high frequencies [Dixon 86]. Feedforward linearization reduces spectral regrowth by canceling the distortion components at the output of the nonlinear power amplifier. A block diagram of a feedforward system is shown in Figure 4. The linearizer is composed of two loops, the first being the signal cancellation loop and the second being the distortion cancellation loop. In the first loop, the input signal is split into two paths; the upper path consists of the main power amplifier and the lower path consists of a delay line. The power amplifier in the main path is saturated and has distortion components at its output. A coupler samples part of this output and adds the signal to the lower path. The delay is adjusted to give 180 degrees of phase difference with the upper path such that the signals add out of phase in the coupler. The resulting signal at the coupler is only the distortion from the main amplifier. The second loop consists of upper and lower paths also. The upper path contains another delay line to shift the signal 180 degrees out of phase with respect to the lower path. The lower path contains an attenuator and amplifier to properly adjust the amplitude of the distortion components. This adjustment is made such that when the two signals combine at the output, the distortion components cancel and leave only the desired signal [Eid 95].

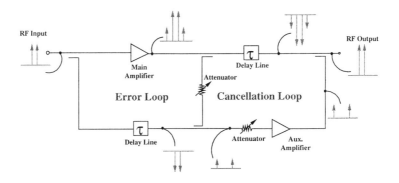

Figure 4. General Feedforward Block Diagram

Feedforward usually has impressive results, over 20dB of cancellation of the third order IMD's is commonly reported [Kenington 91, Stewart 88]. The one downfall to the general configuration is that it is not adaptive. If the either amplifier in Figure 4 were to change due to temperature or bias, complete cancellation of the signal in loop one, or the distortion in loop two would not occur. Phase and amplitude balance have shown to be very sensitive in this design [Cavers 95].

Feedforward definitely has applications to the base station amplifiers. Size of feedforward is a problem since there are two amplifiers and usually the delay lines are implemented with coax cables. This could not be implemented in the hand held mobile unit due to cost and size. It may have applications to the end user equipment in fixed wireless loop depending on the size of the units. Adaptivity is a problem for this configuration. Zhao [Zhao 96], however, has modified this configuration to make it adaptive, but this modification greatly increased the complexity without any decrease in cost.

2.2 Feedback

The previous section demonstrated that the feedforward technique is not inherently adaptive, but any feedback design is adaptive. Figure 5 shows a simple block diagram of a feedback network. $Q(\omega)$ is the filter in the forward path, A is the amplifier gain, τ is the amplifier delay, and β is the feedback amplitude adjustment. Basic operation of feedback is to sample a small portion of the output signal and feed this back to the input 180 degrees out of phase. Now the signal at the input of the amplifier is not only the main signal but also a nonlinear distortion component that is out of phase with respect to the main signal. The distortion cancels with the distortion produced by the amplifier. This is a type of predistortion whereas feedforward is a postdistortion technique since cancellation happens after the amplifier output. This system is essentially a classic feedback control loop; thus, before looking into a specific design, stability needs to be addressed in a feedback system.

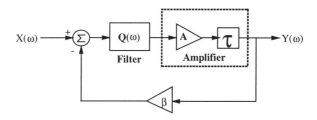

Figure 5. First Order Linear Loop

$Q(\omega)$ is a first order lowpass filter that models the bandwidth of the linearizer. The filter transfer function is given by

$$Q(\omega) = \frac{K\omega_o}{\omega_o + \omega},\tag{1}$$

where ω_o is the 3dB cutoff frequency for the lowpass filter. Stability is usually analyzed by looking at the open loop frequency response. Assuming the open loop transfer function is given by

$$H(\omega) = AQ(\omega) \cdot e^{j\omega\tau} = \frac{AK\omega_n}{\omega_o + j\omega} \cdot e^{-j\omega\tau},\tag{2}$$

where a time delay of τ is seen as a phase shift in the frequency domain expressed as $e^{j\omega\tau}$. The limits of stability are determined by looking at the gain and phase margin of equation 2. Gain margin is the difference between unity gain, $|H(\omega)| = 1$, and the magnitude of $H(\omega)$ when the phase is $180°$. Phase margin is the difference in phase from $180°$ to the phase at the point where $|H(\omega)| = 1$. With AK=G, it can be shown that both the phase margin and gain margin criteria reduce to a single inequality given by

$$G\omega_o\tau < \frac{\pi}{2}. \tag{3}$$

Equation 3 reveals that the product of loop gain, bandwidth, and delay needs to be less than $\pi/2$ to ensure stability. The main contributor to the loop delay is the power amplifier and this delay cannot be decreased; hence, there is a trade off between loop gain and bandwidth. A reduction of the loop gain degrades the performance of the linearizer in order to improve bandwidth. Similarly, if the intermod performance were to be improved, the bandwidth would have to decrease [Boloorian 96].

The other point to note with closed loop feedback systems is that once the loop is closed, the overall gain of the system is degraded by the loop gain. This can be a costly problem especially for the higher frequency applications. In order to compensate for the gain reduction caused by the feedback loop, a larger amplifier is needed. As operating frequency increases, the cost of this larger amplifier can increase exponentially.

2.2.1 Cartesian Feedback

Figure 6 shows a simplified block diagram of a Cartesian feedback linearizer. The figure shows digital data at the input to a modulator with in-phase and quadrature components. Both I and Q channels are lowpass filtered before the quadrature modulator. Some quadrature modulators can directly convert to RF, (as with PCS modems), but for systems like LMDS there would be another upconverter after the quad-modulator to get to 28GHz. The signal is then amplified at RF. The directional coupler at the output feeds a small portion of the amplifier's output signal for the feedback path. The feedback signal is phase and amplitude adjusted, and quad-demodulated to retrieve I and Q bit streams. These I and Q signals are then subtracted from the forward path I and Q bit streams to produce the loop error signal. Then, as described previously, this signal contains the desired signal to be amplified and distortion, which is 180 degrees out of phase with respect to the input signal [Johansson 91, Saleh 82].

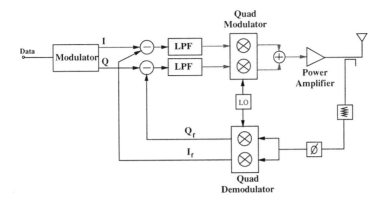

Figure 6. Cartesian Feedback Block Diagram

Cartesian is adaptive and can compensate for drifts in the amplifier characteristics. It is also a complex correction scheme and thus can correct for the IQ impairments of the quadrature modulator. These impairments are the gain imbalance between the I and Q paths, phase imbalance if the quad modulator is not exactly 90 degrees, and DC offsets in the mixers. These non-ideal characteristics of the quad modulator cause spurious signals and degrade the performance of the system [Lohtia 93]. It is noted here that the feedback path is very sensitive to distortion. Cartesian feedback can correct for IQ impairments in the forward path, but not the feedback path. Design on the feedback quadrature modulator is very critical to performance.

Critical design issues like bandwidth and stability need be addressed if Cartesian feedback is to be used for wideband applications. Performance can be limited by the quality of the demodulation circuits. This distortion can be caused from phase noise, carrier leakage, and additional amplitude nonlinearities from the mixers. The bandwidth correction trade off is the most devastation to broadband applications. To date only one Cartesian design reported a bandwidth of 3MHz with 20 dB of correction of the 3rd order distortion and 5MHz for 12dB correction of the 5th order distortion [Johansson 91]. The next generation CDMA systems are 5MHz wide. Linearization bandwidths will need to be on the order of 15 to 25MHz wide.

2.2.2   RF and IF Feedback

Another feedback technique, which is not of the Cartesian type, is a design that uses IF feedback in order to linearize the power amplifier. Figure 7 shows the IF feedback design. This is essentially the same design as RF feedback except the RF cavity filter is replaced with an IF filter and an additional IF gain stage. These components are easier to design and are less expensive at IF. Unlike a cavity filter, this filter can also be of higher order without a significant increase in cost. The IF amplifier can increase the loop gain instead of using the RF amplifier which could decrease performance in either efficiency or spectral regrowth. The phase shifter in the LO path for upconversion is used to ensure negative feedback. The upconverter is a single sideband mixer to take out the unwanted sideband in order to prevent additional distortion from the power amplifier.

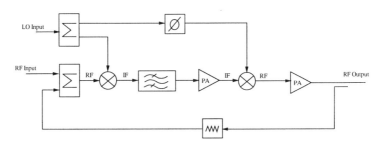

Figure 7. IF Feedback Linearizer

The overall design procedure is first to choose the unity gain frequency, ω_c, based on linearization requirements. Next the zero of the filter is set equal to ω_c and a phase margin is chosen. This phase margin is used to determine the location of the filter poles that gives maximum loop gain for the best linearization performance. This technique reported by Voyce [Voyce 89] was implemented at 450MHz with a 20MHz IF. With a unity gain frequency chosen to be 1.3MHz and a phase margin of 45 degrees, the third order IMD at +/- 500KHz was reduced by 12dB. Although this configuration looks very promising, it is not flexible for the hand held mobile devices and may not be usable for the base station.

Depending on the bandwidth required of the base station, this design may not offer the correction needed because of the gain-bandwidth trade off in feedback designs. There is also need for changing the center frequency of the filter for amplifiers in the handheld mobile devices. Even though the channel bandwidth is usually small, the hand held makes use of all channels. Thus the IF filter needs to be tunable. This would be cheaper than the RF version, but still very costly overall. This does, however, still look promising for fixed wireless loop, where the end user does not change transmit frequencies.

Faulkner proposed using this same technique except to downconvert all the way to baseband where digital techniques can be taken advantage of. Here the filters are digital and controlled by software [Faulkner 95]. This is a significant increase in hardware, a prototype implementation would need to be evaluated to assess whether or not this is practical for a hand held phone and cost-effective for the end user hardware in fixed wireless loop. Even with this modification, this is not practical for a base station as stated earlier.

Other issues that need to be assessed are phase noise introduced by the oscillators and the performance of the SSB mixer at high frequencies. The SSB mixer needs to have good rejection in order to avoid more distortion. Even though this design is adaptive and simple, another downfall common to all of these feedback designs is that gain will be reduced with a closed loop feedback topology. This gain reduction can prove to be costly in the millimeter wave range.

2.2.3   Feedforward + Feedback

One of the drawbacks to feedback is that the overall gain of the system is reduced with a closed loop. The next linearizer combines feedback and feedforward designs so that the gain is not reduced in the process. This design is shown in Figure 8 [McRory 94].

First, two tones are split into upper and lower paths of the loop. The lower path combines with a fed back portion of the output. As a result, only the distortion components are left. The distortion components are filtered and fed into the forward path combiner to form the error signal ε. The cancellation of the main tones in the lower combiner allows an overall reduction of intermodulation distortion at the distortion frequencies without affecting the amplifier gain at the fundamental frequencies.

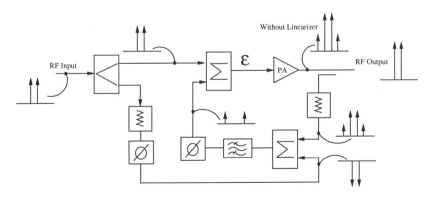

Figure 8. Combined Feedforward and Feedback Design

The filter in the feedback path is used to optimize loop gain and stability. It is not, however, used for cancellation of the main tones. Cancellation of the main tones is done at the combiner. A bandpass Butterworth filter with only one pole was found to provide the best feedback performance.

This topology was demonstrated at 350MHz using two tones that are spaced at 2MHz apart. While maintaining an overall gain of 23dB, the third order IMD's were reduced by 8dB, which was equal to the feedback loop gain. It was mentioned that the design was implemented as proof of concept and was not optimized for performance. Thus, it is expected that reduction of the IMD's can be improved [McRory 94].

This configuration is not only simple, but there are no demodulation circuits or downconverters as with Cartesian feedback and IF feedback. This design doesn't reduce the overall gain of the amplifier, which is by far its greatest advantage. If the performance could be improved, this would be the best choice for implementation in a fixed wireless loop system since it is a fixed bandwidth configuration for the end user. It isn't applicable to the mobile due to bandwidth requirements of the base station and the need to change frequencies at the hand held unit.

2.3   Predistortion

Predistortion is one of the linearizer methodologies that has several different implementations and can be adaptive or not depending upon that implementation.   Figure 9 shows the basic concept behind predistortion. Typically, the predistorter needs to be optimized experimentally.   For example, some predistorters are amplifier stages in which the bias is adjusted until the output of the main amplifier meets some specifications. Predistorters, for which theoretical models are available, take equations that describe the overall AM-AM and AM-PM distortion of the power amplifier and calculate the inverse operation on these equations.   The resulting model is implemented before the power amplifier [D'Andrea 96].

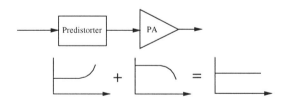

Figure 9. Predistortion Linearizer Block Diagram

2.3.1   General Implementation

In order to make a nonlinear amplifier look linear with a digital predistortion look-up table, only a few simple steps are performed.   The concept is explained first, and then the algorithm implementation follows. Figure 10 shows a typical class A amplifier characteristics. The dotted line shows a corresponding ideal linear response. The goal is to make the predistorter/amplifier combination behave as the linear response. For a given input power (labeled Pin in the figure) the corresponding output power is in the compressed or nonlinear region of the amplifier. At this Pin, the desired output is directly above this point on the dotted line. In order to produce this output power, the predistorter calculates or "looks-up" the input power value needed to operate at the desired output. Thus, for a value Pin as labeled in the figure, the predistorter will change this to Pin_PD. This input power will produce the desired output power from the amplifier [Mingo 97].

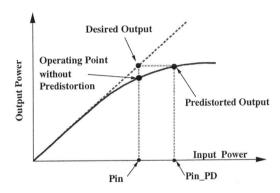

Figure 10. Implementation of Digital Predistortion by LUTs

A predistorter can successfully correct distortion up to the full saturation level of the amplifier. This is the point at which the characteristic levels off and any increase in the input power does not produce an increase in output power. Figure 11 demonstrates the improvement in the amplifiers operating point, as well as the upper limit to linearization. In this figure, the range of input values is represented as a rectangle near the horizontal axis. The range of input power is due to the instantaneous input power of the signal. In any modulation scheme, there is a peak-to-average level that the signal traverses. The probability distribution of the modulating signal will reveal for what percentage of amplitude values the peaks occur. For instance, a CDMA signal amplitude at the base station is Rayleigh distributed with usually a peak-to-average of 12dB.

The first range of input power levels is for an amplifier without a linearizer. It is seen that the peak power cannot go too far into the nonlinear region or the distortion will be too large for transmission. Here, the amplifier must be backed off from its saturation level an amount equal to the peak-to-average level of the signal. With a linearizer, the amplifier is allowed to be at a much higher operating point since the distortion in the peaks can be corrected up to the saturation level. The benefit to this is that a smaller, cheaper amplifier can be built for a given distortion level, and the amplifier's efficiency is much higher.

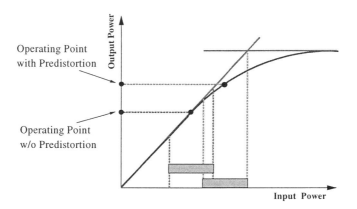

Figure 11. Operating Point Improvement

2.3.2   Digital Implementation

One implementation of digital predistortion that is adaptive is a DSP based design. Figure 12 shows the block diagram of the DSP based linearizer [Nagata 89]. In the forward path, S_i comprises both the complex amplitude and phase of the symbols at the output of the modulator. The RAM block contains predistortion values that compensate for the amplifier's AM-AM and AM-PM distortion. The output is determined from the complex input S_i which are used as memory addresses. The predistorted signal ϕ_i is applied to the forward path where it is quadrature modulated and upconverted to RF. The input signal at the amplifier can be expressed as

$$A(t) = \{S(t) + \phi(t)\}e^{j\omega_c t} . \tag{4}$$

In order for $\phi(t)$ to predistort the amplifier, the following condition of

$$G \cdot S(t) \cdot e^{j\omega_c t} = F[S(t) + \phi(t)] \cdot e^{j\omega_c t} , \tag{5}$$

must be satisfied. G is the amplifier gain and $F[\cdot]$ is the AM-AM and AM-PM characteristics of the amplifier.

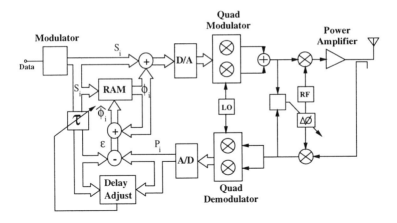

Figure 12. Digital Predistortion

The output of the amplifier is fed back, demodulated, and digitized. The resulting P_i's are compared to the input S_i's. This comparison is for the adaptation of the look up tables, (LUTs). To compensate for the delay through the feedback path, S_i's are delayed by the same amount as in the feedback path. An error signal is formed by the subtraction of the delayed S_i's and the P_i's. The error signal is used for the RAM updates, and ϕ is the new predistortion value. The RAM is updated in an iterative way with the number of iterations depending on the amplifier's nonlinear characteristics and the transmission specifications that have to be met. An interesting point to note here is that the feedback loop to adaptively update the LUT is not a closed loop. If this update is only periodic and not continuous, then stability is no longer an issue. This is one of the benefits this linearizer has over continuous feedback methods.

It was reported that the table update time takes 5 to 10 seconds for 32kbps with a 2bits/symbol modulation rate to achieve a –60dB out-of-band emission specification. This was about a 30dB improvement in the adjacent channel. For changes in the amplifier (due to temperature or bias fluctuations), this update time is acceptable. For changes in the channel, this is too slow. Cavers [Cavers 90] presented a memory reduction version for this configuration and was able to reduce the convergence time to 10 ms.

The drawback to Cavers' solution is that this predistorter is a complex gain predistorter. Here the tables are indexed by the amplitude of the input signal envelope and are thus one-dimensional. Nagata's predistorter is called a mapping predistorter where the LUT is indexed by the complex input signal, hence requiring a 2-D LUT. These tables can correct for both envelope and Cartesian mapped nonlinearities. Cartesian nonlinearities show up in the quadrature modulator when there are gain and phase imbalances between the mixers.   Hence Cavers' technique cannot compensate for quad mixer nonlinearities [Mansell 94, Faulkner 94].

3   SUMMARY AND CONCLUSIONS

Table 1 summarizes the results for each type of linearizer. The results from several other publications are included in the table to give a better representation of the linearizers that have actually been implemented. The performance of the prototypes is of particular importance since they reveal actual results whereas simulations tend to give optimistic results.

Column 6 of this table demonstrates the difference between simulation performance and actual performance. Simulations are usually 10 to 20 dB more optimistic than the reported measured results. The baseband simulations cannot predict some of the non-idealities of high-frequency operation, but are important tools for demonstrating concepts and trends. Analog predistortion and Feedforward could possibly be a short-term solution; but since they are not adaptive, it is not a favorable solution.

The topologies that are adaptive have both advantages and disadvantages for wideband implementation. Cartesian feedback seems to be very popular, but at high frequencies phase noise of oscillators and demodulation problems that are not incorporated into the simulations will inevitably degrade the performance.   The IF feedback approach would be somewhat easier to implement than Cartesian. Here, the IF frequency can be chosen such that downconverting to an intermediate frequency wouldn't have as many implementation problems as downconverting to baseband. The absence of demodulation is definitely an advantage; but the IF frequency cannot be too high as to make the IF filter and amplifier costly. There is also the feedback + feedforward design.   Overall this design looks the best for feedback because it is adaptive, doesn't have any downconversions, doesn't have any demodulation blocks, and doesn't reduce the center frequency gain of the overall linearizer. The latter is probably the strongest advantage to this topology.   Stability is by far the biggest concern for all the feedback topologies; this alone makes it a doubtful candidate for wideband implementation.

Digital predistortion seems to be the best candidate for a wideband adaptive design. It is cost effective since it is a full digital implementation, and digital adaptation is a well-developed field. As for making it wideband, the sample frequency usually is the biggest problem, but recent advances in processors make a wideband design more realizable.

Table 1. Performance Result Summary

Type	Simulation/ Prototype	Center Freq./BW	Stability & Complexity	Two-Tone spacing	Reduction of IMD$_3$	Comments
Cartesian Feedback [Boloorian]	Simulation	Baseband /16 kHz	Cond. Medium to High	4 kHz	40 dB	Stability is a problem for wide-Band.
Cartesian Feedback [Johansson]	Prototype	900 MHz /3 MHz	Cond. Med. To High	1 MHz	20 dB	
IF Feedback [Voyce]	Prototype	450 MHz /1 MHz	Cond. Medium	400 kHz	12 dB	Looks promising
Feedback+ Feed-Forward [McRory]	Prototype	350 MHz /6 MHz	Cond. Medium	2 MHz	8 dB not optimized	Fixed wireless loop a possibility.
Adaptive Feed-Forward [Zhao]	Simulation	1.2 GHz /300 MHz	Cond. High	30 MHz	30 dB	Complex no cost benefit
Feed-Forward [Stewart]	Prototype	950 MHz /1 MHz	Uncond. Low	200 kHz	20 dB	Easy, possibly a short term solution.
Feed-Forward [Kenington]	Simulation	880 MHz /250 kHz	Uncond. Low	22.3 kHz	50 dB	
Digital Predistort. [Nagata]	Simulation	Baseband /64 kHz	Cond. Med. To High	$\pi/4$ QPSK	35 dB	Most promising for wideband applications
Predistort. [Typynamba]	Prototype	6.18GHz /15 MHz	Uncond. Medium	5 MHz	10 dB	Not adaptive, Not a favorable solution

*Cond.=Condtionally, Uncond.=Unconditionally

REFERENCES

Boloorian 96   M. Boloorian and J. McGeehan, "The Frequency-Hopped Cartesian Feedback Linear Transmitter," IEEE Transactions on Vehicular Technology, Vol. 45, No. 4, November 1996, pp. 688 - 706.

Cavers 90   J. Cavers, "Amplifier Linearization Using a Digital Predistorter with Fast Adaptation and Low Memory Requirements," IEEE Transactions on Vehicular Technology, Vol. 39, No. 4, November 1990, pp. 374 - 382.

Cavers 95   J. Cavers, "Adaption Behavior of a Feedforward Amplifier Linearizer," IEEE Transactions on Vehicular Technology, Vol. 44, No. 1, February 1995, pp. 31 - 39.

D'Andrea 96   A. D'Andrea, V. Lottici, and R. Reggianni, "RF Power Amplifier Linearization Through Amplitude and Phase Predistortion," IEEE Transactions on Communications, Vol. 44, No. 11, November 1996, pp. 1477 - 1484.

Dixon 86   J. Dixon, "A Solid-State Amplifier With Feedforward Correction for Linear Single-Sideband Applications," IEEE International Conference on Communications, Toronto, Canada, June 1986, pp. 728 - 732.

Eid 95   E. Eid, F. Ghannouchi, and F. Beauregard, "Optimal Feedforward Linearization System Design," Microwave Journal, November 1995, pp 78 - 84.

Faulkner 94   M. Faulkner and M. Johansson, "Adaptive Linearization Using Predistortion - Experimental Results," IEEE Transactions on Vehicular Technology, Vol. 43, No. 2, May 1994, pp 323 – 332.

Faulkner 95   M. Faulkner, D. Contos, M Johansson, "Linearization of Power Amplifiers Using RF Feedback," IEEE Electronics Letters, Vol. 31, No. 23, Nov. 95, pp. 2023-24.

Johansson 91   M. Johansson and T. Mattson, "Transmitter Linearization Using Cartesian Feedback for Linear TDMA Modulation," 41st IEEE Transactions on Vehicular Technology Conference, St. Louis, MO, May 1991, pp. 155 – 160.

Kenington 91   P. Keningtion, R. Wilkinson, and J Marvill, "Broadband Linear Amplifier Design for PCN Base-Station," 41st IEEE Transactions on Vehicular Technology Conference, St. Louis, MO, May 1991, pp. 155 – 160.

Lohtia 93   A. Lohtia, P. Goud, and C. Englefield, "Adaptive Digital Technique for compensating for analog quadrature modulator / demodulator Impairments," IEEE Pacific Rim Conference on Communications, Computers, and Signal Processing, May 1993, pp.447-450.

Mansell 94   A. Mansell and A. Bateman, "Practical Implementation Issues for Adaptive Predistortion Transmitter Linearization," IEE, London, UK WC2R 0BL, 1994, pp.5/1 – 5.7.

McRory 94   J. McRory and R. Johnston, "An RF Amplifier for Low Intermodulation Distortion," IEEE MTT-S Digest, 1994, pp. 1741 - 1744.

Mingo 97   J. Mingo and A. Valdovinos, "Amplifier Linearization Using a New Digital Predistorter for Digital Mobile Radio Systems," IEEE Transactions, March 1997, pp. 671 – 75.

Nagata 89   Y. Nagata, "Linear Amplification Technique for Digital Mobile Communications," IEEE Vehicular Technology Conference, 1989, pp. 159 - 164.

Saleh 82   A. Saleh and J. Salz, "Adaptive Linearization of Power Amplifiers in Digital Radio Systems," Bell System Technical Journal, Vol. 62, No. 4, April 1983, pp. 1019 - 1033.

Stewart 88   R. Stewart, and F. Tusubira, " Feedforward Linearization of 950 MHz Amplifiers," IEE Proceedings, Vol. 135, Pt. H, No. 5, October 1988, pp. 347 - 350.

Tupynamba 92   R. Tupynamba and E. Camargo, "MESFET Nonlinearities Applied to Predistortion Linearizer Design," IEEE MTT-S Digest, 1992, pp. 955 - 958.

Voyce 89   K. Voyce and J. McCandless, "Power Amplifier Linearization Using IF Feedback," IEEE MTT-S Digest, 1989, pp. 863 - 866.

Zhao 96   G. Zhao, F. Ghannouchi, F. Beauregard, and A. Kouki, "Digital Implementations of Adaptive Feedforward Amplifier Linearization Techniques," IEEE MTT-S Digest, 1996, pp. 543 - 546.

Chapter 15

WIRELESS CHANNEL MODELS- COPING WITH COMPLEXITY

AN MEI CHEN [†] and RAMESH R. RAO
TRW and UCSD

Abstract   In this work we explore two techniques to capture the behavior of wireless channels with mathematically tractable models. The first technique involves state-space aggregation to reduce a large number of states of a Markov chain to a fewer number of states. The property of strong and weak lumpability is discussed. The second technique involves stochastic bounding. These techniques are applied to three different previously published wireless channel models: mobile VHF, wireless indoor, and Rayleigh fading channels. Results show that our stochastic bounding technique can produce simple yet useful upper bounds for the original channel model. We investigate the goodness of these bounds through the performance of higher-layer error control protocols such as stop-and-go and TCP.

[*]This research was supported in part by the National Science Foundation under NSF grant CCR-9714651.

[†]This work was performed while the author was with the Electrical & Computer Engineering Department of the University of California, San Diego.

1.   INTRODUCTION

Errors occur on the wireless channel due to a variety of transmission impairments include fading, mobility, shadowing, interference and noise. Some of these impairments exhibit a significant degree of correlation. Markov models have been widely used in modeling communication channels to characterize this bursty behavior. The quest to develop models that adequately represent real channel behavior and that are mathematically tractable took several directions. The broad range of attempts include an initial two-state Markov model, some modifications to it, models with large number of states, those with restrictions on transitions, a general study on finite-state Markov chains, a subsequent study of infinite state models, higher order Markov models, and other variations [1].

The first attempt at channel modeling was done by Gilbert [2]. Although more general than the i.i.d. model, the two-state channel model is known to be inadequate for the representation of some time-varying channels. One way to overcome this problem is to enlarge the number of states. Fritchman [3] investigated a finite-state Markov chain model with N ($N > 2$) states. The state space was then partitioned into two groups with Group A corresponding to k error-free states and Group B corresponding to $N - k$ error states. Fritchman's model has been applied by many researchers to represent error sequences obtained over the fading channels. Although a large number of states provides a better representation of the channel, the complexity of the model makes subsequent performance analysis intractable. It is therefore worth investigating if we can simplify the representation by reducing the N-state ($N > 2$) Markov chain to a two-state Markov chain and at the same time maintain a better representation of the channel. We shall outline two approaches in this regard, involving the notions of lumpability and hazard rate ordering.

The rest of the paper is organized as follows. In Section II, we briefly summarize the Markov characterization of three different types of digital fading channels that we selected for this study. In particular, we will consider the mobile VHF channel model developed by Swarts [4], wireless indoor channel model by Sivaprakasam [5], and the Rayleigh fading channel by Wang [6]. Section III describes techniques for reducing an N-state ($N > 2$) Markov chain to a two-state Markov chain. In Section IV, we determine if the state space of the three described channel models can be lumped into a two-state Markov chain without any loss of information. When this procedure fails, the channel is stochastically lower and upper bounded by two two-state Markov chains. The goodness of the bounds is assessed by comparing the fraction of slots that the original chain spent in the good and bad states to those of the aggregated chains. We also study the goodness of the bounding process based on higher-layer error control protocols such as stop-and-go and TCP.

2.   WIRELESS CHANNEL MODELS

In this section we briefly describe three previously published channel models. These channel models will be reexamined in section 4. with a view to simplifying their structure.

2.1   MOBILE VHF CHANNEL MODEL

The first model that we will consider for our state-space aggregation study is the digital fading mobile VHF channels developed by Swarts and Ferreira [4]. The model is based on a simplified Fritchman model, where there are $N - 1$ good states and 1 bad state. The model was developed for studying the transmission of data from a motor vehicle traveling through an urban area. Four different modulation schemes, FSK, DPSK, QPSK, and 8-ary PSK, were used to transmit the data, and at the fixed receiving end, the error patterns that occurred were recorded. From these recordings, curves were fitted to the error-free runs to find the parameters of the partitioned Markov chain. In the model setup, the average vehicle speed for city center driving was 40 km/h. An FM transmitter was used to transmit the output of the modems at an RF carrier frequency of 145.2 MHz. The data (baud) rates in bits/s for the four different modulation schemes are FSK: 300 (300), DPSK: 1200 (1200), QPSK: 2400 (1200), and 8-ary PSK: 4800 (1600). The probability transition matrix for the four-state Fritchman channel model for different modulation schemes is shown in Table 15.1.

2.2   WIRELESS INDOOR CHANNEL MODEL

Another channel model which we study is a wireless indoor channel model developed by Sivaprakasam in [5]. Sivaprakasam used a forward-only recursion based hidden Markov modeling (FORM-HMM) technique for modeling burst errors in indoor digital channels. In [5], the Signal Processing Worksystem (SPW) was used to build an end-to-end-communication system to simulate the effects of channel in order to obtain the FORM-HMM models. The simulation measured errors in a QPSK modulation system with a bit rate of 10 kbits/s and sampling frequency of 40 kHz. The number of states in the HMM was six with three good and three bad states. Other simulation parameters include 30dB SNR. The transceivers were separated by 25 meters. The carrier frequency is 1.28 GHz and the mobile unit speed is 1 m/s. The probability transition matrix for the six-state model of the indoor wireless channel is shown in Table 15.2.

2.3 RAYLEIGH FADING CHANNEL MODEL

The two-state Gilbert-Elliot Markov model is not adequate for characterizing the dramatic and continuous channel quality changes in the Rayleigh fading channel. A Markov channel model with more than two states has to be formed to better represent the fading channel. In the work by Wang and Moayeri [6], a finite-state Markov channel (FSMC) was built by partitioning the received instantaneous signal-to-noise ratio (SNR) into K intervals. The channel is represented by K-state stationary Markov chain with state space of $\mathbf{S} = s_1, s_2, \ldots, s_K$. The state space $\mathbf{S}$ is the set of K different channel states with corresponding bit-error-rate ϵ_i, $i \in \{1, 2, \ldots, K\}$. Let S_n be the state at time n, $n = 0, 1, 2, \ldots$. The transition probability $P(i, j)$ and the steady state probability π_i are

$$P(i,j) = P[S_{n+1} = s_j | S_n = s_i], \quad i, j \in \{1, 2, \ldots, K\}, \tag{15.1}$$

$$\pi_i = P[S_n = s_i], \quad i = 1, 2, \ldots, K, \tag{15.2}$$

and

$$P(k) = \sum_{i=1}^{K} P(k, i), \quad k = 1, 2, \ldots, K. \tag{15.3}$$

The model allows the transitions only between adjacent states,

$$P(k, i) = 0, \quad \text{if } |k - i| > 1. \tag{15.4}$$

In a typical multipath propagation environment, the received signal envelope r has the Rayleigh probability density function given by

$$p(r) = \frac{r}{\sigma_r^2} \exp(-\frac{r^2}{2\sigma_r^2}), \quad r \geq 0, \tag{15.5}$$

where σ_r^2 is the variance of the in-phase and quadrature components of the received signal. With additive Gaussian noise of variance σ_0^2, the received instantaneous SNR $\gamma = r^2/2\sigma_0^2$ is distributed exponentially with probability density function

$$p(\gamma) = \frac{1}{\gamma_0} \exp(-\frac{\gamma}{\gamma_0}), \quad \gamma \geq 0, \tag{15.6}$$

where $\gamma_0 = \sigma_r^2/\sigma_0^2$ is the average SNR. The fading characteristics of the signal envelope are determined by the Doppler frequency. The level crossing rate of the instantaneous SNR process γ is the average number of times per unit interval that a fading signal crosses a given signal level Γ. For a random distribution of direction of motion providing a maximum Doppler frequency f_m, the level crossing rate of level Γ for the SNR process is

$$N(\Gamma) = \sqrt{\frac{2\pi\Gamma}{\gamma_0}} f_m \exp(-\frac{\Gamma}{\gamma_0}). \tag{15.7}$$

A finite-state Markov channel model can be built to represent the time-varying behavior of the Rayleigh fading channel.
Let $\Gamma = \{\Gamma_1, \Gamma_2, \ldots, \Gamma_{K+1}\}^T$ be the received SNR thresholds in increasing order with $\Gamma_1 = 0$ and $\Gamma_{K+1} = \infty$. The channel is in state k if the SNR is between Γ_k and Γ_{k+1}. For our study, we consider an example from the paper [6]. In this model, an eight-state Markov channel model was used with the maximum Doppler frequency f_m equals to 10 Hz. The transmission was 10^5 symbols/s and the SNR thresholds are chosen such that the steady state probability of being in any state is equally likely. Table 15.3 shows the transition probability $P(i,j)$ from Wang's analytical model.

3.   STATE-SPACE AGGREGATION TECHNIQUES

In this section, we describe two broad techniques for reducing the complexity of channel models. The first technique for aggregation is applicable to Markov chains while the second technique for bounding is applicable to a larger class of processes.

3.1   LUMPABLE CHAINS

Let X be a homogeneous, irreducible, discrete-time Markov chain on a finite-state space denoted by S, which without loss of generality we assume to be a subset of the natural numbers $\mathcal{N}$, $S = \{1, 2, \ldots, N\}$. The process X has an aperiodic transition matrix P with an equilibrium distribution denoted by a row vector π and initial probability vector $\alpha \in A$ where A is the set of all probability vectors. Let $\Omega = \{\Omega(1), \Omega(2), \ldots, \Omega(M)\}$ with $M < N$ be a fixed partition of the state space S.

With the given process X and the partition Ω, we can associate an aggregated stochastic process Y with values on $\hat{S} = \{1, 2, \ldots, M\}$, defined by $Y_n = m \iff X_n \in \Omega(m), \forall n \geq 0, m = 1, 2, \ldots, M$. Conditions under which the aggregated stochastic process $agg(\alpha, P, \Omega)$ is a homogeneous Markov chain $\forall \alpha \in \mathcal{A}$ were studied by Burke and Rosenblatt in 1958 [8],

Hachgian in 1963 [9], and Kemeny and Snell in 1967 [10]. A chain that has this property is called *strongly lumpable* with respect to the partition Ω. A more general problem is to determine if there exists some initial distributions α such that $agg(\alpha, P, \Omega)$ is a homogeneous Markov chain but not necessarily for every vector in $\mathcal{A}$. The Markov chain, in this case, is called *weakly lumpable* with respect to the partition Ω.

Weakly lumpability was first studied by Kemeny and Snell in 1976 [10], where they showed that it is possible that there exists a proper subset $\mathcal{A}_{\mathcal{M}}$ of the set of all initial probability vectors in $\mathcal{A}$ such that the aggregated process Y is a homogeneous Markov process if and only if $\alpha \in \mathcal{A}_{\mathcal{M}}$. Kemeny and Snell also provided a simple but strong sufficient condition for weak lumpability. Rubino and Sericola in 1991 [12] obtained a characterization of weak lumpability by means of an algorithm which computes the set $\mathcal{A}_{\mathcal{M}}$ of initial distributions and gave necessary and sufficient conditions for weak lumpability. Lumpability of a Markov chain with a denumerable state space was first discussed by Hachgian in 1963 [9]. Rubino and Sericola in [12] gave an algorithm to compute the set $\mathcal{A}_{\mathcal{M}}$.

In the following sections, we will present the algorithm for determining the existence of the proper subset $\mathcal{A}_{\mathcal{M}}$ and give some examples to illustrate the notions of strong and weak lumpability. In our application, the initial distribution vector is always the steady-state distribution vector, i.e. $\alpha = \pi$. Hence, proving that there exists a proper subset $\mathcal{A}_{\mathcal{M}}$ is sufficient, since $\mathcal{A}_{\mathcal{M}} \neq \emptyset \implies \pi \in \mathcal{A}_{\mathcal{M}}$.

3.2   ALGORITHM TO DETERMINE EXISTENCE OF A_M

Before presenting the algorithm for determining the existence of the set $\mathcal{A}_M$, we need the following notation. Let $\pi^{\Omega(l)}$ be obtained from the steady-state distribution, π, by setting to zero all entries corresponding to states which are not in $\Omega(l)$. We define

$$\hat{P}(l, m) = \sum_{i \in \Omega(l)} \pi^{\Omega(l)}(i) P(i, \Omega(m)) \quad l, m \in S, \qquad (15.8)$$

and $\hat{P}_l$ be the lth row of the transition matrix $\hat{P}$, where $\hat{P}$ is the probability transition matrix of the aggregated chain if the original chain is lumpable. For each $l \in \hat{S}$, we define the following matrices

$$\tilde{P}_l = (P(i, \Omega(k)) \quad i \in \Omega(l), k \in \hat{S}, \qquad (15.9)$$

$$H_l = \tilde{P}_l - 1^T \hat{P}_l, \qquad (15.10)$$

with 1 as a row vector where the dimension id defined by context and T denotes the transposition. Let P_l is the submatrix of P constituted by the transition probabilities from the states of $\Omega(l)$ to the states of S,

$$P_l = (P(\Omega(l), \Omega(0)) \ldots P(\Omega(l), \Omega(k)) \ldots). \qquad (15.11)$$

We define the block matrices

$$
\begin{aligned}
H^{[1]} &= H \\
H^{[j+1]} &= Diag(P_l H^{[j]}), \qquad j \geq 1,
\end{aligned}
\qquad (15.12)
$$

and the convex sets, for all $j \geq 1$, $\mathcal{A}^j = \{\alpha \in \mathcal{A} | \alpha H^{[k]} = 0, \text{ for } 1 \leq k \leq j\}$, where $\mathcal{A}^j$ is known as a polytope of $\Re^N$ [14].

To determine the existence of the set $\mathcal{A}_\mathcal{M}$, we can proceed as follows. If the chain is *not* strongly lumpable with respect to the partition $\mathbf{\Omega}$, then $\mathcal{A}_\mathcal{M} = \mathcal{A}$. If not, we first verify whether for all $l \in \hat{S}$ the vector $\pi^{\Omega(l)} P$ is in $\mathcal{A}^1$. If there exists $l \in \hat{S}$ such that $\pi^{\Omega(l)} P \notin \mathcal{A}^1$, then $\pi \notin \mathcal{A}^2$ and $\mathcal{A}_\mathcal{M} = \emptyset$. The loop terminates if the condition of $\mathcal{A}_\mathcal{M} = \emptyset$ is encountered. Otherwise, the loop is continued for N iterations. Note that verifying $\mathcal{A}_\mathcal{M} \neq \emptyset$ at each stage is equivalent to checking $\pi^{\Omega(l)} P H^{[j]} = 0$, for $l \in \hat{S}$ and $j = 1, 2, \ldots, N$. See [12] for additional details. Hence, the algorithm can be expressed as follows:

if X is strongly lumpable **then** $\mathcal{A}_\mathcal{M} := \mathcal{A}$
   stop
else
   $\mathcal{A}_\mathcal{M} :\neq \emptyset$
   for $j = 1$ **to** N
     if $\exists l \in \hat{S} : \pi^{\Omega(l)} P H^{[j]} \neq 0$
       empty := true
       $\mathcal{A}_\mathcal{M} := \emptyset$
       break
     endif
   endfor
endif

3.3   EXAMPLES OF LUMPABLE CHAINS

In this section, we provide some numerical examples to illustrate the concept of strong and weak lumpability.

Example 1-Strong Lumpability [10]

Let the probability transition matrix P be

$$P = \left(\begin{array}{cc|c} 1/2 & 1/4 & 1/4 \\ 1/4 & 1/2 & 1/4 \\ \hline 1/2 & 1/2 & 0 \end{array} \right) \tag{15.13}$$

and the fixed partition $\Omega = \{\Omega(1), \Omega(2)\}$, $\Omega(1) = \{1, 2\}$ and $\Omega(2) = \{3\}$. We note that the probability of moving from either of states 1 or 2 to state 3 is the same, i.e. $P(1,3) = P(2,3) = 1/4$. Hence, the strong lumpability condition is satisfied with the given partition Ω. The probability transition matrix of the aggregated chain is

$$\hat{P} = \left(\begin{array}{cc} \frac{3}{4} & \frac{1}{4} \\ 1 & 0 \end{array} \right). \tag{15.14}$$

Note that the condition of strong lumpability is *not* satisfied for the partition $\mathcal{B} = \{B(1), B(2)\}$ with $B(1) = \{1\}$ and $B(2) = \{2, 3\}$, since $P(2, B(1)) = P(2, 1) = 1/4$ and $P(3, B(1)) = P(3, 1) = 1/2$.

Example 2-Weak Lumpability [11]
Let the probability transition matrix P be

$$P = \left(\begin{array}{c|cc} 1/4 & 1/4 & 1/2 \\ \hline 0 & 1/6 & 5/6 \\ 7/8 & 1/8 & 0 \end{array} \right) \tag{15.15}$$

and $\Omega = \{\Omega(1), \Omega(2)\}$, $\Omega(1) = \{1\}$ and $\Omega(2) = \{2, 3\}$. Applying the algorithm in the previous section, it can be shown that $\mathcal{A}_{\mathcal{M}}$ exists, where $\mathcal{A}_{\mathcal{M}} = \{\lambda(1, 0, 0) + (1 - \lambda)(0, 1/3, 2/3), 0 \le \lambda \le 1\}$.

3.4   STOCHASTIC BOUNDS

When a chain fails to be strongly or weakly lumpable, an alternative approach is to stochastically bound (from above and below) the given chain with geometrically distributed on-off processes. It is of course possible that in some instances no such bounding processes may exist. Fortunately, in [15] Sonderman provides sufficient conditions for the existence of such bounding processes. Before presenting Sonderman's result some definitions on stochastic ordering are necessary.

Let F_x, f_x, F_y and f_y be the cumulative distribution and density functions of two renewal processes X and Y respectively. The process X is said to be greater than the process Y in the sense of *hazard rate* ordering if and only if $\forall x$, $\frac{f_X(x)}{1-F_X(x)} = r_X(x) > r_Y(x) = \frac{f_Y(x)}{1-F_Y(x)}$. Consider now two

On-Off processes X^1 and X^2 characterized by the absolutely continous cumulative distribution functions $F_1, G_1, F_2,$ and G_2 with conditional failure rate functions r_1, s_1, r_2, s_2 respectively and initial conditions p^1 and p^2. Sonderman in [15] showed that if $p^1 \leq p^2$ and $r_1(u) \leq r_2(v)$ and $s_1(u) \leq s_2(v), \forall u, v$, then there exist two semi-Markov processes $\tilde{X}^1$ and $\tilde{X}^2$ on the same probability space such that $\mathcal{L}(\tilde{X}^i) = \mathcal{L}(X^i)$ for $i = 1, 2$ and $P[\tilde{X}^1(t) \leq \tilde{X}^2(t), \forall t \geq 0] = 1$.

To use this result, we would like to set the hazard rates of the bounding processes to be a constant. This corresponds to an exponential or geometric distribution for the bounding on and off periods. If it turns out that the hazard rates of the on and off periods of the actual process are bounded away from zero and infinity, then the upper and lower bounds to the hazard rate will specify the parameters of the geometrically distributed (a.s.) bounding On-Off processes. So the task reduces to (1) ensuring that the actual process is indeed an on-off process and (2) computing the sojourn time probability distributions in the aggregated states. The hazard rates are then computed for the sojourn time probability distributions and the max and min values of the hazard rate curves are identified.

In general, aggregated process may not always evolve as an On-Off process even if the underlying chain is Markovian. Nonetheless, it can be shown that if either one of the two aggregated states contains only a single state or if there exists a single gateway (state) to and from the two aggregated states then the evolution of the aggregated process satisfies the requirements of an On-Off process. A chain is said to have a gateway state if all transitions from the good aggregated state to the bad aggregated state must pass through the singleton good gateway state and similarly for transitions from the bad aggregated states to the good. The proof of this result requires us to establish that good and bad dwell periods are IID. The proof, which is omitted for brevity, follows from conditioning the underlying chain on the singleton gateway states and exploiting the renewal nature of the subsequent evolution.

If these conditions do not exist in the original Markov chain, we develop a looser bounding technique by selective reassignment of the states in each of aggregated sets and finding those state-space partitions that would provide the tightest bounds from above and below of the original process.

Computation of Sojourn Times. Without loss of generality, let $\hat{S} = \{1, 2\}$ and the partition $\boldsymbol{\Omega} = \{\Omega(1), \Omega(2)\}$ of the state space S induces a decomposition of probability transition matrix P into four submatrices and a decomposition of steady state probability distribution vector π into two subvectors:

$$P = \begin{pmatrix} P_{\Omega(1)\Omega(1)} & P_{\Omega(1)\Omega(2)} \\ P_{\Omega(2)\Omega(1)} & P_{\Omega(2)\Omega(2)} \end{pmatrix} \qquad (15.16)$$

and

$$\pi = (\pi^{\Omega(1)} \pi^{\Omega(2)}). \qquad (15.17)$$

We define the *sojourn* of stochastic process X in $\Omega(1)$ as any sequence $X_m, X_{m+1}, \ldots, X_{m+k}$ where $k \geq 1$, $X_m, X_{m+1}, \ldots, X_{m+k-1} \in \Omega(1)$, $X_{m+k} \notin \Omega(1)$ and if $m > 0, X_{m-1} \notin \Omega(1)$. This sojourn begins at time m and finishes at time $m + k$. It lasts k steps. We denote $N_{\Omega(1)}$ as a random variable taking values in $\mathcal{N}$ to represent the sojourn time of the process X in $\Omega(1)$. In [13], the following explicit expression of the sojourn time distribution of $N_{\Omega(1)}$ can be obtained. For any $k \geq 1$, the sojourn time distribution of $N_{\Omega(1)}$ is

$$P(N_{\Omega(1)} = k) = v P_{\Omega(1)\Omega(1)}^{k-1}(I - P_{\Omega(1)\Omega(1)})1^T \quad \text{for any } k \geq 1, \qquad (15.18)$$

where v is a vector given by: $v = (1/K)\pi^{\Omega(1)}(I - P_{\Omega(1)\Omega(1)})$ with K as a normalization constant equal to $\pi^{\Omega(1)}(I - P_{\Omega(1)\Omega(1)})1^T$, I is the identity matrix. The sojourn time distribution, $P(N_{\Omega(2)} = k)$, can be obtained in a similar manner by replacing $\Omega(1)$ with $\Omega(2)$.

4. NUMERICAL RESULTS

We apply the state-space aggregation techniques presented in the previous section to the three different wireless channel models to reduce an N-state $(N > 2)$ Markov chain to a two-state Markov chain. The two-state aggregated chain composes of a *good state* G and a *bad or burst state* B and with the following probability transition matrix

$$P = \begin{pmatrix} p_{gg} & p_{gb} \\ p_{bg} & p_{bb} \end{pmatrix}, \qquad (15.19)$$

where $p_{gb} = 1 - p_{gg} = P[G \rightarrow B]$ and $p_{bg} = 1 - p_{bb} = P[B \rightarrow G]$.

4.1 PERFORMANCE OF BOUNDS

Our analysis indicated that none of the channel models is strongly or weakly lumpable with the specified partition. However, the model developed by Wang for Rayleigh channel is nearly weakly lumpable. We then applied the stochastic bounding technique to all channel models. To proceed with the stochastic bounding technique, we first compute the sojourn time probability distribution. It can be easily shown that the sojourn probability distribution $P[N = k]$ and its corresponding hazard rate $\beta(k)$ in the aggregated state have the following forms:

$$P[N = k] = c_1 a_1^k + c_2 a_2^k + c_3 a_3^k \qquad (15.20)$$

and

$$\beta(k) = \frac{c_1 a_1^k + c_2 a_2^k + c_3 a_3^k}{c_1 a_1^k \frac{1}{1-a_1} + c_2 a_2^k \frac{1}{1-a_2} + c_3 a_3^k \frac{1}{1-a_3}}. \qquad (15.21)$$

The coefficients c_i and a_i are constants. Note that for the VHF channel model, the sojourn time probability distribution of the aggregated *bad* state is geometrically distributed since the aggregated *bad* state contains only a single state of the original chain. Our analysis indicated that the hazard rates of all the considered channel models are bounded away from 0 and ∞. Hence, the original channel models can be stochastically upper and lower bounded by two simpler Markov processes.

Fig. 15.1 shows the hazard rate as a function of time slot k for aggregated *good* states for the mobile VHF channel model for the city environment. The hazard rate of the aggregated *good* state has the maximum and minimum values of 0.382 and 0.001 for all values of k, respectively. Note that these values correspond to $1 - p_{gg}$. The aggregated *bad* state contains a single state of the original chain. The optimistic (upper bound) estimation of the channel can be formed by taking the minimum value of the hazard rate function of the aggregated *good* state. Similarly, the pessimistic (lower bound) estimation of the channel is formed by taking the maximum value of the hazard rate function of the aggregated *good* state. Table 15.4 shows the transition probabilities for both bounds for all channel models.

To cross check the optimistic and pessimistic channel models derived above, we verify if the actual channel model falls between these two channel bounds through simulation as well. The numerical values Table 15.5 indicate that the stochastic bound of the optimistic channel model is very tight since the fractions of good and bad slots are very close to the actual channel model while the lower stochastic bound is looser.

We had to apply the gateway-state stochastic bound technique to the wireless indoor channel model because neither the good nor the bad states are singletons or there exists a gateway state going one aggregated state to another. The original process has the following fixed partition $G = \{1, 2, 3\}$ and $B = \{4, 5, 6\}$. We create the following new partitions for the aggregated *good* and *bad* state respectively: $G_a^1 = \{1\}, G_a^2 = \{2\}, G_a^3 = \{3\}, B_a^1 = \{2, 3, 4, 5, 6\}, B_a^2 = \{1, 3, 4, 5, 6\}$ and

$B_a^3 = \{1, 2, 4, 5, 6\}$. The resulting processes $\tilde{X}_a^1, \tilde{X}_a^1$, and $\tilde{X}_a^3$ corresponding to these partitions are alternating renewal processes which bound the original process from below. Similarly, we create the following partitions: $G_b^1 = \{1\}, G_b^2 = \{2\}, G_b^3 = \{3\}, B_b^1 = \{2, 3, 4, 5, 6\}, B_b^2 = \{1, 3, 4, 5, 6\}$ and $B_b^3 = \{1, 2, 4, 5, 6\}$. The resulting processes $\tilde{X}_b^1, \tilde{X}_b^2$, and $\tilde{X}_b^3$ corresponding to these partitions are alternating renewal processes which bound the original process from above. Table 15.6 shows the numerical result when the gateway-state stochastic bounding technique is applied to the wireless indoor channel model. Note that the tightest upper bound to the original process has the state-space partition of $G = \{1, 2, 3, 4, 6\}$ and $B = \{5\}$, where the tightest lower bound has the following partition: $G = \{1\}$ and $B = \{2, 3, 4, 5, 6\}$.

4.2   GOODNESS OF BOUNDS FROM HIGHER-LAYER PERSPECTIVE

The issue of the quality of the channel bounds can be addressed at different levels. One could develop a measure of the goodness at the physical layer. But, given our interest in data transmission, we have explored via simulations the goodness of the bounding process as seen by stop-and-go ARQ protocol and Transmission and Control Protocol (TCP). In both cases we make the assumption that the channel error models are applicable at the packet level. In principle one could take the bit level error models and extract a packet level error model based on further details of the system. We have not attempted to do so at this time but see [16] for additional details on this aspect.

For the stop-and-go simulation model, the source attempts to transmit one packet at a time to the destination. If the source receives an acknowledgment of the transmitted packet, the next packet in the queue will be transmitted. We assume that the source has an infinite number of packets to transmit and the retransmission timeout is set to be five times the round-trip transmission/propagation delay. Table 15.7 shows the comparison of the stop-and-go protocol performance for different channel models. We can see that the bounds are preserved. We also observed that the performance of the stop-and-go protocol over the optimistic channel model is very close to that of the actual model.

Table 15.8 shows the performance of TCP for the three different channel models. In our TCP simulation model, we assume the following TCP parameters: deviation gain, 0.25; initial Retransmission Timeout (RTO), 1.0 sec.; Karns algorithm, enabled; maximum RTO, 240 secs; minimum RTO, 0.5 sec.; maximum acknowledgment delay, 0.001; Nagel silly window syndrome, enabled; persistence timeout, 1.0 sec; Round Trip Time

(RTT) deviation coefficient, 4.0; RTT gain, 0.125; and receiver buffer capacity, 65 KBytes. The performance of TCP degrades significantly as the channel has longer burst period. This is resulted from both retransmission and TCP's reduction of its congestion window. We also observe that the bounds are still preserved. However, the upper bound is not as tight as observed in the stop-and-go protocol performance.

5. CONCLUSIONS

In this paper, we studied techniques for developing tractable channel models. The first step was to identify accurate channel models. A number of published results that involved large order Markov channel models were identified. The next step was to identify conditions under which the models could be simplified. We showed that if the Markov chain is either strongly or weakly lumpable relative to a particular partition, the evolution of the original channel model can be exactly described by a 2-state Markov chain. When the channel model fails to be strongly or weakly lumpable, a hazard rate ordering technique can be applied to stochastically upper and lower bound the original channel model with a 2-state model. We also study the goodness of the bounding process based on higher-layer error control protocols, stop-and-go and TCP. We found that the bounds were preserved and that in both instances the upper bound was very tight.

Figure 15.1 Hazard rate function for aggregated GOOD state for VHF channel model using DPSK modulation scheme for the city environment.

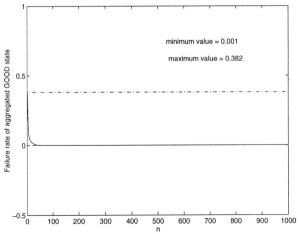

Table 15.1 Transition probabilities for the four-state Fritchman model for the city environment using FSK, DPSK, QPSK, and 8-ary PSK modulation scheme [4].

Prob.	FSK	DPSK	QPSK	PSK
P(1,1)	0.515789	0.467321	0.530715	0.594808
P(2,2)	0.995066	0.900030	0.999489	0.999573
P(3,3)	0.999624	0.999464	0.999950	0.999961
P(4,4)	0.453882	0.385774	0.466127	0.439773
P(4,1)	0.437937	0.416705	0.480427	0.493091
P(4,2)	0.070908	0.124011	0.041837	0.055990
P(4,3)	0.037272	0.073510	0.011609	0.011146
P(1,4)	0.484211	0.532679	0.469284	0.405192
P(2,4)	0.004934	0.099970	0.000511	0.000427
P(3,4)	0.000376	0.000536	0.000050	0.000039

Table 15.2 Transition probabilities for six-state indoor wireless channel model [5].

	1	2	3	4	5	6
1	0.999	0.000	0.000	0.000	0.001	0.000
2	0.000	0.951	0.000	0.000	0.047	0.003
3	0.000	0.000	0.695	0.100	0.023	0.182
4	0.000	0.001	0.514	0.485	0.000	0.000
5	0.077	0.760	0.112	0.000	0.052	0.000
6	0.000	0.045	0.717	0.000	0.000	0.238

Table 15.3 Transition probabilities for eight-state Markov channel with the maximum Doppler frequency $f_m = 10Hz$ [6].

	P(k,k-1)	P(k,k)	P(k,k+1)
$k = 1$	-	0.999359	0.000641
$k = 2$	0.000641	0.998552	0.000807
$k = 3$	0.000807	0.998334	0.000859
$k = 4$	0.000859	0.998306	0.000835
$k = 5$	0.000835	0.998420	0.000745
$k = 6$	0.000745	0.998665	0.000590
$k = 7$	0.000590	0.999048	0.000361
$k = 8$	0.000361	0.999639	-

Table 15.4 Stochastic bounds of hazard rates for different channel models.

	Optimistic		Pessimistic	
	p_{gg}	p_{bb}	p_{gg}	p_{bb}
FSK	0.999624	0.453882	0.611000	0.453882
DPSK	0.999464	0.385774	0.697412	0.385774
QPSK	0.999950	0.466127	0.577654	0.466127
PSK	0.999961	0.439773	0.643322	0.439773
Rayleigh	0.999901	0.997249	0.999168	0.999686

Table 15.5 Comparison of optimistic and pessimistic channel models with the actual channel models using hazard rate ordering technique.

	Optimistic		Actual		Pessimistic	
	% good	% bad	%good	% bad	% good	% bad
FSK	0.9993	6.88e-4	0.9913	0.0087	0.5840	0.4160
DPSK	0.9991	8.71e-4	0.9929	0.0071	0.6700	0.3300
QPSK	0.9999	9.36e-5	0.9970	0.0030	0.5583	0.4417
PSK	0.9999	6.96e-5	0.9976	0.0024	0.6110	0.3890
Indoor	0.9987	0.0013	0.9843	0.0157	0.7219	0.2781
Rayleigh	0.9653	0.0347	0.9168	0.0757	0.2740	0.7260

Table 15.6 Comparison of optimistic and pessimistic channel models with the actual channel models using an exit-gateway bounding technique.

State Decomposition		Percentage of Good Slots		
Good	Bad	Optimistic	Actual	Pessimistic
$\{1,2,3,4,5\}$	$\{6\}$	0.999000	*	*
$\{1,2,3,4,6\}$	$\{5\}$	**0.998800**	*	*
$\{1,2,3,5\}$	$\{4\}$	0.998920	*	*
$\{1,2,3\}$	$\{4,5,6\}$	*	**0.984300**	*
$\{1\}$	$\{2,3,4,5,6\}$	*	*	**0.699600**
$\{2\}$	$\{1,3,4,5,6\}$	*	*	0.018100
$\{3\}$	$\{1,2,4,5,6\}$	*	*	0.002000

Table 15.7 Comparison of Stop and Go protocol performance for optimistic, pessimistic, and actual channel models (d =delay, γ = throughput).

	Optimistic		*Actual*		*Pessimistic*	
	d	γ	d	γ	d	γ
FSK	1.0009	0.9991	1.0092	0.9909	1.7125	0.5839
DPSK	1.0011	0.9988	1.0071	0.9930	1.4925	0.6700
QPSK	1.0003	0.9996	1.0016	0.9984	1.7921	0.5580
PSK	1.0003	0.9997	1.0021	0.9980	1.6359	0.6113
Indoor	1.0014	0.9986	1.0091	0.9908	1.3851	0.7220
Rayleigh	1.0134	0.9877	1.0000	0.9009	1.4599	0.7654

Table 15.8 Comparison of TCP performance for optimistic, pessimistic, and actual channel models (d =delay, γ = throughput).

	Optimistic		*Actual*		*Pessimistic*	
	d	γ	d	γ	d	γ
FSK	0.3283	0.9999	24.20	0.9494	122.30	0.7959
DPSK	0.3294	0.9960	17.98	0.9959	77.33	0.7345
QPSK	0.3258	0.9991	50.33	0.9161	139.80	0.7578
PSK	0.3258	0.9999	29.25	0.9606	125.12	0.8114
Indoor	0.3276	0.9995	4.438	0.9888	15.84	0.9356
Rayleigh	0.3227	0.6000	0.3209	0.4379	0.8977	0.2146

References

[1] L. Kanal and A. Sastry. Models for Channels with Memory and Their Applications to Error Control. *Proc. of the IEEE*, vol. 66, pp. 724-744, July 1978.

[2] E. Gilbert. Capacity of a burst-noise channel. *Bell Systems Tech. Journal*, vol. 39, pp. 1253-1266, Sept. 1960.

[3] B. Fritchman. A Binary Channel Characterization Using Partitioned Markov Chains. *IEEE Trans. on Info. Theory*, vol. IT-13, pp. 221-227, Apr. 1967.

[4] F. Swarts and H. Ferreira. Markov Characterization of Digital Fading Mobile VHF Channels. *IEEE Trans. on Comm.*, vol. COM-43, pp. 997-985, Mov. 1994.

[5] S. Sivaprakasam and K. Shanmugan. An Equivalent Markov Model for Burst Errors in Digital Channels. *IEEE Trans. on Comm.*, vol. COM-43, pp. 1347-1354, 1995.

[6] H. Wang and N. Moayeri. Finite-State Markov Channel- A Useful Model for Radio Communication Channels. *IEEE Trans. on Veh. Tech.*, vol. VT-44, pp. 163-171, Feb. 1995.

[7] M. Zorzi, R. Rao, and L. Milstein. On the Accuracy of a First-Order Markov Model for Data Block Transmission on Fading Channels. *Proc. ICUPC'95*, pp. 221-215.

[8] C. Burke and M. Rosenblatt. A Markovian function of a Markov chain. *Ann. of Math. Stat.*, vol. 29, 1958, pp. 1112-1122.

[9] J. Hachgian. Collapsed Markov chain and the Chapman-Kolmogorov equation. *Ann. of Math. Stat.*, vol. 34, 1963, pp. 233-237.

[10] J. Kemeny and J. Snell. *Finite Markov Chains*. D. Van Nostrand Company, Inc., 1960.

[11] G. Rubino and B. Sericola. On Weak Lumpability in Markov Chains. *J. Appl. Prob.*, No. 26, pp. 446-457, 1989.

[12] G. Rubino and B. Sericola. A finite characterization of weak lumpable Markov processes: Part I- The discrete-time case. *Stochastic Processes Appl.*, vol. 38, pp. 195-204, 1991.

[13] G. Rubino and B. Sericola. Sojourn Times in Finite Markov Processes. *J. Appl. Prob.*, No. 27, pp. 744-756, 1989.

[14] R. T. Rockafellar. *Convex Analysis*, Princeton Univ. Press, Princeton, NJ, 1970.

[15] D. Sonderman. Comparing Semi-Markov Processes. *Mathematics of Operations Research*, Vol. 5, No. 1, Feb. 1980.

[16] M. Zorzi and R. R. Rao On the Statistics of Block Errors in Bursty Channels. *IEEE Transaction on Communications*, Vol. 45, No. 6, Jun. 1997.

Chapter 16

A NEW FRAMEWORK FOR POWER CONTROL IN WIRELESS DATA NETWORKS: GAMES, UTILITY, AND PRICING

D. FAMOLARI, N. MANDAYAM, D. GOODMAN, and V. SHAH
Wireless Information Network Laboratory (WINLAB), Rutgers University

Abstract   We develop a new framework for distributed power control for wireless data based on the economic principles of utility and pricing. Utility is defined as the measure of satisfaction that a user derives from accessing the wireless data network. Properties of utility functions are introduced and a specific function, based on throughput per terminal battery lifetime including forward error control, is presented and shown to conform to those properties. Users enter into a non-cooperative game to maximize their individual utilities by adjusting their transmitter powers. A unique Nash equilibrium for the above game is shown to exist but is not Pareto efficient. A pricing function is then introduced which leads to Pareto improvements for the non-cooperative game.

1.   INTRODUCTION

The technology and business of cellular telephony have made spectacular progress since the first systems were introduced fifteen years ago. With new mobile satellites coming on line, business arrangements, technology and spectrum allocations make it possible for people to make and receive telephone calls anytime anywhere. The cellular telephone success story prompts the wireless communications community to turn its attention to other information services, many of them in the category of "wireless data" communications. One lesson of cellular telephone network operation is that effective radio resource management (power control, channel assignment, and handoffs) is essential to promote quality and efficiency of a system. Radio resource management will be equally, if not more, critical in systems that include high speed data applications. The next generation of wireless systems promise to offer high speed delivery of multimedia information integrated with voice and data. As deployment of these next generation systems draws near, there is an increased urgency to develop and investigate advanced radio resource management techniques centered on wireless data transmission.

This present work focuses on developing a new framework for radio resource management in wireless data networks. The approach proposed here relies on using microeconomic theories that take into account notions of utility and pricing in developing distributed radio resource management algorithms for wireless data services. In particular, a framework for cellular radio power control is developed and investigated, whereby individual users adjust their transmitter powers in order to maximize their utility. It is then shown that system wide increases in utility can be achieved by introducing a penalty, or pricing, function that attempts to police the resource allocations of each user. This type of radio resource management paradigm can be expressed and formulated as a profit maximizing game that attempts to adjust resource levels of each user to maximize the difference between the utility derived and the price exacted.

2.   UTILITY AND WIRELESS DATA

In economics, utility is defined as the level of satisfaction that a person (user) derives from consuming a good or undertaking an activity. The physical channel conditions play different roles in affecting the perceived user quality, and hence the user's level of satisfaction, when it comes to voice and data applications. The factors that affect the perceived quality of voice signals include delay and clarity. Weak signals and interference in the wireless channel cause errors to occur in the speech signal and thus degrade the level of clarity and user satisfaction. Further, as telephone conversations are real time applications, information must be delivered in a timely manner. This condition often precludes the reliance on retransmissions to improve voice quality when frames are received in error. This dependence of voice quality on low and consistent delay differentiates it from data type services. For delay tolerant data applications the commodity is information in the form of bits, and there exist no subjective quality metrics for bits as there are for speech signals. This inherent difference in perceived quality and end user satisfaction points to conceptually different utilities for voice and data systems, and warrants new investigation of the radio resource problem for wireless data.

We believe an explicit economic model can serve as a useful guide in developing a good radio resource mechanism for data users by understanding the factors that affect quality of data communications. To achieve this, we develop a model to quantify the level of satisfaction, or utility, experienced by a data user. This measure, as discussed above, differentiates voice from data traffic. For voice communications the acceptable quality of a connection is often specified by some maximum tolerable BER (usually 10^{-3}). This required BER could be mapped directly to some minimum SIR requirement. Above this required SIR, the voice user has an acceptable connection (utility is "one") and below this required SIR value the voice user does not have satisfactory voice quality (utility is "zero"). Because of this zero-one property of voice utility, a voice user is usually indifferent to small changes in its SIR. A data user has more stringent BER requirements (of the order of 10^{-6} or lower). Further, data users generally communicate information in the form of packets. Due to their intolerance of errors, a packet containing errors will have to be retransmitted, since error correction coding can correct only a few errors. A higher SIR will lead to lesser number of retransmissions, and hence reduce the traffic delay experienced by the data user. Thus for a data user, a higher SIR generally implies a better throughput. This implies that the utility function for a data user is a continuous function of the SIR obtained by the user. Figure 1 shows a conceptual plot of the utilities of a voice user and a data user. Our aim is to formulate a meaningful utility function for data users keeping in mind the dependence of the utility on SIR.

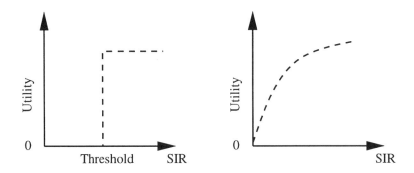

Figure 16.1 Conceptual Utility for Voice and Data

3. DEVELOPING A UTILITY FUNCTION

The definition of utility described above is extremely general and the question is to relate this notion of utility to a wireless data system. In a wireless communication system each user transmits its information over the air using some multiple access system. Since air is a common medium for all the signals, each user's signal acts as an interference to the other users' signals. Further, because of fading, multipath, and other impairments, the radio signal gets distorted by the time it travels from the transmitter to the receiver. A common denominator to account for all these impediments is the SIR of the received signals. Therefore, we can consider the SIR as one of the commodities that a user desires. Further, wireless users typically have a limited battery energy and hence transmitter power is another valuable commodity and users are reluctant to part with it. As the SIR is proportional to the user's transmit power, we see an implicit tradeoff developing. On the one hand, a user desires a strong received signal and the high SIR that can be achieved with a large transmit power. However due to limited battery life a user must also exercise parsimony with its transmitter power. Therefore depending upon the transmit power and the SIR obtained by the user, we would like to formulate an expression to determine the user's satisfaction from using the network.

3.1 PROPERTIES OF A UTILITY FUNCTION

We now stipulate some desirable properties of the utility function with respect to the SIR and transmitter power of a user. We denote the SIR achieved by user j as γ_j, and the transmitter power of user j as P_j. Ideally, a user would prefer to transmit as little power as possible, since battery life of a mobile terminal is limited. At the same time, the user would like its SIR to be as high as possible. Thus we would like the utility function to be a monotonically increasing function of the SIR for a fixed radiated power. However, as in typical communicaiton channels, when the SIR increases beyond large values, the throughput approaches the asymptote of an error free channel. These two facts can be mathematically summarized in the following two properties.

Property 1: The utility function, U_j, is a monotonically increasing function of the user's SIR (γ_j) for a fixed transmitter power (P_j) .

$$\frac{\partial U_j(P_j, \gamma_j)}{\partial \gamma_j} > 0, \quad \forall \gamma_j, P_j > 0. \tag{16.1}$$

Property 2: The utility function, U_j, obeys the law of diminishing marginal utility for high values of γ_j.

$$\lim_{\gamma_j \to \infty} \frac{\partial U_j(P_j, \gamma_j)}{\partial \gamma_j} = 0, \quad P_j > 0. \tag{16.2}$$

We further assume the utility to be a non-negative funciton of SIR, and when $\gamma_j = 0$ we assume that that the utility U_j is zero. In terms of transmit power, the desirable attribute is to conserve battery energy. At the same time, we would like to avoid degenarate situations such as zero transmit power. These attributes can be specified in the form of the following three properties.

Property 3: The utility function, U_j is a monotonically decreasing function of the user's transmitter power, for a fixed SIR γ_j.

$$\frac{\partial U_j(P_j, \gamma_j)}{\partial P_j} < 0, \quad \forall \gamma_j, P_j > 0 \tag{16.3}$$

Property 4: In the limit that the transmitter power tends to zero, the utility value also tends to zero, or:

$$\lim_{P_j \to 0} U_j = 0 \tag{16.4}$$

Property 5: In the limit that the transmitter power tends to infinity, the utility value tends to zero, or:

$$\lim_{P_j \to \infty} U_j = 0 \tag{16.5}$$

The properties that we have specified above will aid us in developing a meaningful utility function.

3.2 DEVELOPING A GENERAL UTILITY FUNCTION

For data communications, information is usually transmitted in the form of packets or frames. The channel introduces errors in the received signal with a probability function that is determined by the received SIR. We will assume that we are able to detect all errors in the received signal through the use of CRC check codes, that we can correct up to and including t errors per frame by employing error correcting codes, and that all data that is unable to be corrected must be retransmitted. The achieved throughput T, in terms of information bits per second, can then be expressed as:

$$T = R_i f(\gamma), \tag{16.6}$$

where the information rate R_i is the rate at which information (total data minus overhead for error correction coding and CRC) is transmitted and $f(\gamma)$ is a measure of the efficiency of the transmission protocol. Note that the efficiency of the protocol should depend on the SIR achieved over the channel and the power of the error correction code t, and its value varies from zero to one (i.e., $f(\gamma) \in [0,1]$). However, we have explicitly dropped t from the notation of the efficiency function because we assume that the error coding is chosen on a system wide basis and that every user employs the same level of error correction. A higher SIR means a lower BER and hence the efficiency of the protocol should be better, while a greater value of t implies that more errors can be corrected without retransmissions, also increasing the efficiency of the protocol.

The measure of satisfaction derived from using a mobile data system clearly is related to the amount of information that a user can transfer in the lifetime of its battery. Therefore, we choose the throughput per battery life as the utility function, as this brings a strong practical and meaningful metric to serve as the definition of utility. If we denote the energy content of user j's battery as E Joules, we can express the utility a user derives from the network as the total number of information bits that it can transmit correctly in the lifetime of its battery, as below:

$$U_j(P_j, \gamma_j) = \frac{E}{P_j} R_i f(\gamma_j). \tag{16.7}$$

Note that the unit of the above utility function is bits. Since $f(\gamma)$ is an increasing nonnegative function of the SIR, it can be shown that the utility function satisfies property (1) through (3). This formulation is very general and applicable to a variety of communication systems. In what follows, however, we illustrate the radio resource allocation concepts with a specific example.

3.3   EXAMPLE UTILITY FUNCTION

In this section we detail a specific utility function that corresponds to a particular CDMA transmission system that employs BCH error correcting codes and interleaving. The specific efficiency function for such a system can be derived as follows. Users are distributed at different distances throughout the cell and access the system through a battery powered device. All users share the same bandwidth and chip rate of W chips per second, and encoded data rate of R (information plus overhead). The SIR of user j can then be written as:

$$\gamma_j = \frac{W}{R} \frac{h_{jk} P_j}{\sum_{i \neq j} h_{ik} P_i + \sigma_k^2} \tag{16.8}$$

where W/R is the processing gain of the spread spectrum system, σ_k^2 is the background noise at the receiver at base station k, and h_{jk} is the path gain of user j to base station k. The path gains, and consequently the location of each user, become significant in determining that user's SIR and therefore its utility.

The SIR value determines the bit error rate seen by user j. Consider the simple case of a Gaussian channel, in which all users share the same frequency band. The bit error rate P_e for a Gaussian channel in terms of the SIR is given as:

$$P_e = Q(\sqrt{\gamma}) \approx \frac{1}{2} e^{-\frac{\gamma}{2}}. \tag{16.9}$$

We incorporate the effects of forward error control into the efficiency function by encoding all information into data frames of length L encoded bits. Forward error control is achieved through the use of error correcting codes that are capable of correcting up to and including t bit errors per frame, and a C bit CRC check is included in each frame for error detection.

The benefits of forward error control come at the expense of a reduction in the amount of information bits present in each frame. In order to provide error correction it is necessary to insert a number of parity bits into each data frame, increasing the overhead. A upper bound for the number of parity bits, m, necessary to correct t corrupted bits per frame can be found, as in [Lin and Costello, 1983], to be $m = t \log_2 (L - C + 1)$. Including the C CRC check bits we can express the ratio of information bits to encoded bits, or code rate $\Gamma(t)$, of the system by

$$\Gamma(t) = \frac{L - C - m}{L} = \frac{L - C - t\log_2(L - C + 1)}{L}. \tag{16.10}$$

Therefore we can represent the information rate of the system R_i as the total data rate R multiplied by the code rate of the system as follows:

$$R_i = R\Gamma(t). \tag{16.11}$$

By assuming sufficient interleaving and hard decision decoding [Proakis, 1989], we can express the frame-success-rate (FSR) experienced by user j of the CDMA system as,

$$FSR_j(\gamma_j, t) = \sum_{i=0}^{t} \binom{L}{i} \left(\frac{e^{-\frac{\gamma_j}{2}}}{2}\right)^i \left(1 - \frac{e^{-\frac{\gamma_j}{2}}}{2}\right)^{L-i}. \tag{16.12}$$

It is very tempting to use the FSR as the efficiency function $f(\gamma)$, however this approach leads to some degenerate solutions as will be shown now. The function does not satisfy Property (4), since at zero SIR the FSR function is not zero, instead it attains a value of $\sum_{i=0}^{t} \binom{L}{i} 2^{-L}$. Therefore, theoretically at zero power a user can obtain infinite utility since it obtains a fixed value of $f(\gamma)$, but is using zero power. This case arises from the fact that when the channel is extremely poor, the worst the receiver can do is to randomly guess at the transmitted bits causing the worst bit error performance to be $P_e = 1/2$. As such there is an extremely small, but non-zero probability that the receiver will guess the right frame. Therefore given enough time the user can get an infinite amount of information through the system at no cost to its battery life. To prevent this degenerate case we would like the efficiency function to approach zero as the power is reduced to zero. One way to do this is to use a probability of bit error function of the form:

$$P_e = e^{-\frac{\gamma}{2}}. \tag{16.13}$$

This will yield a modified frame-success-rate that can be used as the efficiency function and satisfy Properties (1) through (5). So that the utility now becomes

$$U_j(P_j, \gamma_j) = \frac{ER}{P_j} \cdot \frac{L - C - \log_2(L - C + 1)}{L} \cdot \tag{16.14}$$

$$\sum_{i=0}^{t} \binom{L}{i} \left(e^{-\frac{\gamma_j}{2}}\right)^i \left(1 - e^{-\frac{\gamma_j}{2}}\right)^{L-i}$$

$$U_j(P_j, \gamma_j) = \frac{ER\Gamma(t)}{P_j} \cdot f(\gamma_j). \tag{16.15}$$

A plot of this utility function is shown in Figure (16.2). In this figure the level of interference is kept constant, and the user's utility is plotted versus the transmitter power. This utility function still has practical

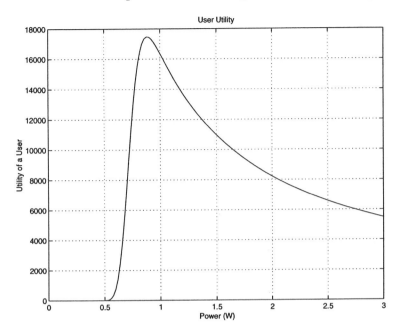

Figure 16.2   Utility function of a data user for fixed interference

resonance for a wireless communications system, in that it serves as an underestimate of the number of information bits that can be transmitted during the lifetime of a terminal's battery. It also is an unbiased modification because the modified function applies equally well to all users of the system regardless of distance, power, or SIR. We feel that although this function does not have an exact practical definition, valuable insights can be derived from its approximate physical significance[1]. We now describe the proposed resource allocation method based on maximizing this measure of utility.

4.   DISTRIBUTED POWER CONTROL BASED ON UTILITY

In this section we present a distributed power control scheme based on the economic framework that we developed in the previous section. We formulate the power control scheme as a distributed utility maximization problem in which each user updates its power to maximize its utility. This is an example of a non-cooperative game, since each user behaves in a selfish way to maximize its own utility. Mathematically, we can formulate this non-cooperative power control game as:

$$\max_{P_i} U_i(P_1, P_2, \ldots, P_i, \ldots, P_K) \quad P_i \in S_i, \quad \forall i = 1, \ldots, K \quad (16.16)$$

where S_i is the strategy space of user i and is the set of values that the user i's power is restricted to. We will assume that the strategy space S_i of each user i is continuous in its power P_i. The joint strategy space $S = S_1 \times S_2 \times \ldots \times S_K$, is the cartesian product of the individual strategy sets for the K users. We will restrict the limits of the strategy space of each user to $S_i = [0, P_{\max}], \forall i = 1, \ldots, K$, where $P_{\max}$ is some maximum power constraint. Equation (16.16) implies that each user maximizes its utility by assuming the power of the other users as exogenously given. The iterations will converge to an equilibrium point (if one exists) where all the (K) users are satisfied with the utility that they are getting. Mathematically we can represent the necessary condition for equilibrium as:

$$\frac{\partial U_j}{\partial P_j} = 0 \quad j = 1, \ldots, K \quad (16.17)$$

which reduces to the following:

$$f(\gamma^*) = \gamma^* \cdot f'(\gamma^*). \quad (16.18)$$

Observe that in the final equation we have dropped the subscript purposely to point to the fact that the first order condition is the same for each user and is only a function of the user's SIR. Each user's utility reaches an (equilibrium) extreme point at some value of SIR given by γ^*. The solution to the non-cooperative equilibrium reduces to finding a power vector $\mathbf{P}^* = [\mathbf{P}_1^*, \mathbf{P}_2^*, \ldots, \mathbf{P}_K^*]$ that satisfies:

$$\left[\mathbf{I} - \gamma^* \frac{R}{W} \mathbf{F}\right] \mathbf{P} = \boldsymbol{\mu}, \quad \mathbf{P} \geq \mathbf{0}, \quad (16.19)$$

where $\mathbf{I}$ is the $K \times K$ identity matrix, and $\mathbf{F}$ is the following nonnegative matrix:

$$F_{ij} = \begin{cases} 0 & i = j \\ h_{ji}/h_{ii} & i \neq j \end{cases} \quad (16.20)$$

and $\boldsymbol{\mu}$ is the vector with elements $\mu_i = \frac{\gamma^* R \sigma_i^2}{W h_{ii}}$. Note that there may be more than one SIR or no SIR value that satisfies the first order equilibrium condition (16.18). We will investigate this issue shortly. For now, assume that there is only one such γ^*.

Theorem 1 *There exists a power vector* $\mathbf{P}^*$ *that is feasible (i.e., it is a solution to equation (16.19)) if the Perron-Frobenius eigenvalue of the matrix F satisfies:*

$$e_{\mathbf{F}} < \frac{W}{R}\frac{1}{\gamma^*}, \qquad (16.21)$$

where $e_{\mathbf{F}}$ *is the Perron-Frobenius eigenvalue defined as the largest eigenvalue of matrix F.*

The proof of the above theorem on the feasibility of power control follows using standard arguments based on the Perron-Frobenius theorem such as those used for fixed target type power control schemes [Yates, 1995]. While equation (16.18) gives only the necessary conditions for the existence of an equilibrium, we will now prove the existence and uniqueness of such an equilibrium.

5. EXISTENCE AND UNIQUENESS OF THE EQUILIBRIUM

In this section we describe the solution to the power control problem. We begin with a definition of the equilibrium for the non-cooperative power control game.

Definition 1: *A Nash equilibrium [Nash, 1950] for the non-cooperative power control game is a power vector* $\mathbf{P}^*$ *such that no single user can improve its utility by a unilateral change in its power.*

Mathematically we can represent a Nash equilibrium as a power vector $\mathbf{P}^*$ such that $U_i(\mathbf{P}^*) \geq \mathbf{U_i}(\mathbf{P'_i}, \mathbf{P^*_{-i}}), \forall P'_i \in S_i, i \in 1, 2, \ldots, K$, where $\mathbf{P_{-i}}$ is a power vector that contains the powers of all the users except the ith user.

Theorem 2 *There exists a unique Nash equilibrium to the non-cooperative power control game in (16.16).*

The proof of the above theorem follows from the fact that the utility function in equation (16.15) is quasi-concave [Roberts and Varberg, 1973] and the use of Debreu's theorem [Debreu, 1952]. This theorem is also equally credited to Fan [Fan, 1952] and Glicksberg [Glicksberg, 1952]. We can see that the Nash equilibrium dictates that every user achieve the same received SIR at the base station. It is interesting to note that the utility maximization approach to power control yields a solution similar to the fixed target type of power control. However, the difference here is that the target, γ^*, is determined by the efficiency function $f(\gamma)$ and the solution to equation (16.18). Further, it can also be shown that a distributed power control scheme using SIR balancing (as in [Yates,

1995]) would also converge to the same unique Nash equilibrium. This result is easily shown by using Yates' framework (see [Yates, 1995]) for power control that is based on the notions of positivity, monotonicity, and scalability of interference functions.

5.1   PROPERTIES OF THE EQUILIBRIUM

Equal Received Powers at the Base Station.   It can be shown, as in [Shah et al., 1998], that in a single cell system the solution to the non-cooperative power control game with forward error correction achieves equal received powers at the base station. That is, at the equilibrium solution all user's have chosen transmitter powers such that their received signals at the base station are all equal. This implies that users who are further from the base station must transmit more power to achieve the same SIR target than those that are closer to the base station.

Pareto Efficiency.   A classical measure of the efficiency of an equilibrium solution is Pareto efficiency [Varian, 1996], which is defined for our wireless system below.

Definition 2: *A power vector P^* is Pareto efficient if and only if there exists no such vector P' such that at least one user achieves higher utility while the other users' utilities remain the same or improve. In other words there exists no power vector $\mathbf{P}'$ such that $U_j(\mathbf{P}') > U_i(\mathbf{P}^*)$, $\forall i = 1, \ldots, K, i \neq j$, where K is the number of users in the cell.*

Alternatively, we can say that if a power vector $\mathbf{P}^*$ is Pareto efficient then for any other allocation $\mathbf{P}'$, if one user improves its utility, then there is at least one other user whose utility is decreased, i.e., for any $U_j' > U_j^*$ there exists at least one i such that $U_i' < U_i^*$. Further, if no such Pareto improvement can be found, then $\mathbf{P}^*$ is Pareto efficient.

Theorem 3 *There exists some $0 < \alpha < 1$ such that for all i, $U_i(\alpha \mathbf{P}^*) > U_i(\mathbf{P}^*)$.*

The proof follows from the strict negativity of the derivative of the utility function with respect to α at the equilibrium solution. Details of the proof for the above theorem can be found in [Famolari, 1999]. The implication of the above theorem is that using utility maximization (or equivalently fixed target type power control) for data users (characterized by non 0-1 utility functions) results in an operating point that is inefficient in power usage.

6. DISTRIBUTED POWER CONTROL BASED ON PRICING

The equilibrium analysis for a non-cooperative game involving only the utility function showed that the equilibrium is Pareto inefficient. In order to improve upon this inefficient equilibrium we would like to influence user behaviour by imposing upon the users a pricing function which we believe will lead to a better equilibrium point. We will first analyze the equilibrium for the non-cooperative power control game that we defined in the last section. This will help us develop a pricing mechanism for the wireless network. A comparison of equilibrium parameters for different users, with and without pricing will give more insight into the effect of the pricing function.

6.1 DEVELOPING A PRICING FUNCTION

In a typical wireless communication system, there may be many users and each will have its own utility, which is dependent on SIR at the base station and its transmitter power. It is evident that as a user tries to increase its power in order to increase its SIR, it causes more intereference to other users thereby degrading their performance. One approach would be to charge a user some price for using the network, and in order to maintain a healthy wireless environment, to charge it for causing harm to other users. We would like to have a pricing function that is a monotonically increasing function of transmitted power. We will now give a more rigorous explanation to support this formulation of the pricing function.

It should be noted that our main intention of introducing pricing is to bring about a Pareto improvement in the utilities of the users and not necessarily as a way of generating revenue for the system. We will now analyze the non-cooperative equilibrium point from the previous section to find out which user is creating the maximum harm in a single cell system with K users. We will denote the base station as y. We define the harm (in terms of the degradation in utility of other users) caused by user j by acting as an interference to user i as follows:

$$C_{ij} = -\frac{\partial U_i}{\partial P_j} P_j. \tag{16.22}$$

Note that C_{ij} has the same units as U_i; in this case bits. The differential term in equation (16.22) is similar to the shadow price concept used in [Mackie-Mason and Varian, 1995] for pricing wireline networks. The total harm done by user j to the rest of the users in cell y is:

$$C_j = \sum_{i \neq j, i=1}^{K} C_{ij}. \tag{16.23}$$

We define $C_{jj} = 0$, i.e., the user does not cause any harm to itself, since it can not act as its own source of interference. Based on the above notion of harm caused by a user we state the following proposition regarding the equilibrium of the non-cooperative power control game.

Proposition 1: *If the path gains of the K users in a cell are in the order $h_{iy} > h_{2y} > \ldots, > h_{Ky}$, then it can be shown that the following is true:*

$$P_1^* < P_2* < \ldots < P_K^* \tag{16.24}$$
$$U_1^* > U_2^* > \ldots > U_K^* \tag{16.25}$$
$$C_1^* < C_2^* < \ldots < C_K^*. \tag{16.26}$$

The proof of the above proposition follows from the fact that all users are received at the same power level at the base station and the definition in equation (16.23).

From Proposition 1, we can draw a number of conclusions which will help us in developing a pricing function. Firstly, a user x with the worst path gain (h_{xy}) causes the most amount of harm to other users (since its C_x is the highest) and vice-versa. Further, the same user transmits the highest power. We would like to price the users according to the amount of harm they are causing to others. Hence we need to develop a pricing function which increases as the transmit power of the user increases. Note that there are a myriad of different functions that have this property. However, as a first step to understanding the effects of pricing, we choose a simple linear pricing function which is proportional to the transmit power of the user. We define the price that we will charge the user as:

$$F_j = \beta P_j, \tag{16.27}$$

where β is a positive constant, and P_j is the power radiated by the mobile. To incorporate the fact that the user is transmitting at a raw bit rate of R bits per second, we would like to factor the constant β in equation (16.27) as $\beta = qR$, where q is a positive constant. The unit of β is bits per watt since we want the units of F_j to be the same as that of U_j. It should be noted that the pricing function that we have formulated is by no means the optimum for our problem. Note that the pricing function her is not in terms of money but rather in terms of abstract credits that are redeemable for network services [Peha, 1997].

Each user would like to have as large a utility as possible, but at the same time would want to pay the smallest possible price for it. This implies that each user would like to maximize the difference between its utility and pricing function as a function of its own power. Since the pricing function is a monotonically increasing function of the user's power, we would expect that if the user maximizes the difference between its utility and price, the equilibrium point would shift to a lower power value than before. Conceptually we illustrate this in Figure 16.3. This figure is somewhat misleading since we consider the effect of the pricing function on only one user, assuming the other users transmit at a constant power level. However, all the users should be using lower powers as well in the

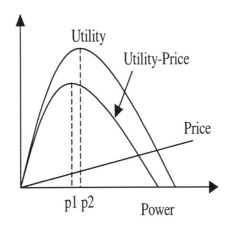

Figure 16.3 Effect of a linear pricing function on the non-cooperative power control game

presence of the pricing function. Recall that from the Pareto optimality discussion for the non-cooperative power control of the utility function we observed that all users would be better off if they reduced their power by some fixed factor α. We can now formulate a distributed power control game based on utility and pricing, where a user's goal is to maximize the difference between utility achieved and price charged as below:

$$\max_{P_i} U_i - F_i \quad P_i \in S_i, \quad \forall i = 1, \ldots, K. \tag{16.28}$$

We show that a pricing function encourages the use of lower transmit powers and thus can lead to Pareto improvements over the non-cooperative game based solely on utility. If the utility-price difference objective function is quasi-concave we can argue as before and show that there exists a Nash equilibrium, but showing the uniqueness of such a solution may be analytically difficult. We do, however, present numerical results that show that an equilibrium does exist and that it does provide Pareto improvements with respect to the the utility only power control game. In the next section we provide some numerical results for the non-cooperative power control game based both on utility and that based on utility and pricing. We then go on to further elaborate the differences in the achieved equilibrium of both games.

7. NUMERICAL RESULTS

The non-cooperative power control game was played for a cell containing users at random distances from the base station. The channel gains of the users was determined by the PCS extension of the Hata propagation model for an urban environment [Rappaport, 1996]. The resulting utilities and powers for a fixed level of error correction equal to four correctable errors per frame are shown in Figures 16.4 and 16.5 respectively. The improvements due to pricing, in terms of increased user utility and reduced transmitter power are clearly evident.

Pricing can be used to achieve a more Pareto efficient equilibrium solution for a non-cooperative power control game for a fixed level of error correction. We can also notice the effects that the non-cooperative game based on pricing has on users located at different distances. When no pricing is applied, we know that an equilibrium solution must provide equal received powers at the base station. This means that users that are located further away must transmit at higher powers to achieve the same SIR, and consequently are expending more battery energy for the same quality channel than the users who are closer to the base station. Thus we see that utility is inversely proportional to user distance.

When pricing is applied to the game the situation for all users improves because pricing provides a more Pareto efficient solution. However the improvements are more pronounced for those users who are closer to the base station. This is due to the fact that the distant users are charged a very high price and reduce their powers more than the closer users. This effect can be seen in Figure 16.6 where after pricing is applied, the SIR's of the users are no longer equal. Those users at the edge of the cell are experiencing much poorer channel conditions then they had enjoyed previously. However, this is compensated for by the drastic reduction in their expended battery energy and the net result is a slightly higher utility. For the users that are closer to the base station, this reduction in interference translates into a conservation of battery power without a subsequent degradation in channel quality. Therefore more information can be sent in the lifetime of the battery and the utility increases.

We also investigate the effect that a moving user has on the non-cooperative game. Figure 16.7 shows a plot of equilibrium utility values of stationary users plotted against the distance of the moving user "1" from the base and distance of the stationary users from the base respectively. In this figure there is no level of error correction, a frame length of 80 bits, and a total of 9 users in the system. As user "1" moves closer to the base station, the stationary users transmit more power and their SIR decreases. As a result, their utility is reduced. The utility of user "1" on the other hand is monotonically increasing as it moves towards the base station. The two effects that are responsible for the utility increase of the moving user are a reduction in transmitter power and an improvement in SIR. The implication of the above behavior is that system concepts like INFOSTATIONS [Goodman et al., 1997] may be a natural solution for providing wireless data services.

8.   CONCLUSION

We have presented a distributed radio resource management scheme for wireless data networks that is based on microeconomic theories. A utility function based on throughput per battery lifetime including forward error control was developed and shown to yield a unique solution to the non-cooperative power control game. This solution, was shown to be equivalent to a fixed-target type power control solution. However, it was shown that the equilibrium solution was Pareto inefficient. A pricing function was developed to achieve Pareto improvements. The effect of pricing applied to the non-cooperative power control game for wireless data networks was shown to yield favorable utility solutions at reduced transmitter powers.

Notes

1. For a general formulation of efficiency and utility functions for BPSK, QPSK, DPSK, and FSK modulation techniques see [Fang et al., 1999]

Acknowledgments

N. Mandayam is supported by the NSF under a CAREER award CCR-9874976

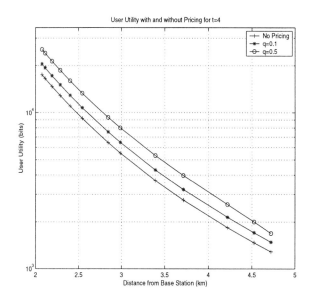

Figure 16.4 User Utility versus Distance with and without Pricing for $t = 4$ Correctable Errors

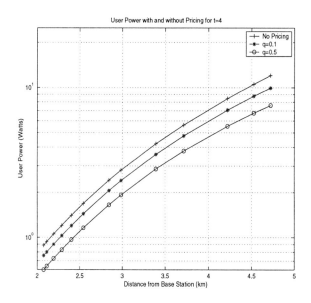

Figure 16.5 User Power versus Distance with and without Pricing for $t = 4$ Correctable Errors

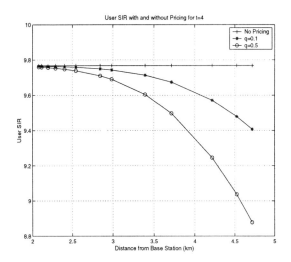

Figure 16.6 User SIR with and without Pricing for $t = 4$ Correctable Errors

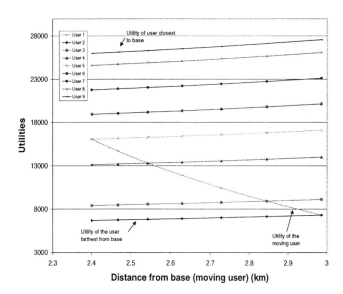

Figure 16.7 Effect a moving user has on the equilibrium utility values

References

[Debreu, 1952] Debreu, G. (1952). A Social Equilibrium Existence Theorem. In *Proceedings of the National Academy of Sciences*, volume 38, pages 886–893.

[Famolari, 1999] Famolari, D. (1999). Parameter Optimization for CDMA Systems. In *Technical Report WINLAB-TR-173*.

[Fan, 1952] Fan, K. (1952). Fixed Point and Minimax Theorems in Locally Convex Topological Linear Spaces. In *Proceedings of the National Academy of Sciences*, volume 38, pages 121–126.

[Fang et al., 1999] Fang, N., Mandayam, N., and Goodman, D. (1999). Joint Power and Rate Optimization for Wireless Data Services based on Utility Functions. In *Proceedings of CISS'99*.

[Glicksberg, 1952] Glicksberg, I. L. (1952). A Further generalization of the Kakutani Fixed Point Theorem with Application to Nash Equilibrium Points. In *Proceedings of the National Academy of Sciences*, volume 38, pages 121–126.

[Goodman et al., 1997] Goodman, D. J., Borras, J., Mandayam, N. B., and Yates, R. D. (1997). INFOSTATIONS : A New System Model for Data and Messaging Services . In *Proceedings of IEEE VTC'97*, volume 2, pages 969–973.

[Lin and Costello, 1983] Lin, S. and Costello, D. (1983). *Error Control Coding: Fundamentals and Applications*. Prentice Hall, Englewood Cliffs, NJ.

[Mackie-Mason and Varian, 1995] Mackie-Mason, J. K. and Varian, H. R. (1995). Pricing congestible network resources. *IEEE Journal on Selected Areas in Communications*, 13(7):1141–1149.

[Nash, 1950] Nash, J. F. (1950). Equilibrium Points in N-Person Games. In *Proceedings of the National Academy of Sciences*, volume 36, pages 48–49.

[Peha, 1997] Peha, J. (1997). Dynamic Pricing as Congestion Control in ATM Networks. In *Proceedings of IEEE GLOBECOM'97*, pages 1367–1372.

[Proakis, 1989] Proakis, J. G. (1989). *Digital Communication*. McGraw-Hill, Inc., USA.

[Rappaport, 1996] Rappaport, T. (1996). *Wireless Communications Principles and Practice*. Prentice-Hall, New York, NY.

[Roberts and Varberg, 1973] Roberts, A. and Varberg, D. (1973). *Convex Functions*. Academic Press.

[Shah et al., 1998] Shah, V., Mandayam, N., and Goodman, D. (1998). Power control for wireless data based on utility and pricing. In *Personal, Indoor and Mobile Radio Communications (PIMRC '98)*, Boston, MA.

[Varian, 1996] Varian, H. (1996). *Intermediate Microeconomics, A Modern Approach*. W. W. Norton and Co., New York, NY.

[Yates, 1995] Yates, R. (1995). A Framework for Uplink Power Control in Cellular Radio Systems. *IEEE Journal on Selected Areas of Communications*, 13(7):1341–1347.

Chapter 17

A UNIFIED APPROACH TO PERFORMANCE EVALUATION OF DIVERSITY SYSTEMS ON FADING CHANNELS

A. ANNAMALAI, C. TELLAMBURA and V. K. BHARGAVA*
Dept. of Electrical and Computer Engineering, University of Victoria, Canada
Computer Science and Software Engineering, Monash University, Australia
Dept. of Electrical and Computer Engineering, University of Victoria, Canada

Abstract   Two theoretical frameworks for computing the average bit or symbol error probability (ASER) of a broad class of coherent, differentially coherent and noncoherent communication systems with microdiversity reception are outlined. We restrict our analyses to four basic predetection diversity schemes, although this method applies to other diversity combining techniques as well. Our novel derivations rely upon the properties of the moment generating function (MGF) or the characteristic function (CHF) of the signal-to-noise ratio at the diversity combiner output, the use of alternative exponential forms for the complementary error functions or the knowledge of the Fourier transforms for the conditional error probabilities, and the application of a Gauss-Chebychev quadrature (GCQ) rule. The analytical expressions are sufficiently general to allow for arbitrary fading parameters as well as dissimilar mean signal strengths across the diversity branches. Moreover, our numerical method is computationally efficient, stable and it approximates the true value of ASER within any degree of accuracy.

*This research was supported in part by a Strategic Project Grant form the Natural Sciences and Engineering Research Council (NSERC) of Canada.

1.   INTRODUCTION

Diversity reception is a classical yet powerful communication receiver technique that provides wireless link improvements at a relatively low cost. The underlying premise is that if two or more statistically independent (or at least highly uncorrelated) replicas of a signal are received over multiple diversity branches with comparable strengths, then it is improbable that all these signals will fade simultaneously. Besides mitigating the deep fades experienced in wireless channels, diversity methods play a crucial role in minimizing the transmit power requirements, particularly in the reverse link, because the battery capacity of handheld subscriber units is usually limited. It also reduces the penalty in signal-to-noise ratio (SNR) due to co-channel interference.

To take advantage of the improvement in signal statistics due to diversity, several combining techniques have been proposed in the research literature, and they can be classified into two groups, namely switched combining and gain combining (see [1] and [2]). Four such techniques are considered here: maximal-ratio combining (MRC — the best known linear combining method), equal-gain combining (EGC), ideal selection diversity (SDC — perhaps the simplest combining method) and one type of switch-and-stay selection diversity (SWC) system.

Let us consider an L-branch diversity combiner as shown in Figure 17.1. Each of the L antennas is receiving a locally coherent signal with statistically independent random amplitudes and random phases. Therefore, the received signal at the k-th antenna, $s_k(t)$, may be written as,

$$s_k(t) = \alpha_k e^{j\varepsilon_k} e^{j[2\pi f_c t + \theta(t)]} \tag{17.1}$$

where f_c is the carrier frequency, $\theta(t)$ is the information signal (desired phase modulation), ε_k is the random phase uniformly distributed between $[0, 2\pi)$, and α_k is the random fading amplitude process.

The additive zero-mean noise component, n_k, is assumed to be independent of the signal and uncorrelated with the noise in any other branch. The composite signal plus noise in each branch is then multiplied by a voltage gain factor g_k and then summed in a linear combiner with replicas of the signal from all the other diversity branches. Thus, the resultant signal amplitude at the predetection diversity output is

$$y_0(t) = \sum_{k=1}^{L} g_k[s_k(t) + n_k(t)] \tag{17.2}$$

The values of the gain factors depend on the type of combining that is employed. For instance, the SDC measures the SNR at each branch (i.e. antenna) and selects the branch with the highest SNR value. However, the ideal SDC may not be practical for radio links that use continuous transmission

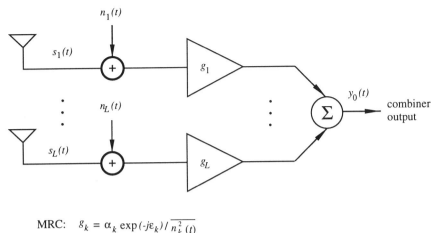

$$\text{MRC:} \quad g_k = \alpha_k \exp(-j\varepsilon_k)/\overline{n_k^2(t)}$$

$$\text{EGC:} \quad g_k = \exp(-j\varepsilon_k)$$

$$\text{SDC:} \quad g_k = \begin{cases} 1 \text{ if } k = l \\ 0 \text{ if } k \neq l \end{cases} \quad \text{where} \quad \frac{\alpha_l^2}{\overline{n_l^2(t)}} = max\left\{\frac{\alpha_k^2}{\overline{n_k^2(t)}}\right\}$$

$$\text{SWC:} \quad g_k = \begin{cases} 1 \text{ if } k = l \\ 0 \text{ if } k \neq l \end{cases} \quad \text{and} \quad \frac{\alpha_l^2}{\overline{n_l^2(t)}} \geq \xi$$

Figure 17.1 Predetection diversity systems.

(e.g., FDMA systems) because it requires continuous monitoring of all the diversity branches. This problem can be circumvented by adopting a suboptimal switched diversity scheme so that the rate of branch switching is reduced (which translates into a reduction of transient effects caused by switching). In [3], Abu-Dayya and Beaulieu proposed a variation of the switch-and-stay diversity (SWC) strategy where the antenna switch is activated in the next switching instant as long as the measured local power in the current antenna is below the threshold level (i.e., envelope of the received signal need not necessarily cross the threshold in the negative direction). Therefore, the SWC scheme does not require comparison of present samples with past samples. In MRC, rather than selecting the single strongest signal, all the diversity branches are first co-phased, and then weighted in proportion to their SNR before summing. Different from MRC, the co-phased signals in an EGC combiner are simply added without having to weight the current SNR level of each signal. Therefore, MRC is known to be optimum in the sense that it yields the best statistical reduction of fading of any linear diversity combiner as well as provides the highest average output SNR.

While there are several excellent papers on the subject of fading channels and diversity reception, with many cases having been thoroughly analyzed, the unified approach adopted in this chapter results in "clean" derivations for the error probability expressions, and they are also numerically efficient. Two of the important techniques recurring throughout this chapter is the use of moment generating functions (MGFs) or the characteristic functions (CHFs), and the application of Gauss-Chebychev sampling to obtain rapidly converging series expressions for the average bit or symbol error rates (ASER).

The organization of this chapter is as follows. Section 17.2 details the derivation of a generic conditional error probability (conditioned on the instantaneous SNR at the output of the predetection combiner) for a broad class of binary and M-ary (multi-level) modulation formats. The average error probability performance of various coherent, differentially coherent and noncoherent communication systems with predetection MRC, SDC and SWC are derived in Section 17.3 using the MGF method. The final expressions are sufficiently general to allow for arbitrary fading parameters as well as dissimilar mean signal strengths across the diversity branches. Yet another general approach for unified analysis of diversity systems on fading channels (CHF method) is discussed in Section 17.4. In particular, exact analytical expressions for the predetection EGC diversity receivers are derived. Finally, the main points are summarized in Section 17.5.

2. ERROR PROBABILITY FOR BINARY AND M-ARY SIGNALLING CONSTELLATIONS IN AN AWGN CHANNEL

Table 17.1 summarizes the instantaneous symbol error rate for a wide range of modulation schemes in an AWGN channel. One can immediately recognize that these expressions may be categorized into one of the four general forms: (a) function of only the erfc $(\sqrt{\gamma})$; (b) functions of both erfc $(\sqrt{\gamma})$ and erfc$^2(\sqrt{\gamma})$; (c) exponential form; and (d) finite-range integral with exponential integrand. Using the alternative exponential forms for the complementary error functions, i.e., erfc $(\sqrt{\gamma}) = \frac{2}{\pi} \int_0^{\pi/2} \exp[-\gamma \csc^2(\theta)]d\theta$ [4]–[8] and erfc$^2(\sqrt{\gamma}) = \frac{4}{\pi} \int_0^{\pi/4} \exp[-\gamma \csc^2(\theta)]d\theta$ [9]–[11], we can show that the conditional error probability for a broad class of binary and two-dimension signalling constellations (for coherent, differentially coherent and noncoherent modulation formats) as a special case of the following generic form [12],

$$P_S(\varepsilon|\gamma) = \sum_u \int_0^{\eta_u} a_u(\theta) \exp[-\gamma b_u(\theta)]d\theta \qquad (17.3)$$

where $a_u(\theta)$ and $b_u(\theta)$ are coefficients independent of the instantaneous SNR per symbol γ at the output of the predetection deversity combiner, but may be

Table 17.1 Instantaneous SER of several common modulation schemes.

Modulation Scheme	Conditional Error Probability $P_S(\varepsilon|\gamma)$
Coherent binary signalling:	
(a) Coherent PSK	$\frac{1}{2}\mathrm{erfc}\left(\sqrt{\gamma}\right)$
(b) Coherent detection of differentially encoded PSK	$\mathrm{erfc}(\sqrt{\gamma}) - \frac{1}{2}\mathrm{erfc}^2\left(\sqrt{\gamma}\right)$
(c) Coherent FSK	$\frac{1}{2}\mathrm{erfc}\left(\sqrt{\gamma/2}\right)$
Noncoherent binary signalling:	
(a) DPSK	$\frac{1}{2}\exp(-\gamma)$
(b) Noncoherent FSK	$\frac{1}{2}\exp(-\gamma/2)$
Quadrature signalling:	
(a) QPSK	$\mathrm{erfc}\left(\sqrt{\gamma}\right) - \frac{1}{4}\mathrm{erfc}^2\left(\sqrt{\gamma}\right)$
(b) MSK	$\mathrm{erfc}\left(\sqrt{\gamma}\right) - \frac{1}{4}\mathrm{erfc}^2\left(\sqrt{\gamma}\right)$
(c) $\pi/4$-DQPSK [13]	$\frac{1}{2\pi}\int_0^\pi \frac{\exp[-\gamma(2-\sqrt{2}\cos\theta)]}{\sqrt{2}-\cos\theta}d\theta$
Multilevel signalling:	
(a) Square QAM	$2q\,\mathrm{erfc}\left(\sqrt{p\gamma}\right) - q^2\mathrm{erfc}^2\left(\sqrt{p\gamma}\right)$ where $q = 1 - 1/\sqrt{M}$, $p = 1.5\log_2 M/(M-1)$
(b) MPSK	$\frac{1}{\pi}\int_0^{\pi-\pi/M} \exp\left(\frac{-\gamma\sin^2(\frac{\pi}{M})\log_2 M}{\sin^2\theta}\right)d\theta$
(c) MDPSK [14]	$\frac{\sin(\frac{\pi}{M})}{\pi}\int_0^{\pi/2} \frac{\exp\{-\gamma\log_2 M[1-\cos(\frac{\pi}{M})\cos\theta]\}}{1-\cos(\frac{\pi}{M})\cos\theta}d\theta$ or $\frac{1}{\pi}\int_0^{\pi-\pi/M} \exp\left(\frac{-\gamma\sin^2(\frac{\pi}{M})\log_2 M}{1+\cos(\frac{\pi}{M})\cos\theta}\right)d\theta$
(d) Two-dimension M-ary signal constellations [4][15]	$\frac{1}{2\pi}\sum_{k=1}^{S} Pr(S_k)\int_0^{\eta_k} \exp\left(\frac{-\gamma c_k\sin^2(\Psi_k)}{\sin^2(\theta+\Psi_k)}\right)d\theta$ where S is the number of signal points, and $Pr(S_k)$ is the *a priori* probability that the kth signal point is transmitted.

dependent on θ. Alternatively, we can express the conditional error probability in terms of the combiner output envelope $\vartheta = \sqrt{\gamma}$ as

$$P_S(\varepsilon|\vartheta) = \sum_u \int_0^{\eta_u} a_u(\theta) \exp[-\vartheta^2 b_u(\theta)] d\theta \tag{17.4}$$

3. UNIFIED ANALYSIS OF ASER USING MGF

The ASER in the fading channels with an L-branch diversity may be derived by averaging the conditional error probability over the probability density function (PDF) of SNR at the output of the diversity combiner $p_\gamma(\cdot)$ in a specified fading environment, i.e.,

$$P_S(\varepsilon) = \int_0^\infty P_S(\varepsilon|\gamma) p_\gamma(\gamma) d\gamma \tag{17.5}$$

Substituting (17.3) into (17.5), interchanging the order of the integration, and then recognizing that the Laplace transform (LT) integral of the PDF of γ yields the MGF of the resultant SNR at the combiner output, we get

$$P_S(\varepsilon) = \sum_u \int_0^{\eta_u} a_u(\theta) \phi_\gamma[b_u(\theta)] d\theta \tag{17.6}$$

where $\phi_\gamma(\cdot)$ is the MGF of SNR at the output of the diversity combiner. If the conditional error probability is in the exponential form, $P_S(\varepsilon|\gamma) = a \exp(-b\gamma)$, then (17.6) reduces to a closed-form expression. For instance, the average bit error rate performance for binary DPSK and noncoherent FSK with dual-branch SC is given by

$$P_S(\varepsilon) = a\phi_\gamma(b) \tag{17.7}$$

where $\{a = 1/2,\ b = 1\}$ for binary DPSK and $\{a = 1/2,\ b = 1/2\}$ for binary orthogonal FSK.

If the conditional error probability is of the form $P_S(\varepsilon|\gamma) = a\,\mathrm{erfc}\,(\sqrt{b\gamma})$ (e.g., coherent binary PSK or FSK), then the ASER can be expressed as

$$P_S(\varepsilon) = \frac{2a}{\pi} \int_0^{\pi/2} \phi_\gamma\left(b \csc^2 \theta\right) d\theta \tag{17.8}$$

Using variable substitution $t = \cos(2\theta)$ in (17.8) and then applying GCQ formula [16, pp. 889], we obtain a rapidly converging series expression for the ASER:

$$\begin{aligned} P_S(\varepsilon) &= \frac{a}{\pi} \int_{-1}^1 \phi_\gamma\left[b \csc^2\left(\frac{1}{2}\cos^{-1} t\right)\right] \frac{dt}{\sqrt{1-t^2}} \\ &= \frac{a}{\pi} \sum_{i=1}^n \phi_\gamma\left[b \csc^2\left(\frac{(2i-1)\pi}{4n}\right)\right] + R_n \end{aligned} \tag{17.9}$$

where n is a small positive integer, and the remainder term R_n can be bounded using procedure outlined in [6] or [17].

Similarly for $P_S(\varepsilon|\gamma) = a\operatorname{erfc}(\sqrt{b\gamma}) - \operatorname{cerfc}^2(\sqrt{b\gamma})$ (e.g., square QAM, quaternary PSK, coherent detection of differentially encoded PSK), the ASER is given by

$$
\begin{aligned}
P_S(\varepsilon) &= \frac{2a}{\pi} \int_0^{\pi/2} \phi_\gamma \left(b\csc^2\theta\right) d\theta - \frac{4c}{\pi} \int_0^{\pi/4} \phi_\gamma \left(b\csc^2\theta\right) d\theta \\
&= \frac{1}{n} \sum_{i=1}^{n} a\phi_\gamma \left[b\csc^2\left(\frac{(2i-1)\pi}{4n}\right)\right] - c\phi_\gamma \left[b\csc^2\left(\frac{(2i-1)\pi}{8n}\right)\right] + R_n
\end{aligned}
$$

$$(17.10)$$

Notice that the ASER formulas [see (17.6)-(17.10)] are expressed in terms of only the MGF of the resultant SNR, $\phi_\gamma(\cdot)$, which is dependent of the type of the diversity combining employed and the fading channel model. Therefore, in order to evaluate the error performance of any modulation formats listed in Table 17.1 in conjunction with a specific diversity combining scheme (MRC, SDC or SWC), we now need to determine only the $\phi_\gamma(\cdot)$ for each type of the diversity combiners. Hence, in the following we will directly derive the MGF of SNR at the MRC, SDC and SWC combiner outputs over generalized fading channels.

3.1   MAXIMAL-RATIO DIVERSITY

It should be highlighted that the direct evaluation of (17.5) involves an L-fold convolution integral. Therefore, it is apparent that the MGF method is an attractive proposition for a rapid calculation of the ASER with MRC diversity receivers, especially when L is large. From the definition of the MGF, we can easily show that

$$
\phi_\gamma^{(MRC)}(s) = E\left[\exp\left(-s\sum_{k=1}^{L}\gamma_k\right)\right] = \prod_{k=1}^{L} \phi_k(s) \qquad (17.11)
$$

where $E[x]$ denotes the expectation (statistical average) of the random variable x, L is the diversity order, and $\gamma_k = (E_S/N_0)\alpha_k^2$ where E_S/N_0 is the symbol energy-to-Gaussian noise spectral density ratio. In other words, the MGF of SNR at the MRC combiner output is simply the product of the MGF of SNR at each of the statistically independent diversity branches.

At this juncture, it is important to highlight that we do not impose any restrictions on the mean received signal strength and/or fading severity index for each of the L diversity branches. In fact, the fading statistics for different diversity branches may even be modelled using different families of distributions. The MGFs of the signal power $\phi_k(\cdot)$ for all common fading distributions are listed in the Appendix at the end of this chapter. The approach presented here can be easily extended to analyze the performance of MRC in arbitrarily correlated Nakagami-m and Rayleigh fading channels. Details can be found in [18].

3.2   SELECTION DIVERSITY

The cumulative distribution function (CDF) of SNR at the SDC combiner output is simply the product of the CDF of each of the individual diversity branches, i.e.,

$$F_\gamma^{(\text{SDC})}(y) = \prod_{k=1}^{L} \text{Prob}\,[\gamma_k \le y] = \prod_{k=1}^{L} F_k(y) \qquad (17.12)$$

where $F_k(\cdot)$ denotes the CDF of the received signal power in the k-th branch. Exploiting the relation between the MGF and CDF (through the use of the Laplace transform of a derivative property), the MGF of γ is given by

$$\begin{aligned}
\phi_\gamma^{(\text{SDC})}(s) &= \int_0^\infty e^{-sy} \frac{d}{dy} F_\gamma^{(\text{SDC})}(y)\,dy \\
&= s \int_0^\infty e^{-sy} F_\gamma^{(\text{SDC})}(y)\,dy - F_\gamma^{(\text{SDC})}(0) \qquad (17.13)
\end{aligned}$$

For instance, the MGF of SNR with SDC in Rician and Nakagami-m fading are given by (17.14) and (17.15), respectively:

Rician:

$$\phi_\gamma^{(\text{SDC})}(s) = \sum_{i=1}^{N} \omega_i \prod_{k=1}^{L} \left\{ 1 - Q\left[\sqrt{2K_k},\, \sqrt{\frac{2(1+K_k)\chi_i}{s\bar{\gamma}_k}} \right] \right\} + R_N$$

$$(17.14)$$

Nakagami-m:

$$\phi_\gamma^{(\text{SDC})}(s) = \sum_{i=1}^{N} \omega_i \prod_{k=1}^{L} \left\{ \frac{1}{\Gamma(m_k)} \Psi\left[m_k,\, \frac{m_k \chi_i}{s\bar{\gamma}_k} \right] \right\} + R_N \qquad (17.15)$$

where $Q(\sqrt{2a}, \sqrt{2b}) = \int_b^\infty \exp(-t-a)I_0(2\sqrt{at})\,dt$ is the Marcum-Q function [2], χ_i and ω_i are the i-th abscissa and weight of the N-th order Laguerre polynomial [16], and $\Psi(a, x) = \int_0^x t^{a-1} \exp(-t)\,dt$ denotes the incomplete Gamma function [19].

3.3   SWITCHED DIVERSITY

The analysis of the switch-and-stay selection (SWC) diversity system is more involved compared to the ideal SDC scheme. In [20], the MGF for the SNR at the output of a dual-branch SWC combiner (statistically independent but with dissimilar fading statistics) is derived using a discrete-time model,

$$
\begin{aligned}
\phi_\gamma^{(SWC)}(s) &= A(\xi)[\lambda_1(\xi, s) + \phi_2(s)F_1(\xi)] \\
&\quad + [1 - A(\xi)][\phi_1(s)F_2(\xi) + \lambda_2(\xi, s)] \qquad (17.16)
\end{aligned}
$$

where ξ is the fixed switching threshold and $A(\xi) = \frac{F_2(\xi)}{F_1(\xi)+F_2(\xi)}$. Notation $f_k(\cdot)$, $F_k(\cdot)$ and $\phi_k(\cdot)$ correspond to the PDF, CDF and MGF of the received signal power in the k-th antenna. The marginal MGF $\lambda_k(\xi, s) = \int_\xi^\infty \exp(-su)f_k(u)du$ can be written in a closed-form for the Rician, Rayleigh and Nakagami-m channels:

Rician:

$$
\begin{aligned}
\lambda_k(\xi, s) &= \frac{1 + K_k}{s\Omega_k + K_k + 1} \exp\left(\frac{-sK_k\Omega_k}{s\Omega_k + K_k + 1}\right) \\
&\quad \times Q\left(\sqrt{\frac{2K_k(K_k + 1)}{s\Omega_k + K_k + 1}}, \sqrt{\frac{2(s\Omega_k + K_k + 1)\xi}{\Omega_k}}\right) \qquad (17.17)
\end{aligned}
$$

Rayleigh:

$$
\lambda_k(\xi, s) = \frac{1}{s\Omega_k + 1} \exp\left[-\frac{\xi}{\Omega_k}(s\Omega_k + 1)\right] \qquad (17.18)
$$

Nakagami-m:

$$
\lambda_k(\xi, s) = \frac{1}{\Gamma(m_k)}\left(\frac{m_k}{m_k + s\Omega_k}\right)^{m_k} \Gamma[m_k, \xi(s + m_k/\Omega_k)] \qquad (17.19)
$$

where $\Gamma(a, x) = \Gamma(a) - \Psi(a, x) = \int_x^\infty \exp(-t)t^{a-1}dt$.

When both the diversity branches are assumed to be independent and identically distributed, then (17.16) reduces to:

$$
\phi_\gamma^{(SWC)}(s) = \phi_1(s)F_1(\xi) + \lambda_1(\xi, s) \qquad (17.20)
$$

While the SWC strategy has the ability to reduce the transient effects compared to the ideal SDC scheme, its operation and the attainable diversity gain are largely dependent on the selection of the fixed threshold level. For instance, there will be constant switching between two antennas if a large value is assigned to the switching threshold because the probability of the received signal power exceeding ξ will be small. In this case, the performance of SWC will be equivalent to the performance of a diversity branch selected in random, which resembles the behavior for no diversity case. On the other extreme (i.e., the value of ξ is set to be very small), the SWC combiner will be stuck in one of the diversity branches because the likelihood of received signal power staying above the specified threshold increases. Once again, the performance of the SWC will be close to the single diversity branch case. It is evident that the proper choice of ξ will minimize the average error probability.

In [20], an analytical expression to compute the optimal switching threshold ξ^* in a correlated fading channel with dissimilar signal strength across the two diversity branches is derived. For the special case of statistically independent and identically distributed diversity branches, the general expression collapses to

$$P_S(\varepsilon|\xi) = P_S^{(1)} = P_S^{(2)} \qquad (17.21)$$

where $P_S^{(k)}$ corresponds to the ASER of the k-th diversity branch. Next we will identify three special cases of the conditional error probability $P_S(\varepsilon|\gamma)$ which lend themselves to closed-form formulas for the calculation of ξ^* in all common fading environments:

Case (a): If $P_S(\varepsilon|\gamma) = a \exp(-b\gamma)$, then $\xi^* = \dfrac{-\ln[\phi_x(b)]}{b}$.

Case (b): If $P_S(\varepsilon|\gamma) = a\mathrm{erfc}(\sqrt{b\gamma})$, then $\xi^* = \dfrac{1}{b}\left[\mathrm{erfcinv}\left(\dfrac{P_S^{(1)}}{a}\right)\right]^2$.

Case (c): If $P_S(\varepsilon|\gamma) = a\mathrm{erfc}(\sqrt{b\gamma}) - c\mathrm{erfc}^2(\sqrt{b\gamma})$,

　　　　then by solving a quadratic equation formed using (17.21),

　　　　we get $\xi^* = \dfrac{1}{b}\left\{\mathrm{erfcinv}\left[\dfrac{a-\sqrt{a^2-4cP_s^{(1)}}}{2c}\right]\right\}^2$.

4.　　UNIFIED ANALYSIS OF ASER USING CHF

In Section 17.3, we tried to unify the performance analysis of the diversity receivers on fading channels by putting the conditional error probability in a desirable form (i.e., a finite range integral with exponential integrand) so that one can directly apply the Laplace transform to perform the averaging over the PDF of γ. Determining a desirable exponential form for an arbitrary $P_S(\varepsilon|\gamma)$ may not be a trivial task and in some cases it is impossible to partition the integral (17.5) into a product form. In fact, one cannot analyze the EGC receiver performance using the MGF approach. In light of the above considerations, we develop yet another general approach for calculating the error probability of all common diversity combining techniques over fading channels in a single common framework. Our novel derivation relies on the use of the Parseval's theorem to transform the product integral into the frequency domain, thereby circumventing the task to find and express the conditional error probability in a desirable form. But then we also need the Fourier transform (FT) of the conditional error probability, which surprisingly turns out to be very easily computed. As an application background, we will first derive exact analytical expressions for the EGC diversity receivers for both binary and M-ary modulation formats. Subsequently, we will show how this approach can be easily extended to other diversity combining techniques.

4.1   EQUAL-GAIN DIVERSITY

In an equal gain combiner, the outputs of different diversity branches are first co-phased, equally weighted, and then summed to give the resultant output. The instantaneous SNR at the output of the EGC combiner is $\gamma = \vartheta^2$ where ϑ is defined as

$$\vartheta = \sqrt{\frac{E_b}{LN_0}} \sum_{l=1}^{L} \alpha_l \tag{17.22}$$

where α_l is the fading amplitude random variable (RV). Let $\bar{\gamma}_k = \Omega_k(E_S/N_0)$ denote the average SNR for the k-th branch, which is consistent with our definition for the MRC case.

The average bit or symbol error probability in fading channels can be obtained by averaging the conditional error probability over the PDF of the combined signal amplitude at the output of the EGC combiner, namely

$$P_S(\varepsilon) = \int_0^\infty P_S(\varepsilon|\vartheta) p_\vartheta(\vartheta) d\vartheta \tag{17.23}$$

where $p_\vartheta(\cdot)$ denotes the PDF of RV ϑ.

If the fading amplitudes are assumed to be independent, then the evaluation of ASER using the classical solution in the form of (17.23) will require L-fold convolution integrals. It is more insightful if we transform the PDF into frequency domain since the CHF of ϑ (i.e., sum of L fading amplitudes) is simply the product of the individual CHFs. However, it is difficult (or impossible) to invert the CHF $\psi_\vartheta^{(\mathrm{EGC})}(\cdot)$ to get a closed-form expression for the PDF of ϑ. Therefore, a Fourier series approach has previously been used [25], [26].

Using the inverse Fourier transform representation for the PDF, and then rearranging the order of integration, (17.23) can be restated as

$$
\begin{aligned}
P_S(\varepsilon) &= \int_0^\infty P_S(\varepsilon|\vartheta) \left[\frac{1}{2\pi} \int_{-\infty}^\infty \psi_\vartheta^{(\mathrm{EGC})}(\omega) \exp(-j\omega\vartheta) d\omega \right] d\vartheta \\
&= \frac{1}{2\pi} \int_{-\infty}^\infty \psi_\vartheta^{(\mathrm{EGC})}(\omega) \left[\int_0^\infty P_S(\varepsilon|\vartheta) \exp(-j\omega\vartheta) d\vartheta \right] d\omega \\
&= \frac{1}{2\pi} \int_{-\infty}^\infty \mathrm{FT}[P_S(\varepsilon|\vartheta)] \psi_\vartheta^{*(\mathrm{EGC})}(\omega) d\omega \tag{17.24}
\end{aligned}
$$

where notation $\psi_\vartheta^{*(\mathrm{EGC})}(\cdot)$ denotes the complex conjugate of the CHF of ϑ. In fact, the final result of (17.24) follows directly from the application of Parseval's theorem [27, pp. 371] to transform the product integral in (17.23) into the frequency domain, thereby circumventing the need to find the PDF of ϑ. But we now need to compute the FT of $P_S(\varepsilon|\vartheta)$.

The FT of the generic conditional error probability (i.e., Eq. (17.4)) is given by

$$
\begin{aligned}
G_\vartheta(\omega) &= \frac{1}{2} \sum_u \int_0^{\eta_u} \frac{a_u(\theta)}{b_u(\theta)} \left\{ \sqrt{\pi b_u(\theta)} \exp\left[\frac{-\omega^2}{4 b_u(\theta)}\right] \right. \\
&\quad \left. + j\omega \Phi\left[1, \frac{3}{2}; \frac{-\omega^2}{4 b_u(\theta)}\right] \right\} d\theta \\
&= \sum_u \int_0^{\eta_u} \frac{a_u(\theta)}{\sqrt{b_u(\theta)}} \left\{ \frac{\sqrt{\pi}}{2} \exp\left[\frac{-\omega^2}{4 b_u(\theta)}\right] + j F\left[\frac{\omega}{2\sqrt{b_u(\theta)}}\right] \right\} d\theta \quad (17.25)
\end{aligned}
$$

where $F(\cdot)$ denotes the Dawson integral,

$$
F(x) = \exp[-x^2] \int_0^x \exp(t^2) dt = x \Phi\left(1, \frac{3}{2}; -x^2\right) \quad (17.26)
$$

and $\Phi(\cdot, \cdot; \cdot)$ is the confluent hypergeometric series. Dawson's integral can be computed more efficiently using a direct method (based on the sampling theorem) suggested by Rybicki [28] instead of evaluating sufficiently large number of terms in the series representation of $\Phi(1, \frac{3}{2}; -x^2)$. For this reason, we have expressed $G_\vartheta(\omega)$ in terms of Dawson's integral.

Substituting (17.25) into (17.24), and realizing that the imaginary part of this integral is zero (since the ASER is real), we get an exact analytical ASER expression for binary and M-ary modulation formats with predetection EGC:

$$
\begin{aligned}
P_s(\varepsilon) &= \frac{1}{\pi} \int_0^\infty \text{Real}\{G_\vartheta(\omega) \psi_\vartheta^{*(\text{EGC})}(\omega)\} d\omega \\
&= \frac{2}{\pi} \int_0^{\pi/2} \frac{\Lambda_\vartheta(\tan\theta)}{\sin(2\theta)} d\theta \quad (17.27)
\end{aligned}
$$

where $\Lambda_\vartheta(\omega) = \text{Real}\{\omega G_\vartheta(\omega) \psi_\vartheta^{*(\text{EGC})}(\omega)\}$, and the CHF of ϑ in Nakagami-m, Rician and Rayleigh channels are given by (17.28), (17.29) and (17.30), respectively:

Nakagami-m:

$$
\begin{aligned}
\psi_\vartheta(\omega) &= \prod_{k=1}^L \left\{ \Phi\left(m_k, \frac{1}{2}; \frac{-\omega^2}{4 L \lambda_k}\right) \right. \\
&\quad \left. + j\omega \frac{\Gamma(m_k + 1/2)}{\Gamma(m_k)\sqrt{L\lambda_k}} \Phi\left(m_k + \frac{1}{2}, \frac{3}{2}; \frac{-\omega^2}{4 L \lambda_k}\right) \right\} \quad (17.28)
\end{aligned}
$$

Rician:

$$\psi_\vartheta(\omega) = \prod_{k=1}^{L} \left\{ \exp(-K_k) \sum_{i=0}^{\infty} \frac{K_k^i}{i!} \Phi\left(i+1, \frac{1}{2}; \frac{-\bar{\gamma}_k \omega^2}{4L(1+K_k)}\right) \right.$$

$$+ j\omega \sqrt{\frac{\bar{\gamma}_k}{L(1+K_k)}} \exp(-K_k) \sum_{i=0}^{\infty} \frac{\Gamma(i+3/2)K_k^i}{(i!)^2}$$

$$\left. \times \Phi\left(i+\frac{3}{2}, \frac{3}{2}; \frac{-\bar{\gamma}_k \omega^2}{4L(1+K_k)}\right) \right\} \qquad (17.29)$$

Rayleigh:

$$\psi_\vartheta(\omega) = \prod_{k=1}^{L} \left\{ \Phi\left(1, \frac{1}{2}; \frac{-\bar{\gamma}_k \omega^2}{4L}\right) + j\omega \sqrt{\frac{\pi \bar{\gamma}_k}{4L}} \exp\left(\frac{-\bar{\gamma}_k \omega^2}{4L}\right) \right\} \quad (17.30)$$

Notice that the evaluation of (17.27) for the most general case involves two-fold integrals. Next we will identify three special cases of the conditional error probability $P_S(\varepsilon|\vartheta)$ which allow the evaluation of the generic expression given in (17.27) to be further simplified into a single finite-range integral. This simplification is attributed to the availability of closed-form formulas for the FT of $P_S(\varepsilon|\vartheta)$:

Case (a): If $P_S(\varepsilon|\vartheta) = a\exp[-b\vartheta^2]$, then

$$G_\vartheta(\omega) = \frac{a}{\sqrt{b}} \left[\frac{\sqrt{\pi}}{2} \exp\left(\frac{-\omega^2}{4b}\right) + jF\left(\frac{\omega}{2\sqrt{b}}\right) \right] \qquad (17.31)$$

Case (b): If $P_S(\varepsilon|\vartheta) = a\,\mathrm{erfc}(\sqrt{b}\vartheta)$, then

$$G_\vartheta(\omega) = \frac{a}{\omega} \left\{ frac2\sqrt{\pi}F\left(\frac{\omega}{2b}\right) + j\left[1 - \exp\left(\frac{-\omega^2}{4b^2}\right)\right] \right\} \qquad (17.32)$$

Case (c): If $P_S(\varepsilon|\vartheta) = a\,\mathrm{erfc}(\sqrt{b}\vartheta) - c\,\mathrm{erfc}^2(\sqrt{b}\vartheta)$, then

$$G_\vartheta(\omega) = \frac{2a}{\omega\sqrt{\pi}}F\left(\frac{\omega}{2b}\right) - \frac{4c}{\omega\sqrt{\pi}} \left[F\left(\frac{\omega}{2b}\right) - F\left(\frac{\omega}{2b\sqrt{2}}\right) \exp\left(\frac{-\omega^2}{8b^2}\right) \right]$$

$$+ j\left\{ \frac{a}{\omega}\left[1 - \exp\left(\frac{-\omega^2}{4b^2}\right)\right] - \frac{c}{\omega}\left[1 - \exp\left(\frac{-\omega^2}{4b^2}\right) - \frac{4}{\pi}F^2\left(\frac{\omega}{2b\sqrt{2}}\right)\right] \right\}$$

$$(17.33)$$

Similar to the development in (17.9), we can replace the finite-range integral (17.27) with the series expression using GCQ approximation,

$$P_S(\varepsilon) = \frac{1}{n} \sum_{i=1}^{n} \frac{\Lambda_\vartheta\left[\tan\left(\frac{(2i-1)\pi}{4n}\right)\right]}{\sin\left[\frac{(2i-1)\pi}{2n}\right]} + R_n \qquad (17.34)$$

4.2   MAXIMAL-RATIO DIVERSITY

Following the development in (17.24), it is straight-forward to show that (17.5) can be restated as

$$
\begin{aligned}
P_S(\varepsilon) &= \frac{1}{2\pi} \int_{-\infty}^{\infty} \mathrm{FT}[P_S(\varepsilon|\gamma)] \psi_\gamma^{*(\mathrm{MRC})}(\omega) d\omega \\
&= \frac{2}{\pi} \int_0^{\pi/2} \frac{\Lambda_\gamma(\tan\theta)}{\sin(2\theta)} d\theta
\end{aligned}
\tag{17.35}
$$

where $\psi_\gamma^{(\mathrm{MRC})}(\omega) = \phi_\gamma^{(\mathrm{MRC})}(-j\omega)$ is the CHF of SNR at the MRC combiner output, $G_\gamma(\omega) = \mathrm{FT}[P_S(\varepsilon|\gamma)]$ and $\Lambda_\gamma(\omega) = \mathrm{Real}\left[\omega G_\gamma(\omega)\psi_\gamma^{*(\mathrm{MRC})}(\omega)\right]$.

It is clear that only the knowledge of $G_\gamma(\cdot)$ is further required to investigate the error performance of different modulation formats. Since the FT of the generic error probability illustrated in (17.3) is given by

$$
G_\gamma(\omega) = \sum_u \int_0^{\eta_u} \frac{a_u(\theta)}{b_u(\theta) - j\omega} d\theta
\tag{17.36}
$$

then (17.35) can be rewritten as

$$
\begin{aligned}
P_S(\varepsilon) &= \sum_u \int_0^{\eta_u} a_u(\theta) \left[\frac{1}{2\pi j} \int_{-\infty}^{\infty} \frac{1}{\omega - jb_u(\theta)} \phi_\gamma^{(\mathrm{MRC})}(-j\omega) d\omega \right] d\theta \\
&= \sum_u \int_0^{\eta_u} a_u(\theta) \left[\phi_\gamma^{(\mathrm{MRC})}(-j\omega) \Big|_{\omega = jb_u(\theta)} \right] d\theta
\end{aligned}
\tag{17.37}
$$

The inner integral of (17.37) can be easily evaluated by applying Cauchy's integral theorem. Then (17.37) reduces to

$$
P_S(\varepsilon) = \sum_u \int_0^{\eta_u} a_u(\theta) \phi_\gamma^{(\mathrm{MRC})}[b_u(\theta)] d\theta
\tag{17.38}
$$

which is identical to the result obtained in Section 17.3.1 using the MGF approach.

If the conditional error probability can be expressed in terms of the complementary error functions, then we may derive an alternative expression for computing the ASER by directly evaluating the FT of $P_S(\varepsilon|\gamma)$ and substituting the result in (17.35):

Case (a): If $P_S(\varepsilon|\gamma) = a\,\mathrm{erfc}\left(\sqrt{b\gamma}\right)$, then

$$
G_\gamma(\omega) = \frac{a}{j\omega}\left(\sqrt{\frac{b}{b - j\omega}} - 1\right)
\tag{17.39}
$$

Case (b): If $P_S(\varepsilon|\gamma) = a\operatorname{erfc}(\sqrt{b\gamma}) - c\operatorname{erfc}^2(\sqrt{b\gamma})$, then

$$G_\gamma(\omega) = \frac{a}{j\omega}\left(\sqrt{\frac{b}{b - j\omega}} - 1\right) + \frac{c}{j\omega}\left[1 - \frac{4}{\pi}\frac{\arctan(\sqrt{1 - j\omega/b})}{\sqrt{1 - j\omega/b}}\right] \quad (17.40)$$

However, the final expression for the ASER with MRC diversity obtained using the CHF approach is slightly more complicated compared to the result attained via the MGF method. Nevertheless, the former is attractive because it allows us to unify the performance evaluation for all the common diversity combining techniques under a single common framework. It is also interesting to note the similarities between (17.27) and (17.35).

4.3 SELECTION AND SWITCHED DIVERSITY

Similar to the analysis for the MRC case, the CHFs of the SNR at the output of the SDC and the SWC combiners are related to their MGFs (derived in Sections 17.3.2 and 17.3.3) as $\psi_\gamma^{(\text{SDC})}(\omega) = \phi_\gamma^{(\text{SDC})}(-j\omega)$ and $\psi_\gamma^{(\text{SWC})}(\omega) = \phi_\gamma^{(\text{SWC})}(-j\omega)$, respectively. Once the CHF of SNR at the combiner output is determined, the ASER can be evaluated directly using the framework developed in Section 17.4.2. In fact, the final expression for the selection and/or switched diversity systems will be identical to the MRC case, with the exception that the expression for the $\psi_\gamma^{(\text{MRC})}(\cdot)$ is now replaced with $\psi_\gamma^{(\text{SDC})}(\cdot)$ or $\psi_\gamma^{(\text{SWC})}(\cdot)$, respectively.

5. CONCLUDING REMARKS

Two unified approaches for evaluating the average error probability performance of diversity receivers in generalized fading channels are outlined. In the first approach, our derivation relies upon the properties of the MGF of the SNR at the combiner output, the use of an alternative exponential form for complementary error functions, and the application of a GCQ formula. In the second approach, we require the knowledge of Fourier transform of the conditional error probability, the CHF of SNR, and the application of a GCQ formula. These unified approaches allow previously obtained results to be simplified both analytically and computationally and new results to be obtained for special cases that heretofore resisted solutions in simple form. The exact SER is mostly expressed in terms of a single finite-range integral, and in some cases in the form of double finite-range integrals. Remarkably, the exact SER integrals can also be replaced by a rapidly converging series formulas. This offers a convenient method to perform a comprehensive study of all common diversity combining techniques (MRC, EGC, SDC and SWC) with different modulation formats. The CHF method allows us to unify the above problem under a single common framework. Nevertheless, the MGF method sometimes yields a more concise solution than the CHF approach in the analysis of MRC, SDC and

326

SWC diversity systems. The generality and computational efficiency of the new results presented in this chapter render themselves as powerful means for both theoretical analysis and practical applications.

References

[1] W. C. Jakes, *Microwave Mobile Communications*, IEEE Press, New Jersey, 1974.

[2] M. Schwartz, W. R. Bennett and S. Stein, *Communication Systems and Techniques*, McGraw-Hill, New York, 1966.

[3] A. Abu-Dayya and N. C. Beaulieu, "Analysis of Switched Diversity System on Generalized- Fading Channels," IEEE Trans. Communications, Vol. 42, pp. 2959-2966, Nov. 1994.

[4] J. W. Craig, "A New, Simple and Exact Results for Calculating the Probability of Error for Two Dimensional Signal Constellations," IEEE Milcom'91 Conference Record, pp. 25.5.1- 25.5.5.

[5] C. Tellambura, A. J. Mueller and V. K. Bhargava, "Analysis of M-ary Phase-Shift-Keying with Diversity Reception for Land-Mobile Satellite Channels," IEEE Trans. Vehicular Technology, Vol. 46, pp. 910-922, November 1997.

[6] C. Tellambura, "Evaluation of the Exact Union Bound for Trellis Coded Modulations over Fading Channels," IEEE Trans. Communications, Vol. 44, pp. 1693-1699, December 1996.

[7] C. Tellambura, A. J. Mueller and V. K. Bhargava, "BER and Outage Probability for the Land Mobile Satellite Channel with Maximal Ratio Combining," IEE Electronics Letters, pp. 606-608, April 1995.

[8] A. J. Mueller, *Issues in Diversity and Adaptive Error Control Coding for Wireless Communications*, M. A. Sc Thesis, Dept. of Electrical and Computer Engineering, University of Victoria, September 1995.

[9] M. K. Simon and D. Divsalar, "Some New Twists to Problems Involving the Gaussian Probability Integral," IEEE Trans. on Communications, Vol. 46, pp. 200-210, Feb. 1998.

[10] M. Alouini and A. Goldsmith, "A Unified Approach for Calculating Error Rates of Linearly Modulated Signals over Generalized Fading Channels," Proc. ICC'98, Atlanta, pp. 459-464, June 1998.

[11] M. Simon and M. Alouini, "A Unified Approach to Performance Analysis of Digital Communication over Generalized Fading Channels," Proc. IEEE, Vol. 86, September 1998, pp. 1860-1877.

[12] A. Annamalai, C. Tellambura and V. K. Bhargava, "A Unified Approach to Performance Evaluation of Switched Diversity in Independent and Correlated Fading Channels," submitted to IEEE WCNC'99.

[13] C. Tellambura and V. K. Bhargava, "Unified Error Analysis of DQPSK in Fading Channels," IEE Electronics Letters, Vol. 30, No. 25, pp. 2110-2111, Dec. 1994.

[14] R. F. Pawula, "A New Formula for MDPSK Symbol Error Probability" IEEE Communications Letters, Vol. 2, pp. 271-272, Oct. 1998.

[15] X. Dong, N. C. Beaulieu and P. H. Wittke, "Two Dimensional Signal Constellations for Fading Channels," IEEE GLOBECOM, Communication Theory Mini Conference, pp. 22- 27, 8-12 Nov. 1998.

[16] M. Abramowitz and I. A. Stegun, *Handbook of Mathematical Functions*, National Bureau of Standards, Applied Mathematics Series 55, 1964.

[17] A. Annamalai, C. Tellambura and V. K. Bhargava, "Unified Analysis of MPSK and MDPSK with Diversity Reception in Different Fading Environments," IEE Electronics Letters, Vol. 34, No. 16, pp. 1564-1565, 6th August 1998.

[18] A. Annamalai, C. Tellambura and V. K. Bhargava, "Exact Evaluation of Maximal-Ratio and Equal-Gain Diversity Receivers for M-ary QAM on Nakagami Fading Channels," to appear in the IEEE Trans. Communications.

[19] I. S. Gradshteyn and I. M. Ryzhik, *Table of Integrals, Series and Products*, Academic Press, 5th edition, 1995.

[20] C. Tellambura, A. Annamalai and V. K. Bhargava, "Unified Analysis of Switched Diversity Systems in Independent and Correlated Fading Channels," submitted to the IEEE Trans. Communications.

[21] J. G. Proakis, *Digital Communications*, McGraw Hill, New York, 3rd edition, 1995.

[22] R. S. Hoyt, "Probability Functions for the Modulus and Angle of the Normal Complex Variate," Bell System Tech. Journal, Vol. 26, pp. 318-359, 1947.

[23] A. Annamalai, C. Tellambura and V. K. Bhargava, "New, Simple and Accurate Methods for Outage Analysis in Cellular Mobile Radio Systems - A Unified Approach," submitted to IEEE Trans. Communications.

[24] E. Lutz, D. Cygan, M. Dippold, F. Dolainsky and W. Papke, "The Land Mobile Satellite Communication Channel - Recordings, Statistics and Channel Model," IEEE Trans. Vehicular Technology, Vol. 40, pp. 375-386, May 1991.

[25] N. C. Beaulieu and A. Abu-Dayya, "Analysis of Equal Gain Diversity on Nakagami Fading Channels," IEEE Trans. Communications, Vol. 39, pp. 225 234, February 1991.

[26] A. Abu-Dayya and N. C. Beaulieu, "Microdiversity on Rician Fading Channels," IEEE Trans. Communications, Vol. 42, June 1994, pp. 2258-2267.

[27] P. Z. Peebles, *Probability, Random Variables and Random Signal Principles*, McGraw-Hill, 1993.

[28] G. B. Rybicki, "Dawson's Integral and Sampling Theorem," Computers in Physics, Vol. 3, pp. 85-87, March 1989.

Appendix

The random variable of interest for the analysis of MRC, SDC and SWC is the power in the fading envelope of the received signal. Since our approach for computing the error performance requires the knowledge of the MGF or CHF of SNR at the combiner output, in the following we will summarize the MGF of SNR (without diversity) for some of the commonly used fading channel models.

Rician and Rayleigh Channels

The MGF for the non-centralized chi-squared distribution is given by [21, pp. 44],

$$\phi_k(z) = \frac{1 + K_k}{1 + K_k + z\Omega_k} \exp\left(\frac{-zK_k\Omega_k}{1 + K_k + z\Omega_k}\right) \qquad (17.A.1)$$

where $\Omega_k = E[\gamma_k]$, and K_k is the Rice factor of the k-th diversity branch, defined as the ratio of the power in the line-of-sight (LOS) path to the power in the multipaths.

In a limiting case when the power in LOS path approaches zero, then $K \to 0$ and the channel reverts to the Rayleigh fading channel. Then the corresponding MGF of thecentralized chi-squared distribution is

$$\phi_k(z) = \frac{1}{1 + z\Omega_k} \qquad (17.A.2)$$

Nakagami-m and Nakagami-q Channels

The Nakagami-m distribution is a versatile statistical distribution because it can accurately model a variety of fading environments. It has greater flexibility in matching some empirical data than the Rayleigh, lognormal or Rice distributions owing to its characterization of the received signal as the sum of vectors with random moduli and random phases. As well, this statistical model includes the Rayleigh and the one-sided Gaussian distributions as special cases for the fading figure $m = 1$ and $m = 0.5$, respectively. Moreover, the m-distribution can closely approximate the Rician distribution via relationship $m = (K + 1)^2/(2K + 1)$. The MGF for this fading channel is given by

$$\phi_k(z) = \left(\frac{1}{1 + z/\lambda_k}\right)^{m_k} \qquad (17.A.3)$$

where $\lambda_k = m_k/\Omega_k$. It is evident that (17.A.3) reduces to (17.A.2) when $m = 1$ (i.e., Rayleigh fading).

By computing the Laplace transform of the signal power in Nakagami-q (Hoyt)[22] fading, we can shown that the desired MGF is given by

$$\phi_k(z) = \frac{1}{\sqrt{[z\Omega_k(1 + b_k) + 1][z\Omega_k(1 - b_k) + 1]}} \qquad (17.A.4)$$

for $-1 \le b_k \le 1$, where $b_k = [1 - q_k^2]/[1 + q_k^2]$ and $q_k(0 \le q_k \le \infty)$ is the fading parameter.

Lognormal Rice and Suzuki Channels

The Rician distribution assumes a constant K. In reality, this may not be the case since as the mobile terminal moves through the cell, a variety of topographical surrounding are encountered. This is particularly evident in terrestrial environment where shadowing (physical obstruction of the signal path) is more severe. Hence, it is plausible to consider a combined distribution incorporating the effects of shadowing into the Rician distribution.

Expressing the received fading envelope as the product of independent Rice and lognormal distributions, and then applying Hermitian integration, the desired MGF is given by [8][23]

$$\begin{aligned}\phi_k(z) &= \frac{1}{\sqrt{\pi}} \sum_{i=1}^{H} \frac{w_i(1 + K_k)}{1 + K_k + z\mu_k \exp[\sqrt{2}\sigma_k x_i]} \\ &\times \exp\left(\frac{-K_k z\mu_k \exp[\sqrt{2}\sigma_k x_i]}{1 + K_k + z\mu_k \exp[\sqrt{2}\sigma_k x_i]}\right) + R_H \quad (17.A.5)\end{aligned}$$

where σ_k is the logarithmic standard deviation of shadowing, and μ_k is the local mean power. The abscissas x_i (ith root of an Hth order Hermite polynomial) and weights w_i are tabulated in [16] for $H \le 20$ and R_H is a remainder term.

Suzuki fading characterizes the joint effects of Rayleigh fading and lognormal shadowing and models a shadowed multipath channel without a LOS path. Since Suzuki distribution is a special case of the lognormal Rician distribution, its MGF is readily obtained by setting $K = 0$ in (17.A.5), i.e.,

$$\phi_k(z) = \frac{1}{\sqrt{\pi}} \sum_{i=1}^{H} \frac{w_i}{1 + z\mu_k \exp[\sqrt{2}\sigma_k x_i]} + R_H \qquad (17.A.6)$$

Lognormal Nakagami-m Channel

Similar to the derivation of (17.A.5), the MGF of the received power in a Nakagami-m fading channel with lognormal shadowing can be expressed as,

$$\phi_k(z) = \frac{1}{\sqrt{\pi}} \sum_{i=1}^{H} \frac{w_i}{\left[1 + z\mu_k \exp(\sqrt{2}\sigma_k x_i)/m_k\right]^{m_k}} + R_H \qquad (17.A.7)$$

Mixed Fading Channel

Owing to the time-varying nature of the wireless channels, a practical wireless channel may be more realistically modelled as a combination of different statistical distributions. The lognormal Rice or the lognormal Nakagami distribution may be further refined if additional information of the channel condition is available.

For instance, in [24] Lutz presented a two-state land mobile satellite channel model based on channel measurements. In this model, the channel is assumed to be in the good state for a fraction of the time $1 - \delta$, and modelled as a Rician random process. For the remaining fraction of the time δ, the channel is in the bad state modelled as a lognormally shadowed Rayleigh random process, or equivalently, a Suzuki random process. The net MGF of the received power is thus the weighted sum of the Rician and Suzuki MGFs, respectively:

$$\phi_k(z) = (1 - \delta)\phi_k^{(Rice)}(z) + \delta\phi_k^{(Suzuki)}(z) \qquad (17.A.8)$$

where $\phi_k^{(Rice)}(z)$ and $\phi_k^{(Suzuki)}(z)$ are the MGFs for the Rician and Suzuki fading states given in (17.A.1) and (17.A.6), respectively.

Chapter 18

*m*MFSK Frequency Hopping Modulation for Wireless Ad Hoc Networks

S. GLISIC, Z. NIKOLIC, N. MILOSEVIC, A. POUTTU

Abstract: Ad hoc networks have long been considered for military tactical communications under the more familiar name Packet Radio Networks. Recently, however, use of ad hoc networking technology in commercial systems has been investigated. Ad hoc networks represent a communication concept, which is not dependent on preexisting infrastructure. Thus ad hoc networks can be rapidly deployed, without prior planning and in unknown radio propagation conditions. In this chapter we consider a modification of the FH/MFSK system to include multitone MFSK signal which will be designated as *m*MFSK modulation in conjunction with the ad hoc networking principles. In *m*MFSK case the signal energy is split to *m* separate tones. This will reduce the signal power spectral density by factor *1/m* and make it more attractive for applications such as low probability of intercept systems. Different versions of *m*MFSK approach have been described earlier without in depth analysis of the performance. The main contribution of this chapter is a detailed performance analysis of this system under different channel, network load and jamming conditions. The results demonstrate that under the large range of the signal, channel, network load and interference parameters the new system offers better performance.

1   INTRODUCTION

Two the most often used versions of spread spectrum systems are frequency hopping (FH) and direct sequence (DS) configurations. The optimum DS receiver is difficult to implement unless a cellular network infrastructure is used. In this concept mobile units communicate through centralized entities (base - stations) which serve as point of wireless access to fixed network. In contrast, ad hoc networks represent a different communication concept, which is not dependent on preexisting infrastructure. Thus ad hoc networks can be rapidly deployed, without prior planning and in unknown radio propagation conditions. In ad hoc networks communications between adjacent nodes is allowed, giving rise to multihop routing. The topology of ad hoc networks is distributed. Nodes in an ad hoc network can migrate freely, joining, leaving, and rejoining the network often, without warning, and without disruption to communication between other nodes. The challenges in the design of ad hoc networks stem from the lack of centralized entities (no possibility for centralized processing like multiuser detection in CDMA networks), from the multihop mode of communication, and from the fact that all communication is carried over wireless medium giving rise to fading and jamming issues. Ad hoc networks have long been considered for military tactical communications under the more familiar name Packet Radio Networks. Recently, however, use of ad hoc networking technology in commercial systems has been investigated. Examples are law enforcement, rescue missions, virtual classrooms or wireless local area networks.

A wireless LAN standard developed by IEEE committee P802.11 operates in the unlicensed 2.4GHz ISM band. This band is a very hostile environment due to many unpredictable interference sources, such as microwave ovens, utilizing the same frequency bands. At the same time there is a limitation about the allowed signal power density that can be used in order to minimize interference to other users in the same band. In order to avoid these interference and keep low signal power density the standard supports both direct sequence (DS) and frequency hopping (FH) modulation for these applications. The new test beds for multimedia WLANs also use FH modulation [26].

All the environments described above will demonstrate severe near far effect, and communicator may prefer to use a frequency-hopping spread-spectrum (FH-SS) system rather than CDMA concept based on Direct Sequence modulation which is sensitive to near far effect. To mitigate this effect power control or multiuser detection must be used, both of which are feasible only in cellular networks.

Performance of FH-SS systems, for different applications, has been already discussed in the literature. In order to improve performance in jamming and fading environment FH-SS systems can use diversity. Traditionally, diversity was obtained via multiple hops per information (or coded) symbol, i.e. frequency diversity. Such a fast hopping makes difficult the synchronization of the carrier phase and, consequently, imposes the use of a noncoherent receiver. Thus, a significant loss in error performance results, due to both noncoherent demodulation and noncoherent combining of the received diversity replicas. Taking into account these losses, and using binary frequency-shift keying (BFSK) modulation, an optimum diversity scheme is analyzed in [1] for the worst case jammer and with side information on noise and jamming levels. Since optimum diversity has more analytical value than practical existence, the error probability is much higher in practice. In order to recover these performance losses, some authors have studied a solution that makes coherent reception feasible; see, e.g., [4], [7], [5]. Frequency diversity as used on Rayleigh fading channels, and which differs from the diversity mentioned above, was proposed in [2] to counter band-limited interference. Such a diversity allows one to avoid noncoherent combining loss. In this system, called frequency-diversity spread spectrum (FD-SS), the communicator frequency band is partitioned into N disjoint subbands on which N replicas of the signal are simultaneously transmitted. However, since frequency hopping is considered as mandatory in some applications, some solutions combine both FD-SS and FH-SS systems [1]. The main objective is to guarantee coherent demodulation and to avoid noncoherent combining loss.

If coherency is not feasible then noncoherent solution is the only option. The effect of barrage and partial-band noise jamming on frequency-hopped, M-ary frequency-shift keyed (FH/MFSK) noncoherent receivers, when one or more symbols per hop are transmitted, has been examined both for channels with no fading and for Rayleigh fading channels in [9] and [10], respectively. The effect of partialband noise jamming on fast frequency-hopped (FFH) binary frequency-shift keyed (BFSK) noncoherent receivers with diversity has been examined for channels with no fading [11], as has the effect of partial-band noise jamming on FFH/MFSK for Ricean fading channels [12]. The performance degradation resulting from both band and independent multitone jamming of FH/MFSK, where the jamming tones are assumed to correspond to some or all of the possible FH M-ary signaling tones and when thermal and other wideband noise is negligible, is examined in [13]-[15]. The effect of tone interference on noncoherent MFSK when AWGN is not neglected is examined for channels with no fading in [16], and the effect of independent multitone jamming on noncoherent FH/BFSK when AWGN is not neglected is examined for channels with no fading in [17]. In this paper we consider a modification of the FH/MFSK system to include multitone MFSK signal which will be designated as *m*MFSK modulation. In this case the signal energy is split to *m* separate tones. This will reduce the signal power spectral density by factor *1/m* and make it more attractive for applications such as low probability of intercept systems. This is also very attractive characteristic for wireless LANs in ISM frequency band where the transmitted power density is limited by regulations in order to prevent excessive interference to other systems sharing the same bandwidth. Due to reduced power level of each tone, the system will be more vulnerable to noise and fading but still the overall flow of useful information will be increased. Different versions of this approach are described in [19-25] without in depth analysis of their performance. The main contribution of this chapter is a detailed performance analysis of this system under different channel, network load and jamming conditions. The chapter is organized as follows: In Section 2 we present the signal model and performance criteria. In Section 3 performance analysis is given. A number of numerical results is presented in Section 4. These results demonstrate that under the large range of the signal, channel, network load and interference parameters the new system offers better performance.

2 SYSTEM MODEL

A standard M-ary FSK modulation uses one out of M frequencies each T_s seconds to transmit a block of $n=\log_2 M$ bits. The optimum receiver has a bank of M matched filters and every T_s seconds makes decision based on the largest output filter sample. This configuration requires coherent demodulation and still is widely considered for practical applications [1-7] due to higher gain in a frequency diversity scheme compared with noncoherent combining. In spite of the combining losses, noncoherent schemes are also being considered for practical applications due to its simplicity [8-15].

Let's suppose that now instead of sending one out of M frequencies we send two frequencies simultaneously. If amplitudes are the same we can form

$$M_2 = \sum_{r=1}^{M-1} r = (M-1)M/2 \tag{1}$$

different combinations and send

$$n_2 = \log(M - 1)M / 2 = \log M + \log(M - 1) - 1 \qquad (2)$$

bits. If M is large $n_2 \approx 2\log M - 1 \approx 2n$ is almost twice as much as in the case of simple M-ary modulation. The optimum receiver will now have to find two the largest signals at the output of the bank of matched filters. If now instead of two, m out of M frequencies are simultaneously transmitted we have m**MFSK** modulation.

Table 1. Number of possible signaling combinations for mMFSK with m and M as parameters

m \ M	3	4	5	6	7	8	9
1	3	4	5	6	7	8	9
2	3	6	10	15	21	28	36
3	1	4	10	20	35	56	84
4		1	5	15	35	70	126
5			1	6	21	56	126
6				1	7	28	84
7					1	8	36
8						1	9
9							1

The number of transmitted bits per T_s and the number of possible signaling combinations is now further increased to

$$M_m = \binom{M}{m}, \qquad n_m = \log_2\binom{M}{m}$$

A simple calculus is needed to evaluate capacity improvement in noise free channel. In this paper mMFSK modulation will be considered for frequency hopping systems instead of standard MFSK modulation, hence the name FH/mMFSK modulation. As the performance measure we will be discussing symbol error rate or the system efficiency improvement factor defined as [18]

$$E_{eff} = [(1 - P)n]/[(1 - P_0)n_0] \qquad (3)$$

where parameters with index zero refer to the standard MFSK modulation. In Eq. 3, n is the number of bits per symbol, P is the bit error rate and $(1-P)n$ is the average number of correctly transmitted bits per symbol.

By splitting the available signal power on m frequencies the system will be more error prone but the average number of correctly transmitted bits should be still higher.

3   PERFORMANCE ANALYSIS

In this section we first calculate the performance in the presence of white Gaussian noise and thereafter extend the results to ad hoc networks.

3.1   Error probability for coherent mMFSK

The transmitted signals can be represented as

$$u_k(t) = Ae^{j2\pi f_k t} = Au_{ku}(t)$$
$$s_k(t) = \mathrm{Re}\{u_k(t)e^{j2\pi f_c t}\} = A\,\mathrm{Re}\{u_{ku}(t)e^{j2\pi f_c t}\} = As_{ku}(t),\ \ k=1,2,...,M \tag{4}$$

$$\text{with } \int_0^T u_k(t)u_m(t)dt = \delta_{km} \tag{5}$$

where $u_k(t)$ is the complex envelope of the signal and δ_{km} is the Kronecker delta function. Eq. 5 implies that signals $u_k(t)$ are mutually orthogonal. The energy of these signals is

$$E_k = \int_0^T s_k^2(t)dt = \tfrac{1}{2}\int_0^T |u_k(t)|^2 dt = A^2 E_u = E \ \ \ k=1,2,...,M \tag{6}$$

After frequency down conversion the received signal envelope is

$$r(t) = \alpha e^{-j\phi}u_k(t) + z(t) \quad 0 \le t \le T \tag{7}$$

where α is due to channel attenuation, ϕ is the phase difference between the input and local signal, and $z(t)$ is Gaussian noise. For simplicity and without loosing on generality let us assume that the first m frequencies are transmitted i.e. $s_l(t),\ l=1,2,...,m$. The received low pass equivalent of the signal becomes

$$r(t) = \alpha e^{-j\phi}(u_1(t) + \cdots + u_m(t)) + z(t);\ \ 0 \le t \le T \tag{8}$$

The decision variables are now given as

$$U_k = \mathrm{Re}\left\{e^{j\phi}\int_0^T r(t)u_k^*(t)dt\right\}\ \ k=1,2,...,M \tag{9}$$

An optimum receiver will choose m the largest decision variables. The receiver block diagram for the coherent mMFSK-scheme is shown in Fig.1.

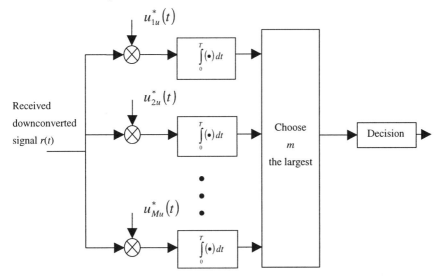

Fig. 1. m**MFSK demodulator for coherent detection**

The decision variables U_k can be represented as

$$U_l = 2\alpha E + N_{lr}, \ l = 1,2,...,m$$
$$U_p = N_{pr}, \ p = m+1,...,M$$

(10)

were N_{kr} are Gaussian zero mean variables with variance $\sigma^2 = 2EN_0$, and N_0 is the noise spectral density. Probability density functions (pdf) for U_k can be represented as

$$p(U_l) = \frac{1}{\sqrt{2\pi}\sigma} e^{-(U_l - 2\alpha E)^2/2\sigma^2} \ l = 1,2,...,m$$

(11)

$$p(U_p) = \frac{1}{\sqrt{2\pi}\sigma} e^{-U_p^2/2\sigma^2} \ p = m+1,...,M$$

Probability of correct decision assumes the form

$$F_c - F(U_1 > U_{m+1}, U_1 > U_{m+2},...,U_1 > U_M) \cdots$$
$$P(U_m > U_{m+1}, U_m > U_{m+2},...,U_m > U_M)$$

Because $P(U_x > U_{m+1}, U_x > U_{m+2}, ..., U_x > U_M)$, $x \leq m$, are the same,

$$P_c = \left[P(U_1 > U_{m+1}, U_1 > U_{m+2}, ..., U_1 > U_M) \right]^m. \tag{12}$$

Inserting the pdfs of the decision variables given in Eq. 11 into Eq. 12 and by performing the necessary calculus we obtain an expression for the probability of correct decision

$$P_c = \left[\frac{1}{2^{M-m} \sqrt{\pi}} \int_{-\infty}^{+\infty} e^{-x^2} \left[1 + \operatorname{erf}\left(x + \sqrt{\frac{\alpha^2 E}{N_0}} \right) \right]^{M-m} dx \right]^m. \tag{13}$$

The symbol error probability is the complement of the probability of correct decision and can be hence written as $P_s = 1 - P_c$.

3.2 Error probability for noncoherent mMFSK with envelope detector

The transmitted signal, its energies and the equivalent baseband form of the received signal are given again by Eqs. 4-8. Due to the envelope detector the receiver will create decision variables

$$U_k = \left| \int_0^T r(t) u_k^*(t) dt \right|, \quad k = 1, 2, ..., M \tag{14}$$

and choose m the largest ones. The receiver block diagram for the noncoherent mMFSK with envelope detector is shown in Fig. 2. If we assume that the transmitted signals are $s_l(t)$, $l = 1, 2, ..., m$, then the received signal is given by Eq. 8. Bearing in mind that signals $u_k(t)$ are mutually orthogonal U_l and U_p are obtained by taking modulo of Eq. 10. Parameters N_{kr} are zero mean Gaussian variables having the variance $\sigma^2 = 2EN_0$, and N_0 is the noise power density. One can show that pdf's for U_k can be expressed as:

$$p(U_l) = \frac{U_l}{2EN_0} \exp\left(-\frac{U_l^2 + 4\alpha^2 E^2}{4EN_0} \right) I_0\left(\frac{\alpha U_l}{N_0} \right), l = 1, 2, ..., m$$

$$p(U_p) = \frac{U_p}{2EN_0} \exp\left(-\frac{U_p^2}{4EN_0} \right) \quad p = m+1, ..., M \tag{15}$$

where $I_0(\cdot)$ is the zero[th] order modified Bessel function of the first kind. Probability of correct decision is defined in general form by Eq. 12 and in the case of noncoherent reception becomes

$$P_c = \left[\sum_{n=0}^{M-m} (-1)^n \binom{M-m}{n} \frac{1}{n+1} e^{-\frac{\alpha^2 E}{N_0} \frac{n}{n+1}} \right]^m \qquad (16)$$

Finally the symbol error probability is again given as $P_s = 1 - P_c$.

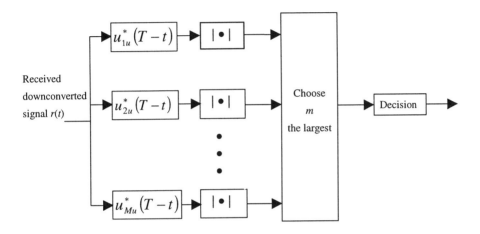

Fig. 2. *m*MFSK demodulator for noncoherent detection

One can show that error probability remain the same if square-law instead of envelope detector is used. In the presence of other users in the network, jamming and fading, expressions for symbol error rate are further modified. Details are shown in the sequel.

3.3 Error probability for coherent FH/*m*MFSK system in ad hoc network

The ad-hoc networking implies that the system is an asynchronous system. This means that the frequency hops of different communicators can overlap. Furthermore the decentralized nature of ad-hoc networking prevents intelligent frequency hop code allocation policies and hence the random hopping code approach is sensible in the analysis. Given these assumptions, in the presence of multiple access interference (MAI) the symbol error rate (SER) for FH/mMFSK system can be represented as

$$P_s = \frac{q}{N} P_s(\text{HJ}) + \frac{N-q}{N} P_s(\text{HNJ}) \tag{17}$$

where q is the number users, N is the number of available FH carriers, $P_s(\text{HJ})$ is SER when a hop is jammed by other users, and $P_s(\text{HNJ})$ is SER when the hop is not jammed. In the sequel we derive expressions for $P_s(\text{HJ})$ and $P_s(\text{HNJ})$.

3.3.1 Hop is not jammed

As a starting point in this derivation we use Eq. 13 which defines SER for coherent mMFSK in the presence of Gaussian noise to obtain

$$P_s(\text{HNJ} \mid a_c) = F_{koh}\left(\frac{a_c^2}{\sigma^2}\right), \text{ where } \frac{a_c^2}{\sigma^2} = \frac{\alpha^2 E}{N_0} \tag{18}$$

$$F_{koh}(z) = 1 - \left[\frac{1}{2^{M-m}\sqrt{\pi}} \int_{-\infty}^{+\infty} e^{-x^2}\left[1 + \text{erf}\left(x + \sqrt{z}\right)\right]^{M-m} dx\right]^m \tag{19}$$

In order to be able to evaluate the system in realistic channel we assume that the channel exhibits Rician (or Rayleigh) fading and hence the pdf for the signal amplitude is

$$f_{A_c}(a_c) = \frac{a_c}{\sigma_c^2}\exp\left[-\frac{a_c^2 + \alpha_c^2}{2\sigma_c^2}\right] I_0\left(\frac{a_c \alpha_c}{2\sigma_c^2}\right) \tag{20}$$

where α_c and σ_c are standard deviation for the steady and diffuse signal component, respectively. The Rayleigh fading case is obtained by setting $\alpha_c = 0$. By averaging Eq. 18 with respect to a_c we get

$$P_s(\text{HNJ}) = \int_0^\infty P_s(\text{HNJ} \mid a_c) \cdot f_{A_c}(a_c) da_c$$

$$= \exp\left(-\frac{\rho_c}{\xi_c}\right) \int_0^\infty F_{koh}\left(x^2 \frac{\xi_c}{2}\right) \cdot x \exp\left[-\frac{x^2}{2}\right] I_0\left(x\sqrt{2\frac{\rho_c}{\xi_c}}\right) dx \tag{21}$$

where $\rho_c = \dfrac{\alpha_c^2}{\sigma^2}$, $\xi_c = \dfrac{2\sigma_c^2}{\sigma^2}$.

3.3.2 Hop is jammed

As there are m jamming tones, when a hop is jammed the jamming signal frequencies may overlap with one useful signal frequency, with two of them, ..., with m of them. So the SER can be expressed as

$$P_s(\text{HJ}) = \frac{1}{\binom{M}{m}} \sum_{k=0}^m \binom{m}{k}\binom{M-m}{m-k} P_s(\text{HJ}/k\,\text{SBJ}) \tag{22}$$

where $P_s(\text{HJ}/k\text{SBJ})$ is SER when the hop and k Signal frequency **B**ins are **J**ammed. Without loosing the generality we can assume that the signal occupies the first m frequency slots and the jamming signal occupies the first k, the $(m+1)$th, ...,$(2m-k)$th. So we have (see Fig. 3.)

$$p(U_l) = \frac{1}{\sqrt{2\pi}\sigma} e^{-\left(U_l - \sqrt{2}\sqrt{a_c^2 + a_J^2 + 2a_c a_J \cos\theta}\right)^2 / 2\sigma^2}, \quad l = 1,...,k$$

$$p(U_l) = \frac{1}{\sqrt{2\pi}\sigma} e^{-\left(U_l - \sqrt{2}a_c\right)^2 / 2\sigma^2}, \quad l = k+1,...,m$$

$$p(U_l) = \frac{1}{\sqrt{2\pi}\sigma} e^{-\left(U_l - \sqrt{2}a_J\right)^2 / 2\sigma^2}, \quad l = m+1,...,2m-k \tag{23}$$

$$p(U_l) = \frac{1}{\sqrt{2\pi}\sigma} e^{-U_l^2 / 2\sigma^2}, \quad l = 2m-k+1,...,M$$

where a_J is the jamming signal amplitude, and θ is the jamming signal phase relative to the useful signal uniformly distributed in $[0,2\pi]$ interval.

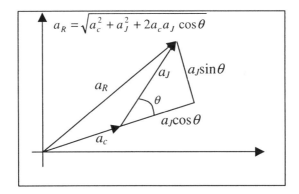

Fig. 3. Signal phasor diagram when the signal and the interferer hit the same detector branch

For a given amplitudes of the useful and jamming signal the conditional SER becomes

$$
\begin{aligned}
P_s\,(HJ/\,k\,SBJ\,|\,a_{c}\,U_1,...,U_m) = \\
1 - [P(U_1 > U_{m+1})P(U_1 > U_{m+2})\cdots P(U_1 > U_M)]\times\cdots\times \\
[P(U_m > U_{m+1})P(U_m > U_{m+2})\cdots P(U_m > U_M)]
\end{aligned}
\tag{24}
$$

$p(U_l), l = 1,2,...,k$, $\quad p(U_l), l = k+1,...,m$, $\quad p(U_l), l = m+1,...,2m-k$ and $p(U_l), l = 2m-k+1,...,M$ are the same and hence we have

$$
P_s\,(HJ/\,k\,SBJ\,|\,a_c,\theta,U_1) = 1 - \left[\left(P(U_{k+1} > U_{m+1})\right)^{m-k}\left(P(U_{k+1} > U_{2m-k+1})\right)^{M-2m+k}\right]^{m-k} \times
\tag{25}
$$

$$
\times\left[\left(P(U_1 > U_{m+1})\right)^{m-k}\left(P(U_1 > U_{2m-k+1})\right)^{M-2m+k}\right]^{k}
$$

where

$$
P(U_{k+1} > U_{m+1}) = \int_{-\infty}^{U_{k+1}} p(U_{m+1})dU_{m+1} = \frac{1}{2}\left(1 + \mathrm{erf}\left(\frac{U_{k+1} - \sqrt{2}a_J}{\sqrt{2}\sigma}\right)\right)
$$

$$
P(U_{k+1} > U_{2m-k+1}) = \int_{-\infty}^{U_{k+1}} p(U_{2m-k+1})dU_{2m-k+1} = \frac{1}{2}\left(1 + \mathrm{erf}\left(\frac{U_{k+1}}{\sqrt{2}\sigma}\right)\right)
\tag{26}
$$

$$
P(U_1 > U_{2m-k+1}) = \int_{-\infty}^{U_1} p(U_{2m-k+1})dU_{2m-k+1} = \frac{1}{2}\left(1 + \mathrm{erf}\left(\frac{U_1}{\sqrt{2}\sigma}\right)\right)
$$

By averaging Eq. 24 with respect to U_1 and U_{k+1} we have

$$P_s(\text{HJ}/k\,\text{SBJ}\,|\,a_c,\theta) = F_{koh1}\left(\frac{a_c^2}{\sigma^2},\frac{a_J^2}{\sigma^2},\theta\right), \text{ where} \tag{27}$$

$$F_{koh1}(y,z,\theta) = 1 - \left[\frac{1}{\sqrt{\pi}}\int_{-\infty}^{+\infty}e^{-x^2}F_{erf}\left(x+\sqrt{y}-\sqrt{z},m-k\right)F_{erf}\left(x+\sqrt{y},M-2m+k\right)dx\right]^{m-k} \times$$
$$\left[\frac{1}{\sqrt{\pi}}\int_{-\infty}^{+\infty}e^{-x^2}F_{erf}\left(x+\sqrt{y+z+2\sqrt{yz}\cos\theta}-\sqrt{z},m-k\right)\times \right. \tag{28}$$
$$\left. F_{erf}\left(x+\sqrt{y+z+2\sqrt{yz}\cos\theta},M-2m+k\right)dx\right]^k$$

and

$$F_{erf}(x,n) = \begin{cases}\left[\frac{1}{2}(1+\text{erf}(x))\right]^n, & n \geq 0 \\ 1, & n < 0\end{cases} \tag{29}$$

After averaging Eq. 27 we get

$$P_s(\text{HJ}/k\,\text{SBJ}) = \frac{1}{2\pi}\int_0^{2\pi}\int_0^{\infty}F_{koh1}\left(\frac{a_c^2}{\sigma^2},\frac{a_J^2}{\sigma^2},\theta\right)f_{A_c}(a_c)da_cd\theta \tag{30}$$

3.4 Error probability for noncoherent FH/mMFSK system in ad hoc network

In the presence of MAI interference the symbol error rate (SER) for FH/mMFSK system can be represented in general again by Eq.17.

3.4.1 Hop is not jammed

As a starting point in this derivation we use Eq. 16 which defines SER for noncoherent mMFSK in Gaussian noise. It can be written as

$$P_s(\text{HNJ}) = 1 - \sum_k b_k \exp\left(-c_k\frac{a_c^2}{\sigma^2}\right) \tag{31}$$

where b_k and c_k depend on m and M. Pdf for the signal amplitude is given by Eq. 20. By averaging Eq. 31 with respect to a_c we have:

$$P_s(\text{HNJ}) = \int_0^\infty f_{A_c}(a_c) \cdot P_s(\text{HNJ} \mid a_c) da_c = \sum_k b_k \frac{1}{1 + c_k \xi_c} \exp\left(-\frac{c_k \rho_c}{1 + c_k \xi_c}\right) \quad (32)$$

3.4.2 Hop is jammed

Similarly to the coherent detection, SER is given by Eq. 22. Pdfs for U_i, $i = 1, ..., M$ can be now represented as

$$p(U_l) = \frac{1}{2\sigma^2} \exp\left[-\frac{1}{2}\left(\frac{U_l + 2(a_c^2 + a_J^2 + 2a_c a_J \cos\theta)}{\sigma^2}\right)\right] \times$$
$$I_0\left(\frac{\sqrt{a_c^2 + a_J^2 + 2a_c a_J \cos\theta}\sqrt{2U_l}}{\sigma^2}\right) \quad l = 1,.., k$$

$$p(U_l) = \frac{1}{2\sigma^2} \exp\left[-\frac{1}{2}\left(\frac{U_l + 2a_c^2}{\sigma^2}\right)\right] I_0\left(\frac{a_c\sqrt{2U_l}}{\sigma^2}\right), l = k+1,..., m \quad (33)$$

$$p(U_l) = \frac{1}{2\sigma^2} \exp\left[-\frac{1}{2}\left(\frac{U_l + 2a_J^2}{\sigma^2}\right)\right] I_0\left(\frac{a_J\sqrt{2U_l}}{\sigma^2}\right), l = m+1,...,2m-k$$

$$p(U_l) = \frac{1}{2\sigma^2} \exp\left[-\frac{U_l}{2\sigma^2}\right], l = 2m-k+1,..., M$$

where a_J is the jamming signal amplitude, and θ is the jamming signal phase relative to the useful signal, uniformly distributed in $[0,2\pi]$ interval. Probability $P(\text{HJ}/k\text{SBJ})$ is given by Eq. 25, where

$$P(U_{k+1} > U_{m+1}) = \int_{-\infty}^{U_{k+1}} p(U_{m+1}) dU_{m+1} = 1 - Q\left(\sqrt{2\frac{a_J^2}{\sigma^2}}, \sqrt{\frac{U_{k+1}}{\sigma^2}}\right)$$

$$P(U_{k+1} > U_{2m-k+1}) = \int_{-\infty}^{U_{k+1}} p(U_{2m-k+1}) dU_{2m-k+1} = 1 - \exp\left(-\frac{U_{k+1}}{2\sigma^2}\right)$$

$$P(U_1 > U_{m+1}) = \int_{-\infty}^{U_1} p(U_{m+1}) dU_{m+1} = 1 - Q\left(\sqrt{2\frac{a_J^2}{\sigma^2}}, \sqrt{\frac{U_{k+1}}{\sigma^2}}\right) \quad (34)$$

$$P(U_1 > U_{2m-k+1}) = \int_{-\infty}^{U_1} p(U_{2m-k+1}) dU_{2m-k+1} = 1 - \exp\left(-\frac{U_{k+1}}{2\sigma^2}\right)$$

where $Q(a,b)$ is Marcum's Q-function. By averaging Eq. 25 with respect to U_1 and U_{k+1} we have

$$P_s(\text{HJ}/\,k\,\text{SBJ}\mid a_c,\theta) = F_{noncoh}\left(\frac{a_c^2}{\sigma^2},\frac{a_J^2}{\sigma^2},\theta\right), \text{ where} \tag{35}$$

$$F_{noncoh}(y,z,\theta) = 1 - \left[\ \frac{1}{2}\int_0^\infty \exp\left(-\frac{x}{2}-y-z-2\sqrt{yz}\cos\theta\right)\times\right.$$

$$\left. I_0\left(\sqrt{y-z-2\sqrt{yz}\cos\theta}\sqrt{2x}\right)\times F_Q(x,z,m-k)F_{\exp}(x,M-2m+k)dx\ \right]^k \times \tag{36}$$

$$\times\left[\ \frac{1}{2}\int_0^\infty \exp\left(-\frac{x}{2}-y\right)I_0\left(\sqrt{y}\sqrt{2x}\right)F_Q(x,z,m-k)F_{\exp}(x,M-2m+k)dx\ \right]^{m-k}$$

in which

$$F_Q(x,z,n) = \begin{cases} \left[1-Q\left(\sqrt{2z},\sqrt{x}\right)\right]^n, & n \ge 0 \\ 1, & n < 0 \end{cases}$$

$$F_{\exp}(x,n) = \begin{cases} \left[1-\exp\left(-\frac{x}{2}\right)\right]^n, & n \ge 0 \\ 1, & n < 0 \end{cases} \tag{37}$$

After averaging Eq. 35 with respect to θ and a_c we get

$$P_s(\text{HJ}/\,k\,\text{SBJ}) = \frac{1}{2\pi}\int_0^{2\pi}\int_0^\infty F_{noncoh}\left(\frac{a_c^2}{\sigma^2},\frac{a_J^2}{\sigma^2},\theta\right)f_{A_c}(a_c)da_c\,d\theta \tag{38}$$

4 NUMERICAL RESULTS

By introducing multitone MFSK signaling the energy of the useful signal is split into m frequency bins and SER is expected to be higher. Still for as long as system efficiency $E_{\text{eff}} > 1$ the new schemes will perform better than the standard MFSK/FH system. In this section we present numerical results to illustrate under what conditions we can expect $E_{\text{eff}} > 1$. The numerical results presented are computed with the number of frequency hopping carriers $N=1000$ and the number of users $q=100$.

Efficiency improvement as a function of SNR in the presence of MAI interference for coherent FH/mMFSK system is shown in Fig. 4. Solid lines represent the case with no MAI interference and dashed lines represent the case with MAI interference with $P_c / P_{mait} = -20$ dB. This corresponds to $q=100$ additional users of the same power. The values of the other signal parameters are: C - $m=3$, F - $m=6$, b - $M=8$, c - $M=16$, d - $M=32$. One can see that for lower m (set C), E_{eff} becomes higher for lower E_s / N_0 but $\max(E_{eff,C}) < \max(E_{eff,F})$. For each set of curves, $\max(E_{eff})$ is higher if M is higher.

In a multihop ad hoc network not all signals will reach a certain point with the same power level. Relative power levels of these signals will depend on the network topology. As an indication of the problem, in Fig. 5 symbol error probability with coherent detection is shown as a function of signal to total interference ratio. The parameters are as follows: A - $m=1$, B - $m=2$, D - $m=4$, a - $M=4$, b - $M=8$, c - $M=16$, d - $M=32$. The thermal noise level was set at $E_b / N_0 = 14$ dB. As it was expected the lower the m the lower the BER and vice versa. Also, for lower M lower BER is obtained.

Efficiency improvement as a function of signal to noise ratio in the presence of fading and interference with coherent detection is shown in Fig. 6. The number of signaling tones $m=2$. The other parameters are as follows: A - no fading, B - Rayleigh signal fading, a - $M=4$, b - $M=8$, c - $M=16$, d - $M=32$. The solid lines represent the cases of no interference and the dashed lines interference with $P_c / P_{Jt} = -20$ dB. One can see that efficiency improvement is worse in fading environment. This is due to the fact that splitting signal power to m frequency bins is more critical in fading channel.

The same set of curves is shown in Fig. 7 for, $m = 4$. One can see that for $E_s / N_0 < 10$ dB, E_{eff} is much worse for $m=4$ than for $m=2$. On the other hand if $E_s/N_0 >10$ dB E_{eff} becomes much better for $m=4$.

Efficiency improvement as a function of SNR in the presence of interference for coherent and noncoherent detection is compared in Fig. 8. The notations and parameters are: B - $m=2$, D - $m=4$, a - $M=4$, b - $M=8$, c - $M=16$. The solid lines represent the cases of no interference and the dashed lines interference with $P_c / P_{Jt} = -20$ dB. The thicker lines refer to coherent detection. If $E_s / N_0 > 12$ dB, improvements are larger than one. Improvements are larger for coherent system. For $E_s / N_0 > 18$ dB, parameter E_{eff} for both coherent and noncoherent systems become the same.

Fig. 9 represents SER as a function of signal to total interference ratio for no fading environment with $M = 4$ and $E_b / N_0 = 14$ dB. The solid lines represent the case of coherent detection and the dashed lines noncoherent detection. Other parameters and notations are: A - $m=1$, B - $m=2$. In a multihop ad hoc network for the same number of users P_c / P_{mait} may increase and the system performance will become better. For $q=100$, P_c / $P_{mait} = -20$dB corresponds to the case when all signals have the same power.

Finally Fig. 10 represents efficiency improvement as a function of signal to noise ratio in the presence of fading for noncoherent detection. The solid lines represent the case of no fading and the dashed lines Rayleigh signal fading. The notation and the signal parameters are as follows: B - $m=2$, D - $m=4$, b - $M=8$, c - $M=16$. In this case also, for $E_s/N_0 > 12$ dB efficiency improvements become larger than one.

As a conclusion one can say that FH/mMFSK modulation offers better performance than standard FH/MFSK signal format. Although for $m>1$ the signal energy is split to m frequency bins, resulting in higher BER, the average number of correctly transmitted bits is increased. To achieve $E_{eff} > 1$, signal to noise ratio is supposed to be higher than a threshold value. Depending on the signal parameters this threshold value is between 9-14 dB.

5   CONCLUSIONS

In this paper we analyze performance of wireless ad hoc networks with the physical layer based on mMFSK Frequency Hopping Modulation. This modulation is a modification of FH/MFSK system that includes multitone MFSK signal designated as mMFSK modulation. In this case the signal energy is split to m separate tones. This will reduce the signal power spectral density by factor $1/m$ and make it more attractive for applications such as low probability of intercept systems. This is also very attractive characteristic for wireless LANs in ISM frequency band where the transmitted power density is limited by regulations in order to prevent excessive interference to other systems sharing the same bandwidth. Due to reduced power level of each tone, the system is more vulnerable to noise and fading but still the overall flow of useful information, characterized by the system efficiency, is increased. The results demonstrate that under the large range of the signal, channel and interference parameter values this system offers better performance. Depending on the signal parameters the efficiency is increased by factor 2-3.5.

REFERENCES

E. Lance and G. Kaleh, "A diversity scheme for phase coherent frequency hopping spread spectrum system," IEEE Transactions on Communications, No. 9, pp. 1123-1130, Sep -97.

G.K. Kaleh, "Frequency-diversity spread spectrum communication system to counter bandlimited jammers," IEEE Trans. Commun., vol. 44, pp. 886-893, July 1996

"Performance comparison of frequency diversity and frequency hopping spread spectrum systems," IEEE Trans. Commun. vol. 45, pp. 910-912, Aug. 1997.

M. Simon and A. Polidoros, "Coherent detection of frequency-hopped quadrature modulation in the presence of jamming - Part I: QPSK and QASK modulations," IEEE Trans. Commun., vol. COM-29, pp. 1644-1660, Nov. 1981.

I. Ghareeb and A. Yongaçoglu, "Performance analysis of frequency-hopped/coherent MPSK in the presence of multitone jamming," IEEE Trans. Commun., vol. 44, pp. 152-155, Feb. 1996.

G. Cherubini and L.B: Milstein, "Performance analysis of both hybrid and frequency-hopped phase-coherent spread-spectrum systems, Part II: An FH system," IEEE Trans. Commun., vol. 37, pp. 612-622, June 1989.

C.M. Su and L.B. Milstein, "Analysis of coherent frequency-hopped spread-spectrum receiver in the presence of jamming," IEEE Trans. Commun. vol. 38, pp. 715-726, May 1990.

R.C. Robertson et al, "Multiple tone interference of frequency hopped noncoherent MFSK signals transmitted over Rician fading channel," IEEE Trans. Commun. vol. 44, No. 7, pp. 867-875, July 1996.

R.L. Peterson, R.E.Ziemer, and D.E.Borth," Introduction to Spread Spectrum Communications", Englewood Cliffs, NJ: Prentice-Hall, 1995

D.L.Nicholson, " Spread Spectrum Signal Design: LPE and AJ Systems" Salt Lake City, UT, W.H.Freeman and Co., Computer Science Press, 1988

J.S.Lee, R.H. French, and L.E. Miller, "Probability of error analyses of a BFSK frequency-hopping system with diversity under partialband jamming interference, Part I: Performance of square-law linear combining soft decision receiver," IEEE Trans. Commun., vol. COM-32, pp. 645-653, June 1984.

R.C. Robertson and K.Y. Lee, "Performance of fast frequency-hopped Rician fading channel with partial-band interference," IEEE J. Select. Areas Commun., pp. 731-741, May 1992.

S.W.Houston, "Modulation techniques for communication, Part I: Tone and noise jamming performance in IEEE NAECON Rec. 1975, pp. 51-58.

B.K. Levitt, "Use of diversity to improve FH/MFSK performance in worst case partial band noise and multitone jamming," in Proc. IEEE MILCOM, 1982, pp. 28.2-1-28.2-5.

J.S. Bird and E.B. Felstead, "Antijam performance of fast frequency-hopped M-ary NCFSK-AN overview, IEEE J. Select. Areas Commun., vol. SAC-4, pp. 216-233, Mar. 1986.

M.J. Massaro, "Error performance of M-ary noncoherent FSK in the presence of CW tone interference," IEEE Trans. Commun., vol. COM-23, pp. 1367-1369, Nov. 1975.

L.B. Milstein, R.L.Pickholtz and D.L. Shilling, "Optimization of the processing gain of an FSK-FH system," IEEE Trans. Commun., vol. COM-28, pp. 1062-1079, July 1980.

S.G.Glisic et al, "Efficiency of digital communication systems," IEEE Transactions on Communications, Vol.Com-35, No.6, pp.679-684, June 1987.

M.Iida and G. Marubayashi, "Multi–Tone Combinatory Frequency Hopping System," Proceedings of ISSSTA'96, Mainz, September 1996, pp. 893-897.

D.J.Goodman, P.S.Henry and V.K.Prabhu, "Frequency-hopped multilevel FSK for mobile radio", Bell Syst.Tech.J., vol.59, no.7, pp.l257-1275, Sept.1980.

Timor.U, "Multitone Frequency-Hopped MFSK System for Mobile Radio", Bell Syst.Tech.J., vol.61, no.10, pp.3007-3017, Dec.1982.

T.Mabuchi, R.Kohno and H.Imai, "Multihopping and Decoding of Error-Correcting Code for MFSK/FH-SSMA Systems", IEICE Trans. comm., vol.E76-B, no.8, pp.874-885, Aug.1993.

G.Einarsson, "Address Assignment for a Time-Frequency-Coded, Spread-Spectrum System", Bell Syst.Tech.J., vol.59, no.7, pp.1241-1255, Sept.1980.

J.Zhu, S.Sasaki and G.Marubayashi, "Proposal of Parallel Combination Spread Spectrum Communication System", IEICE Trans. comm., vol.J74-B-II, no.5, pp.207-214, May.1991

W.Mao, R.Kohno and H.Imai, "MFSK/FH-CDMA System Using Two-Stage Address Coding", Technical report of IEICE SST94-33, pp.l-5, Aug.1994.

P. Agrawal, "Wireless Access Protocols for Mobile Multimedia Networks," PIMRC'98 Tutorial, Boston, September 1998

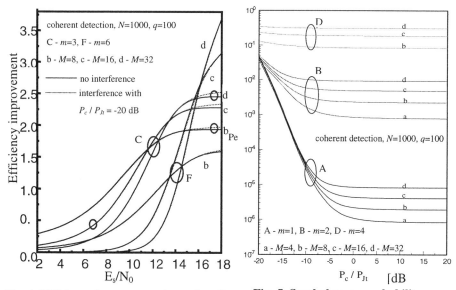

Fig 4. Efficiency improvement as a function of signal to noise ratio in the presence of interference

Fig. 5. Symbol error probability as a function of signal to total interference ratio with $E_b / N_0 = 14$ dB

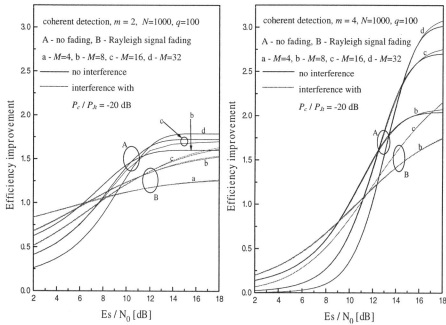

Fig. 6. Efficiency improvement as a function of signal to noise ratio in the presence of fading and interference

Fig. 7. Efficiency improvement as a function of signal to noise ratio in the presence of fading and interference

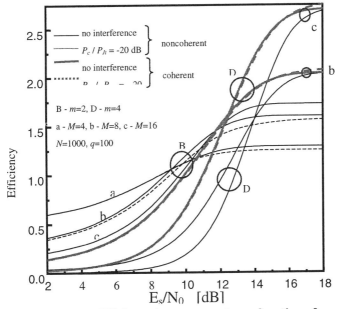

Fig. 8. Efficiency improvement as a function of signal to noise ratio in the presence of interference

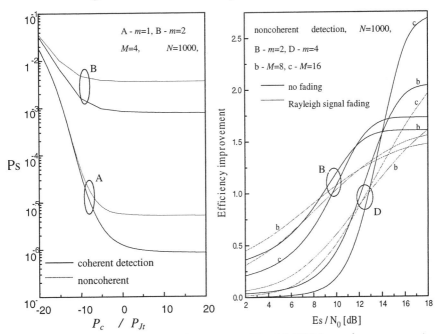

Fig. 9. Symbol error probability as a function of signal to total interference

Fig. 10. Efficiency improvement as a function of signal to noise ratio

Index

A

A/D, *182, 231*
ABR traffic, *207-210, 212, 215*
access long code, *38*
ACTS, *5-7, 15, 19, 20, 46, 54, 57, 58, 60, 65, 66, 68-73, 83, 124*
Ad hoc networks, *331, 332*
adaptive array antenna, *100, 182*
adaptive equalizer, *100, 228*
adaptive receiver technologies, *100*
adaptive zone configuration, *100*
AM-AM distortion, *262, 264*
Amplifier linearity, *252*
AM-PM distortion, *262, 264*
AMPS, *9, 11, 13, 24, 25, 46, 92, 228*
ANSI, *4, 17, 91*
ANSI T1P1, *91*
application specific integrated circuits (ASICs), *228, 236-238, 241-245*
ARIB, *17, 46, 49, 75, 81, 84, 91-93, 116, 134, 175, 176*
ATDMA, *5*
ATM, *57, 58, 64, 66-69, 71-74, 117-135, 158, 160, 166, 171, 175, 179, 184, 206, 214, 221, 222, 241, 248*
ATM Forum's Wireless ATM working group, *134*
ATM network technologies, *117*
ATM switch, *120, 152, 158, 160*
ATM system architecture, *117, 118*

B

bandwidth on demand, *174*
battery capacity, *139*
Base Station Controller (BSC), *154*
Base Terminal Station (BTS), *154, 159*
beamforming, *228*

BER, *101, 104-107, 109, 115, 180, 255, 346, 347*
BiCMOS, *230, 231, 247*
binary frequency-shift keying (BFSK), *333, 348*
BISDN, *206*
BRAN (Broadband Radio Access Networks), *67, 68, 118, 122, 131, 134, 179*
broadband amplifiers, *251*
broadband wireline modems, *175*

C

call control, *155*
Call Origination Percent, *168, 169*
call termination procedures, *164, 168*
canonic signed-digit (CSD), *243*
Cartesian feedback, *255, 258, 259, 262, 266*
Cartesian feedback linearizer, *258*
cascaded integrator-comb (CIC), *244*
CBR, *121, 123, 206-218*
CDG, *13, 175*
CDMA, *5-7, 13, 23-28, 37, 38, 40, 43, 44, 47-50, 54, 55, 84, 88, 91, 97, 98, 101-115, 173-178, 182, 184, 209, 210, 214, 221, 222, 240, 241, 263, 332*
cdma2000, *23, 28-30, 35, 41-43, 47-50*
Cellular Digital Packet Data (CDPD), *37, 127, 145*
Conference of European Post and Telecommunications (CEPT), *5, 16, 17, 20, 53, 68*
channel estimation word (CE), *105*
characteristic function (CHF), *311, 320*
Class A power amplifier, *252*
Complementary Metal Oxide Semiconductor (CMOS), *229-233, 247, 248*

Q

R

S

X

About the Editors

RAJAMANI GANESH received the B.E. degree from Indian Institute of Science, Bangalore, India and the Ph.D. degree in Electrical Engineering from Worcester Polytechnic Institute, Worcester, Massachusetts. From 1991 to 1995 he was part of the Communications Research Laboratory at Sarnoff Corporation in Princeton, New Jersey. While at Sarnoff, he worked on several communication systems design projects including the HDTV transmission system and digital cellular CDMA networks. Since 1995, he has been a part of the Network Planning and Engineering department at GTE Laboratories near Boston. He is actively involved in wireless system design analysis and research and his work has been incorporated in GRANET - a cellular radio planning tool for network optimization which helps GTE Wireless and other GTE business units optimally plan their analog and digital cellular radio networks. He has published several technical papers in many journals and conferences, applied for many patents and has received numerous awards including the prestigious WARNER award, GTE's highest award for outstanding technical achievement. Dr. Ganesh has been an organizing and technical program committee member of many international conferences and is a technical program chairman of the International Conference on Personal Wireless Communications, to be held in December 2000 in India.

KAVEH PAHLAVAN is a Professor of ECE and Director of Center for Wireless Information Network Studies at the Worcester Polytechnic Institute, Worcester, MA. Prior to joining WPI in 1985, he was the Director of Advanced Developments, INFINET Inc., Andover, MA and had taught for four years in the ECE Dept., Northeastern University, Boston, MA. He has served as a consultant for the design and analysis of wireless networks in many companies in different states and countries including Nokia, Elektrobit, LK-Products, TEKES in Finland, GTE Laboratories, Sierra Communications, BBN, DEC, Codex/Motorola, Mercury Computers, WINDATA in Massachusetts, JPL, Savi Technologies, RadioLAN in California, Honeywell Avionics Systems in Arizona, Aironet in Ohio, NTT in Japan. He is the Editor-in-Chief of the International Journal of Wireless Information Networks. He was the technical program chairman and organizer of the third and the ninth IEEE International Symposium on Portable, Indoor, and Mobile Radio Communications, Boston, MA and the organizer and the chairman of the IEEE workshop on wireless local area networks, Worcester, MA in 1991 and 1996. He is the principal author of the Wireless Information Network, John Wiley and Sons, 1995 (co-authored by Dr. Allen Levesque). He has also contributed to numerous book chapters, technical papers and patents. He was the Weston Hadden Professor of ECE at WPI during 1993-1996, became a fellow of the IEEE in 1996 and this year he is a fellow of Nokia. In the past decade he has given invited lectures in more than 10 different countries in 12 major academic institutions (including MIT, Stanford, ETH in Switzerland, and Deft U. in the Netherlands) 8 industrial research institutions (such as Sarnoff Corporation, NTT Laboratories in Japan, Nokia research laboratories in Finland) and 15 international conferences.

ZORAN ZVONAR received the Dipl. Ing. degree in 1986 and the M.S. degree in 1989, both from the Department of Electrical Engineering, University of Belgrade, Yugoslavia, and the Ph.D. degree in Electrical Engineering from the Northeastern University, Boston, in 1993. From 1986 to 1989 he was with the Department of Electrical Engineering, University of Belgrade, Yugoslavia as a faculty member. From 1993 to 1994 he was a Post-Doctoral investigator at the Woods Hole Oceanographic Institution, Woods Hole, MA, where he worked on multiple-access techniques for underwater acoustic communications. Since 1994 he has been with the Analog Devices, Communications Division, Wilmington, USA, where he is currently Systems Development Manager. He was a Guest Editor of the IEEE Transactions on Vehicular Technology, the International Journal of Wireless Information Networks and the ACM/Baltzer Wireless Networks, and a co-editor of the book "GSM: Evolution Towards Third Generation Systems", Kluwer Academic Publishers, 1998. Dr. Zvonar is currently an Associate Editor of the IEEE Communications Letters and a Feature Editor of the series on Software & DSP in Radio in the IEEE Communications Magazine.

CONTRIBUTORS

PART I: OVERVIEW OF EVOLVING 3G TECHNOLOGIES

Chapter 1   *Ensuring the Success of Third Generation*

ROBERT VERRUE is the Director General of DG XIII, European Commission in charge of Information Society: Telecommunications, Markets, Technologies - Innovation and Exploitation of Research. He was until December 1998 responsible for the Advanced Communication Technologies and Services, Telematic Applications, and Innovation programs, and is now responsible for the Innovation and Information Society Technologies programs, the latter bringing together all R&D aspects concerning the Information Society. He previously held several important positions in the European Commission, including Division Head for Co-ordination of Monetary Policies at the DG for Economic and Financial Affairs; Director at the DG for Industrial Affairs and Internal Market; and Director General Adjoint at the DG for External Affairs. He attended the École Supérieure de Commerce et d'Administration des Entreprises, Lille, France; and the College of Europe—International Economics Section, Bruges, Belgium. He also earned an MBA from the European Institute of Business Administration (INSEAD), Fontainebleau, France.

JORGE M. PEREIRA obtained the Engineering and M.Sc. degrees in Electrical and Computer Engineering from Instituto Superior Técnico (IST), Lisboa, Portugal in 1983 and 1987, respectively; and received the Ph.D. in Electrical Engineering-Systems from the University of Southern California, Los Angeles, in 1993. Between 1983 and 1988, taught in the Department of Electrical and Computer Engineering of IST as Full Lecturer; and in 1994 became Assistant Professor. From 1988 to 1990 worked at LinCom Corp., Los Angeles, in a NASA project on space-to-space communications, and on synchronization issues. From 1991 to 1993, worked for Caltrans and PATH on Intelligent Vehicle Highway Systems (IVHS). From 1993 to 1996, at GTE Laboratories Inc., Waltham, MA, was responsible for Communication Analysis and Simulation in the National IVHS Architecture FHWA contract; and represented GTE at the TIA IVHS Section, and in the High Speed Data Systems workgroup of the CDMA Development Group. Since September 1996, he has been with the European Commission, DG XIII, as scientific officer in the area of Mobile and Personal Communications.

Chapter 2   *Third Generation Wireless Communication Technologies*

LIN-NAN LEE is Vice President, Advance Development at Hughes Network Systems (HNS) in Germantown, Maryland, USA. He and his group made many significant contributions to the design and engineering of HNS' wireless products, and are actively working on the architecture and design of HNS' CDMA and 3rd Generation Wireless products. Dr. Lee received his B.S. Degree from National Taiwan University, M.S. and Ph. D. from University of Notre Dame, all in Electrical Engineering. He started his career at Linkabit Corporation, where he was a Senior Scientist working on Packet communications over satellites at the dawn of the Internet age. He then worked at Communication Satellite Corporation (COMSAT) in various research and development capacities with emphasis on source and channel coding technology development for satellite transmission and eventually assumed the position of Chief Scientist, COMSAT Systems Division. Dr. Lee is a Fellow IEEE. He was the co-

recipient of the COMSAT Exceptional Invention Award, and the 1985 and 1988 COMSAT Research Award. He has authored or co-authored more than 20 US patents.

FENG-WEN SUN received the B.S. degree from Heilongjiang University, M.S. from Nankai University in China. He received his Ph.D. degree in Eindhoven University of Technology, the Netherlands in 1994. He was the recipient of the 1995 Canada International Fellowship tenable at McGill University, Montreal. He jointed Hughes Network Systems in 1996, and is a Principal Engineer in the Advance Development Group. He has been actively involved in the US and European third generation wideband CDMA standard activities and served as the chair for the Turbo code ad hoc group in the TIA/EIA TR45.5 cdma2000 Standards development. He has more than 20 papers published in various journals. He was an invited speaker for the 1994 Information Theory Symposium held in Japan. He is the recipient of 1998 Hughes Electronics Patent Award and the co-recipient of the 1998 CDMA Technical Achievement Award. His research area includes coded modulation, convolutional and algebraic coding, signal processing, equalization, receiver design and implementation of wireless communication systems, especially CDMA related technologies.

KHALID KARIMULLAH received his Ph. D in Electrical Engineering from Michigan State University, East Lansing, Michigan in the Year 1980. He started his professional career at Comsat Laboratories, Maryland where he worked on Regenerative Satellite Transponder modem designs. He later joined MA-Com Linkabit, San Diego in April 1987, where he developed his expertise in the areas of communications and signal processing. He joined Hughes Network Systems in 1989 and has since worked on CDMA technology. Currently, he works at HNS Germantown as a Senior Principal Engineer mostly involved in CDMA technology related activities. He has been active in the TIA/EIA TR45.5 Cdma2000 Standards Development, participating in the Physical Layer and enhanced access procedures development. He chaired the TR45.5 cdma2000 enhanced access ad hoc group. He has authored/ co-authored several patents in the CDMA physical layer and enhanced access procedures. In 1998, he received Hughes Electronics Patent Award and was the co-recipient of the 1998 CDMA Technical Achievement Award.

MUSTAFA EROZ received his B.S. from Bilkent University, Ankara, Turkey in 1992 and his Ph.D. from the University of Maryland, College Park in 1996, both in Electrical Engineering. Since then, he has been with the Advanced Development Group of Hughes Network Systems, Germantown, MD. His research interests include trellis-coded modulation, turbo codes and their application to third generation wireless systems. . In 1998, he received Hughes Electronics Patent Award and was the co-recipient of the 1998 CDMA Technical Achievement Award.

A. ROGER HAMMONS, JR. received his B.S. degree with highest honors and distinction in mathematics and the humanities from Harvey Mudd College; the M.A. degree in mathematics from U.C.L.A., where he was a teaching assistant; and the Ph.D. in electrical engineering from the University of Southern California, where he was a Hughes Aircraft Company Ph.D. fellow. His doctoral work in the area of nonbinary sequence designs as families of CDMA signature sequences and as error control codes. For this work, the study of Z4-linear codes is established as a new research area in coding theory. He was awarded the IEEE Information Theory Society 1995 Best Paper Award. Dr. Hammons has been with Hughes Electronics since 1981, first with the Missile Systems Group and since 1993 with Hughes Network Systems. He is currently the manager in the Advanced Development Group for R&D in the areas of transmission technology including channel coding and modem algorithms. He was the co-recipient of the 1998 CDMA Technical Achievement Award.

Chapter 3   *Perspectives on 3G System Development*

TERO OJANPERÄ received the M.Sc. degree from University of Oulu, Finland, in 1991. Thereafter, he worked for Nokia Mobile Phones, Finland, as a research engineer and research group leader. During 1995 and 1997 he was leading development activities of third generation mobile communications systems both in research and standardization within Nokia Research Center, Finland. During 1997 and 1998 he worked as a principal engineer in Nokia Research Center, Irving, TX. Currently, he is with Nokia Telecommunications in Finland as head of research, radio access systems. He is co-author of a book Wideband CDMA for Third Generation Mobile Communications (Artech House, 1998) and author of a chapter in books Wireless Communications TDMA vs. CDMA (Kluwer, 1997), GSM: Evolution towards 3rd generation (Kluwer, 1998), and The Mobile Communications Handbook (CRC Press, 1999). He is member of IEEE and Vice Chair of ICC'2001 conference.

Chapter 4   *An Unified, Open Architecture to Deliver Mobile Multimedia*

JORGE M. PEREIRA (his biography is mentioned under chapter 1).

Chapter 5   *UMTS/IMT-2000 Standardization*

ANTUN SAMUKIC received the M.Sc.E.E. degree from Croatian University, Zagreb, 1976 and the Bachelor of Commerce degree from the Institute for Higher Marketing, Stockholm, in 1984. He is the 3GPP Support Team Program Manager UMTS System Architecture. He was ETSI SMG Program Manager UMTS and a member of ETSI Global Multimedia Mobility Co-ordination Group. He has been with Ericsson since 1976 and has held Manager or Chairman positions within Ericsson, ETSI and ITU related to strategic product planning and standardisation: UMTS and IMT-2000; Private Network Mobility based on DECT and PABX; Mobility between private and public networks based on DECT and/or GSM; and VPN Virtual Private Networks. During 1993 - 1995 he was, at RACE research project, Consensus Manager UMTS and Satellite Systems. He initiated and chaired SMG Group "Evolution towards UMTS" which contributed to the acceptance of the UMTS evolution approach. He was one of the editors for the ITU-R TG8/1 study "Evolution/migration towards FPLMTS". Mr. Samukic was also member of the European Radio Office Study Team "UMTS Frequency Allocations" and its report was used as a basis for the ERC decision on the allocation of frequencies for UMTS.

Part II: TOPICS IN WIRELESS MULTIMEDIA NETWORKS

Chapter 6   *Intelligent and Flexible Radio Access/Transmission Technologies for Wireless Multimedia Communication Systems in the Software Defined Radio Era*

SEIICHI SAMPEI received the B.E., M.E. and Ph.D. degrees in electrical engineering from Tokyo Institute of Technology in 1980, 1982 and 1991 respectively. From 1982 to 1993, he was with the Communications Research Laboratory, Ministry of Posts and Telecommunications, Japan where he was engaged in developing adjacent channel interference rejection, fading compensation and M-ary QAM technologies for land-mobile communication systems. During 1991-1992, he was at the University of California, Davis, as a visiting re-searcher. In 1993, he joined the Department of Communications Engineering,

Osaka University, where he is currently an Associate Professor. Dr. Sampei received the Shinohara Young Engineering Award from the IEICE and the Telecom System Technology Award from the Telecommunication Advancement Foundation. He is a member of the IEICE, IEEE and the Institute of Image Information and Television Engineers of Japan.

NORIHIKO MORINAGA received the B.E. degree in electrical engineering from Shizuoka University, Shizuoka, Japan, in 1963 and the M.E. and Ph.D. degrees from Osaka University, Osaka, Japan in 1965 and 1968, respectively. He is currently a Professor in the Department of Communications Engineering, Osaka University, working in the area of radio, mobile, satellite, and optical communication systems and EMC. Dr. Morinaga received the Telecom Natural Science Award and the Telecom System Technology Award from the Telecommunication Advancement Foundation and the Paper Award from the IEICE. During May 1998 - May 1999, he was the President of the Communication Society of the IEICE, and he is currently the Vice President of the IEICE. He is also a member of the Institute of Image Information and Television Engineers of Japan, and a senior member of the IEEE.

Chapter 7   **Current Topics in Wireless & Mobile ATM Networks: QoS Control, IP Support and Legacy Integration**

DIPANKAR RAYCHAUDHURI received the B.Tech. (Honors) degree from the Indian Institute of Technology, Kharagpur, India in 1976, and the MS and PhD degrees in Electrical Engineering from SUNY, Stony Brook in 1978 and 1979 respectively. He was with the David Sarnoff Research Center (formerly RCA Laboratories), Princeton, NJ, as Member of Technical Staff (1979-87), Senior Member of Technical Staff (1988-89) and Head, Broadband Communications Research (1990-92). Since Jan. 1993, he has been with NEC USA, C&C Research Laboratories, Princeton, NJ, where he is currently Assistant General Manager & Department Head (Systems Architecture), with focus on multimedia networking technologies including IP and ATM switching/protocols, broadband wireless and distributed multimedia software. Dr. Raychaudhuri is a Fellow of the IEEE, and has authored over 100 technical papers and 10 U.S. patents. He is currently a Technical Editor of the IEEE Multimedia Magazine, and has previously served as Editor, IEEE Transactions on Networking (1993-98) and IEEE Communications Society Distinguished Lecturer (1992-96). Since 1996 he has been Vice-Chair of the ATM Forum's Wireless ATM working group.

Chapter 8   **Providing Internet Services to Mobile Phones: A case study with Email**

THOMAS Y.C. WOO received the B.S. (first-class honors) degree in Computer Science from the University of Hong Kong in 1986, and the M.S. and Ph.D. degrees in Computer Science from the University of Texas at Austin, in 1988 and 1994, respectively. He joined AT&T Bell Laboratories in 1994. He is currently a Member of Technical Staff in the Networking Software Research Department at Bell Laboratories, Lucent Technologies. In the last few years, he had worked mostly in the area of Internet protocols, wireless data networking and network security. His current research interests include router design and implementation, Internet protocols, security, wireless networking, and Web technologies. Dr. Woo has served as a program committee member for the 1997 IEEE International Conference on Network Protocols, an area technical program chair for IEEE INFOCOM 1998, and a program committee member for the 1999 ACM SIGCOMM.

KRISHAN SABNANI has made major contributions to the communications protocols' area. He has designed several protocols such as SNR, RMTP, and Airmail. He has also made *significant contributions to conformance test generation, protocol validation, automated converter generation, and reverse engineering. Krishan has also led two impactful wireless networking projects. Krishan is a Bell Labs Fellow. He received the Leonard G. Abraham award from the IEEE Communications Society in 1991. He received the Bell Laboratories Distinguished Technical Staff Award in 1990. He is also a fellow of the IEEE. Krishan, who heads the Networking Software Research Department of the Networking Research Laboratory at Bell Labs, received his Ph.D. in electrical engineering from Columbia University, New York, in 1981. He joined Bell Labs in 1981.*

SCOTT C. MILLER received his B.S. and M.S. degrees in Electrical Engineering from the Cooper Union, New York, NY in 1986 and 1987, respectively. He joined AT&T Bell Laboratories in 1987. He is now a member of the Networking Software Research Department in Bell Laboratories, Lucent Technologies. Scott's interests are in software for communications networks, Internet protocols, and wireless networking.

Chapter 9   ***Design, Implementation, and Performance of a Cluster Mobile Switching Center***

RAMACHANDRAN RAMJEE received his B.Tech in Computer Science and Engineering from the Indian Institute of Technology, Madras in 1992 and his M.S. and Ph.D. in Computer Science from University of Massachusetts, Amherst in 1994 and 1997 respectively. He is currently a Member of Technical Staff at Bell Labs, Lucent Technologies. His research interests are signaling, mobility management and quality of service issues in wireless and high speed networks.

KAZUTAKA MURAKAMI received the B.Eng. and M.Eng. degrees in electrical engineering from University of Tokyo in 1982 and 1984, respectively, and the Ph.D. degree in electrical and computer engineering from Carnegie Mellon University, Pittsburgh, PA, in 1995. He was with the Tokyo Research Laboratory of IBM Japan Ltd. from 1984 to 1991, where he worked in the fields on computer networking and distributed systems. Since 1996, he has been a member of Technical Staff at Bell Labs Research, Lucent Technologies, Holmdel, NJ, where he works on highly-available distributed call processing systems for 3rd generation wireless telecommunications networks with advanced features. His research interests include network optimization, survivable network management, routing control, call processing systems, distributed object-oriented systems, and fault-tolerant systems. Dr. Murakami was a Feature Editor of IEEE Communications Magazine from 1986 to 1991 and a Program Committee Member of IEEE ICON 99. His e-mail address is: kmurakami@bell-labs.com.

RICHARD W. BUSKENS is a member of technical staff in the Networking Software Research Department at Bell Labs in Holmdel, New Jersey. He is interested in distributed and parallel algorithms, networking and network management, and fault tolerance. Mr. Buskens received a B.S. in electrical engineering and an M.S. in computer science, both from the University of Manitoba in Winnipeg, Manitoba, Canada. At Carnegie Mellon University, in Pittsburgh, Pennsylvania, he earned a Ph.D. in electrical and computer engineering.

YOW-JIAN LIN (yjlin@lucent.com) is a member of the technical staff in Bell Labs Research, Lucent Technologies. He received his BS degree in Electrical Engineering from National Taiwan University, MS degree in Electrical and Computer Engineering from University of California, Santa Barbara, and PhD degree in Computer Sciences from University of Texas at Austin. From 1988 to 1995 he was with the Applied Research Area of Bell Communications Research (Bellcore, recently renamed as Telecordia), first as a member of the technical staff, then as a director of Systems Specification and Management Research Group. He was a key member of the Bellcore team that pioneered the feature interaction management research. He has been with Bell Labs Research since 1995. His current research interests include network configuration, network monitoring and control, software systems management, feature modeling and interaction detection.

THOMAS F. LA PORTA received his B.S.E.E. and M.S.E.E. degrees from The Cooper Union, New York, NY, in 1986 and 1987 respectively, and his Ph.D. degree in Electrical Engineering from Columbia University, New York, NY, in 1992. He is currently Head of the Networking Techniques Research Department in Bell Laboratories, Lucent Technologies. He received the Bell Labs Distinguished Technical Staff Award in 1996, and the Eta Kappa Nu Outstanding Young Electrical Engineer Award honorable mention in 1996. His research interests include mobility management algorithms, signaling protocols and architectures for wireless and broadband networks, and protocol design. Dr. La Porta is the Editor-in-Chief of IEEE Personal Communications Magazine and is a technical editor on ACM/Baltzer Journal of Mobile Networking and Applications. He is an adjunct member of faculty at Columbia University where he has taught courses on mobile networking and protocol design.

Chapter 10   *Challenges of Higher Speed Wireless Internet Access*

AHMAD R S BAHAI received his BS degree from Tehran University in 1996, MS degree from Imperial College, University of London in 1988 and Ph.D. degree from University of California at Berkeley in 1993, all in Electrical Engineering. From 1992 to 1994 he worked as a member of technical staff in the wireless communications division of TCSI. He joined AT&T Bell Laboratories in 1994 where he was Technical Manager of Wireless Communication Group in Advanced Communications Technology Labs until 1997. He is currently Chief Technical Officer of ALGOREX Inc. His research interest includes adaptive signal processing and communication theory. He has taught short courses in statistical communication theory and signal processing at several graduate engineering schools. His book on "Multi-carrier Digital Communications" is due to be published by September 1999. Dr. Bahai has five patents in Communications and Signal Processing field and currently serves as editor of IEEE Communication Letters.

Chapter 11   *Energy Efficient Protocols for Wireless Systems*

PRATHIMA AGRAWAL is assistant Vice President and Chief Scientist in the Internet Architecture Research Laboratory at Telcordia Technologies (formerly Bellcore), Morristown, New Jersey. During 1997-98, she was head of the Networked Computing Technology Department at AT&T Labs in Whippany, New Jersey. Prior to that, she headed the Networked Computing Research Department at Bell Labs in Murray Hill, New Jersey. Dr. Agrawal received her BE and ME degrees in electrical communication engineering from the Indian Institute of Science, Bangalore, India, and PhD degree in electrical engineering from the University of Southern California. Her research interests are computer networks, mobile and wireless computing, wireless Internet access systems, parallel processing and VLSI CAD. She has published over 150 papers and has received or applied for more than 30 U.S. patents. Dr. Agrawal is a Fellow of the IEEE and a member of the ACM. Presently, she chairs the IEEE Fellow Selection Committee.

JYH-CHENG CHEN received a B.S. degree in information science from Tunghai University, Taichung, Taiwan, R.O.C. in 1990. In 1992, he received his M.S. degree in computer engineering from Syracuse University, Syracuse, New York. He received his Ph.D. degree in electrical engineering from the State University of New York at Buffalo in 1998. From 1995 to

1996, he worked as a software engineer at ASOMA-TCI, Inc., North Tonawanda, New York. During the summer of 1997, he worked in the Networked Computing Technology Department at AT&T Labs, Whippany, New Jersey. Since August 1998, he has been a research scientist in the Applied Research at Telcordia Technologies (formerly Bellcore), Morristown, New Jersey. At Telcordia Technologies, his research is focused on third generation wireless networking and next generation networks (NGN). He is a member of IEEE and ACM.

DR. KRISHNA M. SIVALINGAM is an Assistant Professor in the School of Electrical Engineering and Computer Science, at Washington State University, Pullman. Earlier, he was an Assistant Professor at University of North Carolina, Greensboro from 1994 until 1997. He has conducted research at Lucent Technologies' Bell Labs in Murray Hill, NJ, and at AT&T Labs in Whippany, NJ, where he also served as a consultant during 1997. He received his Ph.D. and M.S. degrees in Computer Science in 1994 and 1990 respectively from State University of New York at Buffalo where he was a Presidential Fellow from 1988 to 1991. Prior, he received his B.E. degree in Computer Science and Engineering in 1988 from Anna University, Madras, India. His research interests include wireless and mobile networks, optical WDM networks, ATM networks, high-speed communication networks such as Gigabit Ethernet, high performance distributed computing, and performance evaluation.

Chapter 12   ***Guaranteed Quality-of-Service Wireless Access to Broadband Networks***

CHIA-SHENG CHANG received the B.S. and M.S. degrees from the National Tsing Hua University, Taiwan, R.O.C., in 1995 and 1997, respectively, all in electrical engineering. Since 1997, he has been with the Institute of Communications Engineering, College of Electrical Engineering, National Taiwan University, Taiwan, R.O.C., working toward the Ph.D. degree. His research interests include wireless broadband networks and multimedia multiple access.

KWANG-CHENG CHEN received B.S. from the National Taiwan University, M.S. and Ph.D from the University of Maryland, College Park, all in electrical engineering. From 1987 to 1991, Dr. Chen worked with SSE, COMSAT, and IBM Thomas J. Watson Research Center. During 1991 to 1998, he was with the National Tsing Hua University, Taiwan ROC. Since 1998, Dr. Chen is a professor at the Institute of Communications Engineering, National Taiwan University, Taiwan ROC. He was a visiting scientist with Hewlett-Packard Laboratories in California during 1997 and a visiting Professor at the Delft University of Technology, Netherlands, 1998 summer. Dr. Chen actively participates in the technical organization of many IEEE conferences, including as the Technical Program Committee Chair of 1996 PIMRC. He has been technically serving editorial board for the following journals: IEEE Transaction on Communications, IEEE Communications Letters, IEEE Personal Communications Magazine, IEEE Selected Areas in Communications, International Journal of Wireless Information Networks. Dr. Chen has been an IEEE Senior Member since 1993, was elected Outstanding Young Engineer in 1994, one of Ten Outstanding Young Persons in 1996, and listed in the 15th edition Marquis Who's Who in the World.

RAMJEE PRASAD received a B.Sc. degree from Bihar Institute of Technology, Sindri, India and M.Sc and Ph.D. degrees from Birla Institute of Technology (BIT), India, in 1968, 1970 and 1979, respectively. He joined BIT as a Senior Research Fellow in 1970 and became associate professor in 1980. During 1983 to 1988, he was with the University of Dar es Sulaum (UDSM), Tanzania, as a professor of telecommunications at the Department of Electrical Engineering. Since February 1988, he has been with the Telecommunications and Traffic-Control Systems Group of Delft University of Technology (DUT), The Netherlands, where he is actively involved in the area of wireless personal and multimedia communications (WPMC). He is currently involved in the project FRAMES as a Project Leader. He has published over 300 technical papers, authored and co-edited three books on Wireless Personal Communications. As a member of advisory and program committees of several IEEE international conferences he has presented many keynote speeches, invited papers, and tutorials on WPMC. He is a fellow of the IEE, a senior member of IEEE and is listed in the US Who's Who in the World.

PART III: ENABLING COMPONENTS OF WIRELESS NETWORKS

Chapter 13   Integrated Circuit Technologies for Wireless Communications

BABAK DANESHRAD is an assistant professor with the Electrical Engineering Dept. at the University of California, Los Angeles (UCLA). His research interests include the design of systems and VLSI ASICs for wireless data communications. He obtained the BEng. And MEng. degrees with emphasis in Communications from McGill University, in 1986 and 1988 respectively, and the Ph.D. degree from UCLA in 1993 with emphasis in integrated Circuits and Systems. While at UCLA he investigated systems and VLSI circuits for HDSL and ADSL applications. From 1993 to June 1996 he was a member of technical staff with the Wireless Communications Systems Research Dept. of AT&T Bell Laboratories where he was involved in the design and implementation of systems for high-speed wireless packet communications. As a consultant with the TI DSP R&D center from June to Sept. 1996 he was involved with the design of systems for 100Base-T Ethernet. Between 1990 and 1993 he held positions as a consulting engineer in the areas of digital VLSI ASIC design and communication system design with PairGain Technologies, Level One Communications, LinCom Corp. and Raytheon Semiconductors.

AHMED M. ELTAWIL is a graduate student at the Electrical Engineering Department, University Of California, Los Angeles (UCLA) where he is currently working towards his MS./Ph.D. degrees. He received his B.Sc. with honors from the Electronics and Communications Department, Cairo University, Egypt in 1997. In the same year he joined the VLSI wireless research group at Cairo University where he conducted research in low power, low voltage circuits for wireless applications. In summer 1998 he joined the wireless systems research laboratory at UCLA where his primary research interest is the development of highly flexible low power VLSI ASICs for wireless communication systems. As an undergraduate he has held summer intern positions at the British Broadcasting Corporation, Technical Division (BBC).

Chapter 14   Amplifier Linearization - for Broadband Wireless Applications

KATHLEEN MUHONEN is a Ph.D. candidate and a member of the Center for Information and Communications Technology Research (CICTR) at Penn State University. She received her B.S.E.E. from Michigan Technological University in 1991 and M.S.E.E. from Syracuse University in 1994. Between 1991 and 1994, Kathleen worked for GE Aerospace where she was a device engineer working on millimeter-wave solid state amplifiers using PHEMT technology. In 1994, she went to the University of California at Davis for one year conducting research in digital communications. From 1995 she was employed at Hewlett Packard in the Wireless Infrastructure Division as a hardware engineer. At HP she worked on the design of base station amplifiers for the cellular and PCS standards. She also spent one summer at Lucent Bell Laboratories conducting research on linearization. At Penn State, Kathleen's research is on amplifier linearization techniques with emphasis on digital predistortion for wideband applications.

DR. MOHSEN KAVEHRAD, FIEEE, is the first Weiss Chair Professor of Electrical Engineering and the Director of the Center for Information and Communications Technology Research (CICTR) at The Pennsylvania State University. He received his B.Sc. degree in Electronics from Tehran Polytechnic Institute, in 1973, the M.Sc. degree from Worcester Polytechnic Institute (WPI) in Massachusetts, USA, in 1975 and his Ph.D. degree from Polytechnic University (Formerly: Brooklyn Polytechnic Institute), Brooklyn, New York, in November 1977 in Electrical Engineering. His current research interests are broadband fixed and/or mobile wireless communications and optical fiber communications and networking. He has published over 200 papers and holds several issued patents in these areas. He received 3 Exceptional Technical Contributions awards while working at Bell Laboratories, the 1991 TRIO Feedback award for his patent on a "Passive Optical Interconnect" and 5 best paper awards plus a Canada NSERC Ph.D. Thesis Prize, jointly with his graduate students.

Chapter 15   *Wireless Channel Models – Coping with Complexity*

AN MEI CHEN received her Ph.D. degree in Electrical Engineering from the University of California, San Diego, in June 1998. She is currently employed at TRW performing system studies and design for the next-generation aviation platform to support a Tactical Internet. Her present research interests include dynamic resource allocation and mobility management for integrated services in wireless communication network, tactical Internet architecture, and network security. Dr. Chen is a University of California Regents Scholar and a member of Phi Beta Kappa and Tau Beta Pi.

RAMESH R. RAO received his Ph. D. in Electrical Engineering from the University of Maryland, College Park, in Dec.1984. Since then he has been employed at the University of California, San Diego, where he is currently Professor. His research interests are in the area of protocol design, performance analysis and architectures for Wireless and Wireline Networks. He is a member of the industry funded Center for Wireless Communications at UCSD. He is active in IEEE Society affairs, serves on the Information theory Society Board of Governors and is the Editor for Packet Multiple Access of the IEEE Transactions on Communications.

Chapter 16   *A New Framework for Power Control in Wireless Data Networks: Games, Utility, and Pricing*

DAVID FAMOLARI received the Bachelor of Science and Master of Science degrees in electrical engineering from Rutgers University, in 1996 and 1999, respectively. From 1997 to 1999 he was employed as a research assistant at the Wireless Information Network Laboratory (WINLAB) at Rutgers University, where he carried out research in the areas of third generation cellular networks, and dynamic radio resource management and power control for CDMA systems. In 1998 he joined the Applied Research department at Telecordia Technologies (formerly Bellcore) where he is involved in next generation wireless systems, mobile computing, and distributed Internet appliance networking. His current research interests include, high-speed wireless Internet access, QoS provisioning and mobility management for multimedia services, and distributed self-organizing wireless networks.

NARAYAN MANDAYAM received the B.Tech (Hons.) degree in 1989 from the Indian Institute of Technology, Kharagpur, and the M.S. and Ph.D. degrees in 1991 and 1994 from Rice University, Houston, TX, all in electrical engineering. In September 1994, he joined Rutgers University where he is currently an Assistant Professor in the ECE department and also serves as the Research Director for Radio Systems at WINLAB. His research interests are in communication theory, wireless system modeling and performance, multiaccess protocols, software defined radios for multiuser detection, multimedia wireless communications and radio resource management for wireless data networks. Dr. Mandayam was a recipient of the Institute Silver Medal from the Indian Institute of Technology, Kharagpur in 1989. He also received the National Science Foundation CAREER Award in 1998.

DAVID GOODMAN is Professor of Electrical and Computer Engineering and Director of WINLAB, the Wireless Information Network Laboratory at Rutgers, the State University of New Jersey. Prior to joining Rutgers in 1988, he was a Department Head in Communications Systems Research at Bell Labs. Dr. Goodman's research accomplishments are in the fields of speech coding, digital signal processing and wireless communications. He is author of the books "Wireless Personal Communications Systems" and "Probability and Stochastic Processes: A Gentle Introduction for Electrical and Computer Engineers". He is an editor of five books on wireless communications. Three of his papers have won IEEE prize paper awards. In 1997, he received the ACM Sigmobile Outstanding Contributions Award.

VIRAL SHAH received the Master of Science degree in electrical engineering from Rutgers University in 1998. Since then he has been with QUALCOMM Inc. in San Diego. His research interests are in radio resource management for wireless data systems.

Chapter 17   ***A Unified Approach to Performance Evaluation of Diversity Systems on Fading Channels***

A. ANNAMALAI received his B.Eng. degree with honors from the University of Science of Malaysia in 1993, M.A.Sc. and Ph.D. degrees in electrical engineering in 1997 and 1999, respectively, from the University of Victoria, Canada. He is currently a post-doctoral research fellow at the same university. During May 1993 to April 1995, he was with Motorola Inc. as a RF design engineer. From May 1995 till January 1999, he was a research assistant in the Department of Electrical and Computer Engineering at University of Victoria where he was involved in the research of 3rd generation of wireless CDMA systems. His research interests include coding, modulation, communication theory and wireless communications. Dr. Annamalai is the recipient of the 1998 Lieutenant Governor-General's medal from the University of Victoria and the 1998 Daniel E. Noble Fellowship jointly awarded by IEEE Vehicular Technology Society and Motorola Inc.

C. TELLAMBURA received his BSc degree with honors from the University of Moratuwa, Sri-Lanka, in 1986, MSc in electronics from the King's College, UK, in 1988, and PhD in Electrical Engineering from the University of Victoria, Canada, in 1993. He was a post-doctoral research fellow with the University of Victoria (1993-94) and the University of Bradford (1995-96). Currently, Dr. Tellambura is a Senior Lecturer with the Faculty of Information Technology, Monash University, Australia. His research interests include coding, communication theory, modulation, equalization and wireless communications.

VIJAY K. BHARGAVA received the B.Sc., M.Sc., and Ph.D. degrees from Queen's University of Kingston, Canada in 1970, 1972 and 1974 respectively. Currently, he is a Professor of Electrical and Computer Engineering at the University of Victoria. He is a co-author of the book DIGITAL COMMUNICATIONS BY SATELLITE (New York: Wiley, 1981) and co-editor of the IEEE Press Book REED-SOLOMON CODES AND THEIR APPLICATIONS. He is an Editor-in-Chief of WIRELESS PERSONAL COMMUNICATION, a Kluwer Periodical. His research interests are in multi-media wireless communications. Dr. Bhargava is very active in the IEEE and is currently the Vice President of IEEE Information Theory Society. He was co-chair for ISIT'95 and technical program chair for ICC'99. Dr. Bhargava is a Fellow of the

B.C. Advanced Systems Institute, Engineering Institute of Canada (EIC) and the IEEE. He is a recipient of the IEEE Centennial Medal (1984), IEEE Canada's McNaughton Gold Medal (1995) and the IEEE Haraden Pratt Award (1999).

Chapter 18   ***mMFSK Frequency Hopping Modulation for Wireless Ad Hoc Networks***

SAVO GLISIC is a Professor of Electrical Engineering at University of Oulu, and Director of Globalcomm Institute for Telecommunications. He was visiting scientist at Cranfield Institute of Technology, Cranfield, England (1976,77) and University of California at San Diego, San Diego, USA, (1986,87). He has been active in the field of Spread Spectrum and wireless communications for 20 years and has published a number of papers and five books. He is doing consulting in this field for industry and government. He has served as Technical Program Chairman of the Third IEEE ISSSTA'94, the Eighth IEEE PIMRC'97 and IEEE ICC'01. Dr Glisic is Director of IEEE ComSoc MD programs.

ZORICA NIKOLIC was born in 1955 in Nis. She earned BEE, MS and Ph.D. degrees at the University of Nis, Faculty of Electronic Engineering, Nis in 1979, 1985 and 1989, respectively. She is now an associate professor, with Faculty of Electronic Engineering, Nis, where she teaches Basics of Telecommunications and Telecommunication Networks. Her main fields of interest include mobile telecommunications, satellite communications and spread-spectrum systems.

NENAD MILOSEVIC was born in 1973, in Knjazevac. He earned BEE degree at the University of Nis, Faculty of Electronic Engineering, in 1997. He is now with the Faculty of Electronic Engineering, Nis. His fields of interest are spread spectrum systems and mobile communications.

ARI POUTTU was born in 1966, in Kokkola, Finland. He earned M.Sc. degree at the University of Oulu, Faculty of Electronic Engineering, Finland in 1995 and Licentiate of technology degree in 1997. He is now with the Faculty of Electronic Engineering, Telecommunication laboratory, Oulu. His fields of interest are spread spectrum systems and mobile communications with emphasis on spectrally efficient modulation schemes, synchronization, coding and interference suppression.